T0211790

Lecture Notes in Computer Science 12720

Founding Editors

Gerhard Goos
Karlsruhe Institute of Technology, Karlsruhe, Germany

Juris Hartmanis
Cornell University, Ithaca, NY, USA

Editorial Board Members

Elisa Bertino
Purdue University, West Lafayette, IN, USA

Wen Gao
Peking University, Beijing, China

Bernhard Steffen
TU Dortmund University, Dortmund, Germany

Gerhard Woeginger
RWTH Aachen, Aachen, Germany

Moti Yung
Columbia University, New York, NY, USA

More information about this subseries at http://www.springer.com/series/7409

Robert Thomson · Muhammad Nihal Hussain ·
Christopher Dancy · Aryn Pyke (Eds.)

Social, Cultural, and Behavioral Modeling

14th International Conference, SBP-BRiMS 2021
Virtual Event, July 6–9, 2021
Proceedings

 Springer

Editors
Robert Thomson
United States Military Academy
West Point, NY, USA

Muhammad Nihal Hussain
University of Arkansas at Little Rock
Little Rock, AR, USA

Christopher Dancy
Bucknell University
Lewisburg, PA, USA

Aryn Pyke
United States Military Academy
West Point, NY, USA

ISSN 0302-9743   ISSN 1611-3349 (electronic)
Lecture Notes in Computer Science
ISBN 978-3-030-80386-5   ISBN 978-3-030-80387-2 (eBook)
https://doi.org/10.1007/978-3-030-80387-2

LNCS Sublibrary: SL3 – Information Systems and Applications, incl. Internet/Web, and HCI

© Springer Nature Switzerland AG 2021
This work is subject to copyright. All rights are reserved by the Publisher, whether the whole or part of the material is concerned, specifically the rights of translation, reprinting, reuse of illustrations, recitation, broadcasting, reproduction on microfilms or in any other physical way, and transmission or information storage and retrieval, electronic adaptation, computer software, or by similar or dissimilar methodology now known or hereafter developed.
The use of general descriptive names, registered names, trademarks, service marks, etc. in this publication does not imply, even in the absence of a specific statement, that such names are exempt from the relevant protective laws and regulations and therefore free for general use.
The publisher, the authors and the editors are safe to assume that the advice and information in this book are believed to be true and accurate at the date of publication. Neither the publisher nor the authors or the editors give a warranty, expressed or implied, with respect to the material contained herein or for any errors or omissions that may have been made. The publisher remains neutral with regard to jurisdictional claims in published maps and institutional affiliations.

This Springer imprint is published by the registered company Springer Nature Switzerland AG
The registered company address is: Gewerbestrasse 11, 6330 Cham, Switzerland

Preface

Improving the human condition requires understanding, forecasting, and impacting socio-cultural behavior both in the digital and non-digital world. Increasing amounts of digital data, embedded sensors collecting human information, rapidly changing communication media, changes in legislation concerning digital rights and privacy, spread of 4G technology to third world countries, and so on are creating a new cyber mediated world where the very precepts of why, when, and how people interact and make decisions is being called into question. For example, Uber took a deep understanding of human behavior vis-à-vis commuting, developed software to support this behavior, ended up saving human time (and so capital) and reducing stress, and so indirectly created the opportunity for humans with more time and less stress to evolve new behaviors. Scientific and industrial pioneers in this area are relying on both social science and computer science to help make sense of and impact this new frontier. To be successful a true merger of social science and computer science is needed. Solutions that rely only on the social science or only on the computer science are doomed to failure. For example, Anonymous developed an approach for identifying members of terror groups such as ISIS on the Twitter social media platform using state-of-the-art computational techniques. These accounts were then suspended. This was a purely technical solution. The response was those individuals with suspended accounts just moved to new platforms and resurfaced on Twitter under new IDs. In this case, failure to understand basic social behavior resulted in an ineffective solution.

The goal of the International Conference on Social Computing, Behavioral-Cultural Modeling and Prediction and Behavior Representation in Modeling and Simulation (SBP-BRiMS) is to build this new community of social cyber scholars by bringing together and fostering interaction between members of the scientific, corporate, government, and military communities interested in understanding, forecasting, and impacting human socio-cultural behavior. It is the charge of this community to build this new science, its theories, methods, and its scientific culture, in a way that does not give priority to either social science or computer science, and to embrace change as the cornerstone of the community. Despite decades of work in this area, this new scientific field is still in its infancy. To meet this charge, to move this science to the next level, this community must meet the following three challenges: deep understanding, socio-cognitive reasoning, and re-usable computational technology. Fortunately, as the papers in this volume illustrate, this community is poised to answer these challenges. But what does meeting these challenges entail?

Deep understanding refers to the ability to make operational decisions and theoretical arguments on the basis of an empirical based deep and broad understanding of the complex socio-cultural phenomena of interest. Today, although more data is available digitally than ever before, we are still plagued by anecdotal based arguments. For example, in social media, despite the wealth of information available, most analysts focus on small samples, which are typically biased and cover only a small time period, and use them to explain all events and make future predictions. The analyst finds the magic tweet or the unusual tweeter and uses that to prove their point. Tools that can help the analyst to reason using more data or less biased data are not widely used, are often more complex than the average analyst wants to use, or take more time than the analyst wants to spend to generate results. Not only are more scalable technologies needed but so too is a better understanding of the biases in the data and ways to overcome them, and a cultural change to not accept anecdotes as evidence.

Socio-cognitive reasoning refers to the ability of individuals to make sense of the world and to interact with it in terms of groups and not just individuals. Today most social-behavioral models either focus on (1) strong cognitive models of individuals engaged in tasks, and so model a small number of agents with high levels of cognitive accuracy but with little if any social context, or (2) light cognitive models and strong interaction models, and so model massive numbers of agents with high levels of social realisms and little cognitive realism. In both cases, as realism is increased in the other dimension the scalability of the models fails, and their predictive accuracy on one of the two dimensions remains low. In contrast, as agent models are built where the agents are not just cognitive but socially cognitive, we find that the scalability increases and the predictive accuracy increases. Not only are agent models with socio-cognitive reasoning capabilities needed but so too is a better understanding of how individuals form and use these social-cognitions.

More software solutions that support behavioral representation, modeling, data collection, bias identification, analysis, and visualization, support human socio-cultural behavioral modeling and prediction than ever before. However, this software is generally just piling up in giant black holes on the web. Part of the problem is the fallacy of open source; the idea that if you just make code open source others will use it. In contrast, most of the tools and methods available in Git or R are only used by the developer, if that. Reasons for lack of use include lack of documentation, lack of interfaces, lack of interoperability with other tools, difficulty of linking to data, and increased demands on the analyst's time due to a lack of tool-chain and workflow optimization. Part of the problem is the not-invented here syndrome. For social scientists and computer scientists alike it is just more fun to build a quick and dirty tool for your own use than to study and learn tools built by others. And, part of the problem is the insensitivity of people from one scientific or corporate culture to the reward and demand structures of the other cultures that impact what information can or should be shared and when. A related problem is double standards in sharing where universities are expected to share and companies are not, but increasingly universities are relying on that intellectual property as a source of funding just like other companies. While common standards and representations would help, a cultural shift from a focus on sharing to a focus on re-use is as or more critical for moving this area to the next scientific level.

In this volume, and in all the work presented at the SBP-BRiMS 2021 conference, you will see suggestions of how to address the challenges just described. SBP-BRiMS 2021 carries on the scholarly tradition of the past conferences out of which it has emerged like a phoenix: the Social Computing, Behavioral-Cultural Modeling, and Prediction (SBP) Conference and the Behavioral Representation in Modeling and Simulation (BRiMS) Society's conference. A total of 56 papers were submitted as regular track submissions. Of these, 32 were accepted as full papers giving an acceptance rate of 57%. Additionally, there was a large number of papers describing emergent ideas and late breaking results as well as nine tutorials. This is an international group with papers submitted by authors from many countries.

The conference has a strong multi-disciplinary heritage. As the papers in this volume show, people, theories, methods, and data from a wide number of disciplines are represented including computer science, psychology, sociology, communication science, public health, bioinformatics, political science, and organizational science. Numerous types of computational methods are used including, but not limited to, machine learning, language technology, social network analysis and visualization, agent-based simulation, and statistics.

This exciting program could not have been put together without the hard work of a number of dedicated and forward-thinking researchers serving as the Organizing Committee, listed on the following pages. Members of the Program Committee and the Scholarship Committee, along with the publication, advertising and local arrangements chairs, worked tirelessly to put together this event. They were supported by the government sponsors, the area chairs, and the reviewers. Please join me in thanking them for their efforts on behalf of the community. In addition, we gratefully acknowledge the support of our sponsors – the Army Research Office (W911NF-21-1-0102) and the National Science Foundation (IIS-1926691). We hope that you enjoyed the conference and welcome to the community.

July 2021

Kathleen M. Carley
Nitin Agarwal

Organization

Conference Co-chairs

Kathleen M. Carley	Carnegie Mellon University, USA
Nitin Agarwal	University of Arkansas at Little Rock, USA

Program Co-chairs

Christopher Dancy II	Bucknell University, USA
Muhammad Nihal Hussain	Equifax Inc., USA
Robert Thomson	United States Military Academy, USA
Aryn Pyke	United States Military Academy, USA

Advisory Committee

Fahmida N. Chowdhury	National Science Foundation, USA
Rebecca Goolsby	Office of Naval Research, USA
Stephen Marcus	National Institutes of Health, USA
Paul Tandy	Defense Threat Reduction Agency, USA
Edward T. Palazzolo	Army Research Office, USA

Advisory Committee Emeritus

Patricia Mabry	Indiana University, USA
John Lavery	Army Research Office, USA
Tisha Wiley	National Institutes of Health, USA

Scholarship and Sponsorship Committee

Nitin Agarwal	University of Arkansas at Little Rock, USA
Christopher Dancy II	Bucknell University, USA

Industry Sponsorship Committee

Jiliang Tang	Michigan State University, USA

Publicity Chairs

Donald Adjeroh	West Virginia University, USA
Katrin Kania Galeano	University of Arkansas at Little Rock, USA

Local Area Coordination

David Broniatowski   The George Washington University, USA

Proceedings Chair

Robert Thomson   United States Military Academy, USA

Agenda Co-chairs

Robert Thomson   United States Military Academy, USA
Kathleen M. Carley   Carnegie Mellon University, USA

Journal Special Issue Chair

Kathleen M. Carley   Carnegie Mellon University, USA

Tutorial Chair

Kathleen M. Carley   Carnegie Mellon University, USA

Graduate Program Chair

Fred Morstatter   University of Southern California, USA
Kenny Joseph   University at Buffalo, USA

Challenge Problem Committee

Kathleen M. Carley   Carnegie Mellon University
Nitin Agarwal   University of Arkansas at Little Rock, USA

BRiMS Society Chair

Christopher Dancy II   Bucknell University, USA

SBP Steering Committee

Nitin Agarwal   University of Arkansas at Little Rock, USA
Sun Ki Chai   University of Hawaii, USA
Ariel Greenberg   Johns Hopkins University/Applied Physics Laboratory,
                            USA
Huan Liu   Arizona State University, USA
John Salerno   Exelis, USA
Shanchieh (Jay) Yang   Rochester Institute of Technology, USA

SBP Steering Committee Emeritus

Nathan D. Bos	Johns Hopkins University/Applied Physics Lab, USA
Claudio Cioffi-Revilla	George Mason University, USA
V. S. Subrahmanian	University of Maryland, USA
Dana Nau	University of Maryland, USA

SBP-BRIMS Steering Committee Emeritus

Jeffrey Johnson	University of Florida, USA

Technical Program Committee

Tazin Afrin
Nasrin Akhter
Samer Al-Khateeb
Antonio Luca Alfeo
Walaa Alnasser
Georgios Anagnostopoulos
Akshay Aravamudan
Brent Auble
Dmitriy Babichenko
Kiran Kumar Bandeli
Emanuel Ben-David
Nathan Bos
Keith Burghardt
Gian Maria Campedelli
Yutao Chen
Lu Cheng
Yi Han Victoria Chua
Andrew Crooks
Joel Croteau
Iain Cruickshank
Vito D'Orazio
Jacob Dineen
Geoffrey Dobson
Wen Dong
Ma Regina Justina E. Estuar
Laurie Fenstermacher
Juan Fernandez-Gracia
Erika Frydenlund
Rick Galeano
Ozlem Garibay
Ariel Greenberg

Gayane Grigoryan
Ahsan-Ul Haque
Andre Harrison
Yuzi He
Shuyuan Mary Ho
Pedram Hosseini
Aruna Jammalamadaka
Amir Karami
Prakruthi Karuna
Khalid Kattan
Hamdi Kavak
Alex Kelly
James Kennedy
Imane Khaouja
Tuja Khaund
Baris Kirdemir
Ugur Kursuncu
Christian Lebiere
Stephan Leitner
Xiaoyan Li
Larry Lin
Huan Liu
Jiebo Luo
Thomas Magelinski
Chitaranjan Mahapatra
Murali Mani
Esther Mead
Jonathan Morgan
Saddam Mukta
Sahiti Myneni
Andy Novobilski

Salem Othman
Lucas Overbey
Jose Padilla
Xinyue Pan
Rahul Pandey
Hemant Purohit
Rey Rodrigueza
Mitchell Ryan
Mainuddin Shaik
Zhongkai Shangguan

Juliette Shedd
Martin Smyth
Billy Spann
Zachary Stine
Amit Upadhyay
Joshua Uyheng
Alina Vereshchaka
Friederike Wall
Terry Williams
Xupin Zhang

Contents

Social Cybersecurity and Social Networks

Human and Agent Modeling

COVID-Related Focus

Malicious and Low Credibility URLs on Twitter During the AstraZeneca COVID-19 Vaccine Development

Sameera Horawalavithana[1]($\boxtimes$) (iD), Ravindu De Silva[2], Mohamed Nabeel[3],
Charitha Elvitigala[2], Primal Wijesekara[4], and Adriana Iamnitchi[1] (iD)

[1] University of South Florida, Tampa, USA
sameera1@usf.edu, anda@cse.usf.edu
[2] SCoRe Lab, Colombo, Sri Lanka
{ravindud,charitha}@scorelab.org
[3] Qatar Computing Research Institute, Al-Rayyan, Qatar
mnabeel@hbku.edu.qa
[4] University of California, Berkeley, Berkeley, USA
primal@berkeley.edu

Abstract. We investigate the link sharing behavior of Twitter users following the temporary halt of AstraZeneca COVID-19 vaccine development in September 2020. During this period, we show the presence of malicious and low credibility information sources shared on Twitter messages in multiple languages. The malicious URLs, often in shortened forms, are increasingly hosted in content delivery networks and shared cloud hosting infrastructures not only to improve reach but also to avoid being detected and blocked. There are potential signs of coordination to promote both malicious and low credibility URLs on Twitter. Our findings suggest the need to develop a system that monitors the low-quality URLs shared in times of crisis.

Keywords: COVID-19 · URL · Twitter

1 Introduction

During the COVID-19 pandemic, Twitter is being used to spread both high-quality and low-quality information [20]. While information regarding the health, diseases, and vaccines are rapidly shared during this crisis period, there is a high prevalence of health misinformation that often contradicts with the opinions provided by health experts. Many researchers called this an "infodemic" during COVID-19 pandemic [8] that highlights the "safety, efficacy and necessity" concerns around vaccines [21]. Many factors contribute to the spread of health misinformation. A recent report suggested that 20% health misinformation claims are shared by politicians, celebrities and other prominent figures which received 69% total social media engagement [8]. Both social media companies and public

© Springer Nature Switzerland AG 2021
R. Thomson et al. (Eds.): SBP-BRiMS 2021, LNCS 12720, pp. 3–12, 2021.
https://doi.org/10.1007/978-3-030-80387-2_1

health officials are heavily criticized due to the delayed actions of controlling the circulation of these misleading messages [21].

There have also been active efforts to sow seeds of doubt on the efficacy of vaccines in the long term. According to Reuters [7], AstraZeneza vaccine development faces many challenges from the date of its inception. For example, there is a temporary halt in the development of AstraZeneca vaccine on September 2020 due to an unexplained illness that was reported in one of the trial participants [18]. AstraZeneca did not release enough details of this event which lead scientists to question its transparency on the vaccine development efforts [9]. This event also led The speaker of the House of Representatives, Nancy Pelosi to make a public statement regarding the approval of COVID-19 vaccines relying on UK safety tests [5]. Digital media shared various information about this event due to the popularity of AstraZeneca vaccine during the final stages of clinical testing [5,9].

In this paper, we attempt to uncover the patterns of link sharing behavior on Twitter discussions following this event. We discover a strong presence of malicious and low credibility information sources shared on Twitter messages in multiple languages. We also show potential signs of coordination to promote these low-quality information.

2   Data Collection and Processing

We used a publicly available Twitter dataset around AstraZeneca COVID-19 vaccine development released as a part of Grand Challenge, North American Social Network Conference, 2021 [1]. The keywords used to collect this dataset are *AstraZeneca, Astra Zeneca, AZD1222, COVID, vaccine, immunity, herd immunity, Barrington, and focused protection.*

To understand the patterns of sharing *Uniform Resource Locator* (URL) information in Twitter conversations, we only consider the Twitter messages that contain at least a URL, which consist of 25% from the total messages. As shown in Fig. 1, they are shared between August 17 and September 10, 2020. Note that the original Twitter dataset covered a time period between August 17 and October 23, 2020, but include many gaps in the data after September 10, 2020. We filtered out URLs to Twitter itself that typically refer to other tweets. In addition, the external links (e.g., a tweet mentioning a YouTube video, or an external website domain) mentioned in the messages are pre-processed as following. The shortened URLs are expanded, and HTML parameters are removed from the URLs. The YouTube URLs are resolved to the base video URL if they include a parameter referencing a specific time in the video. We represent the URL by the parent domain when multiple child domains exist (e.g., `fr.sputniknews.com`, `arabic.sputniknews.com` etc. are renamed as `sputniknews.com`). This pre-processing code of resolving URLs is publicly available [17]. In this analysis, we do not differentiate among different message types (e.g., tweets, retweets, replies and quotes), but consider all messages as URL mentions. The resulting dataset consists of 85,064 Twitter messages. These messages are shared by 40,287 users citing 26,422 distinct URLs. These URLs span over 6,990 distinct web domains.

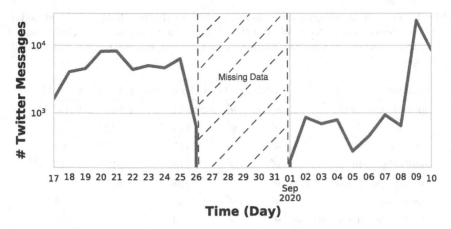

Fig. 1. Number of Twitter URL mentions over time. There is a known gap in the original data collection between August 27–31 [1]

Figure 1 shows a spike in URL mentions on September 9, 2020. There are 23,620 (28%) messages citing distinct 8,566 URLs on this particular day. The most popular URLs point mainstream news articles, while the article published in `statnews.com` received the highest number (1,158) of shares [18]. This article reports the halt in AstraZeneca vaccine development following a suspected adverse reaction of a trial participant.

For additional analysis, we scrape the web page content pointed to an URL using a Python library, Newspaper. We managed to scrape 22,856 articles. The tool failed to extract content from some web domains mainly due to inactive web pages and regulations enforced by the web domain. According to the Langid [14], the majority of the articles are in English (6,453) and Spanish (5,197) languages. Other articles are written in Turkish (1,839), French (1,539), Portuguese (1,277), Russian (784), Italian (772), Greek (708), Japanese (702), Croatian (447), etc. There are 274 and 165 YouTube videos shared on Twitter have the title written in English and Spanish languages, respectively.

3   Low Credibility Information Sources

In this section, we report how Twitter users react to low credibility information sources. We group the web domains according to the classification made by Media Bias/Fact Check (MBFC) [4]. We also cross-check these domains with the list of web sites that publish False COVID-19 information as identified by NewsGuard [6]. We identify 6,860 (26%) URLs from 17 low credibility information sources that are shared on 26,220 (31%) messages (as shown in Table 1). `sputniknews.com` is the most popular web domain by the number of mentions (22,251), the number of users (4,358) and the number of URLs (6,638). This can be expected mainly due to a network of `sputniknews.com` media outlets that publish articles in many languages. For example, 1,377 and

Table 1. Twitter sharing characteristics for low credibility URLs as identified by MBFC and NewsGuard (NG). We use the ✔ mark to reflect whether the domain is being listed as a low credibility information source by the respective fact checking organization.

Domain	MBFC	NG	# Mentions	# Users	# URLs
sputniknews.com	✔	✔	22,251	4,358	6,638
rt.com	✔	✔	3,225	1,922	119
zenith.news	✔	-	117	6	62
zerohedge.com	✔	✔	278	249	15
swarajyamag.com	✔	-	228	105	5
oann.com	✔	✔	50	48	4
childrenshealthdefense.org	✔	✔	3	3	3
globalresearch.ca	✔	✔	16	7	3
torontotoday.net	✔	-	9	5	2
gnews.org	✔	✔	4	4	2
truepundit.com	✔	✔	29	29	1
needtoknow.news	✔	-	5	5	1
oye.news	✔	✔	1	1	1
gellerreport.com	✔	✔	1	1	1
barenakedislam.com	✔	-	1	1	1
wakingtimes.com	✔	✔	1	1	1
thegatewaypundit.com	✔	✔	1	1	1

1,080 `sputniknews.com` articles are published in Turkish and French. However, `sputniknews.com` URLs are not the most popular to be shared in the immediate hours after their first appearance (as shown in Fig. 2b). For example, the median number of `sputniknews.com` URL mentions is 2 in an hour. `sputniknews.com` is frequently described as a Russian propaganda outlet that spreads master narratives in the Russia's disinformation campaign [4].

We noticed that `rt.com` acquires its many mentions from very few URLs: 119 URLs are cited in 3,225 messages during our observation period (Fig. 2a). `rt.com` URLs have the highest number of mentions in the first hour after publication (as shown in Fig. 2b) relative to the rest of popular web sites.

We also note that Twitter messages share the same article heading in the messages when they cite the same URL. These users promoted certain topics through massive repetition of messages via injecting URLs. For example, an article published in `zerohedge.com` was in the Top-10 most popular URLs on the day when AstraZeneza vaccine development halted[1]. However, this article

---

[1] https://www.zerohedge.com/markets/ft-confirms-astrazeneca-covid-19-vaccine-caused-serious-spinal-issues-test-patient.

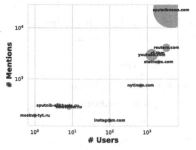

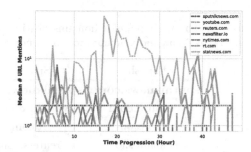

(a) Top-10 shared domains   (b) Twitter lifespan of URLs by domain

Fig. 2. Twitter sharing characteristics of most popular domains. Figure a) shows the Top-10 domains by the number of distinct URLs. The size of the markers in this plot are proportional to the number of URLs associated with the domain. Figure b) shows the number of URL mentions posted at each hour after the first appearance of URL in the respective domain. We count the number of mentions for each URL in the respective hour, and calculate the median number of mentions for the domain of the URL.

tried to build an alternative frame highlighting a statement by the US House Speaker Nancy Pelosi about the issue instead of reporting the details of the main event. We also discovered URLs from two Russian news media domains, `sputnik-abkhazia.ru` and `moskva-tyt.ru` that are shared by very few users. For example, `sputnik-abkhazia.ru` is the fourth most popular web domain by the number of URLs cited on Twitter messages. These URLs are shared solely by the Twitter account controlled by the news media site. However, they gain limited engagement with other users.

4   Malicious URLs

We use VirusTotal (VT) [22] to extract the *maliciousness* of URLs. VT provides the state-of-the-art aggregated intelligence for domains and URLs, and relies on more than 70 third-party updated antivirus (AV) engines. For all distinct URLs in our collection, we extract VT scan reports via querying the publicly available API. Each VT scan report contains of the verdict from every AV engine, information related to the URL such as first and last seen dates of the URL in the VT system, hosting IP address, final redirected URL (if applicable), content length, etc. Each AV engine in a VT report detects if the URL is malicious or not. We use the number of engines that detect a URL as malicious as an indication of the maliciousness of the URL.

In this study, we label a URL as malicious if at least one AV engine, that is $\#VT \geq 1$, detects it as malicious. Such malicious URLs, in general, are either phishing websites that steal user credentials and/or personally identifiable information from victims or malware hosting websites that attempt to install malware on victims' devices. We identify 441 malicious URLs from the collected VT

Table 2. The most popular domains hosting articles written in multiple languages. sputniknews.com and youtube.com are among the most popular domains.

English	Spanish	Turkish
sputniknews.com	**sputniknews.com**	**sputniknews.com**
youtube.com	**youtube.com**	**youtube.com**
reuters.com	rt.com	ntv.com.tr
newsfilter.io	reuters.com	is.gd
nytimes.com	wp.me	sozcu.com.tr

French	Portugese
sputniknews.com	**sputniknews.com**
cvitrolles.wordpress.com	brasil247.com
youtube.com	evsarteblog.wordpress.com
francetvinfo.fr	**youtube.com**
lalibre.be	tvi24.iol.pt

reports. We observe that 25.75% of the malicious URLs utilize URL shortening services with top 4 services being bit.ly, tinyurl.com, ow.ly and goo.su whereas as only 0.97% of benign URLs utilize such services. This observation is consistent with the trend that malicious actors increasingly utilize shortening services to camouflage malicious URLs presenting non-suspecting URLs to users [2]. We find that 30.80% of the domains related to malicious URLs are ranked below 100K by Alexa [3] (The lower the rank value, the higher the popularity). This indicates the concerning fact that malicious actors are able to reach a large user base reaping a high return on investment for their attacks (Table 2).

We further analyze the malicious URLs to identify related malicious URLs. To this end, based on the lexical features in the literature [19] and the hosting features mentioned in Table 3, we cluster the malicious URLs using PCA/OPTICS algorithm.

While lexical features identify characteristics related to URLs themselves, hosting features, extracted from Farsight Passive DNS (PDNS) data [10], capture the characteristics of underlying hosting infrastructure. As shown in Fig. 3, these features collectively identify 10 distinct malicious URL clusters. We manually verified the accuracy of the top 2 clusters by checking the web page content, registration information and domain certificate information. Our observations suggest that attackers launch multiple attacks at the same time. We also analyze the clusters based on the maliciousness of URLs. The maliciousness of a URL can loosely be measured by #VT, the number of VT positives. An interesting observation is that URLs belonging to different maliciousness levels share similar lexical and hosting features. We further analyze these malicious URLs in terms of where they are hosted. To our surprise, we find that 80.04% of these malicious URLs are hosted in CDNs such as Cloudflare and Akamai. While CDNs provide fast delivery of content across the globe through their distributed

Table 3. Details of the hosting features for malicious URL clustering

Feature	Description
VT_Dur	URL duration in VT
PDNS_Dur	Domain duration in PDNS
#IPs	# hosting IPs
#Queries	# times the domain is accessed
#NSes	# Name servers
Is_NS	NS domain matches?
#SOAs	# administrative domains
Is_SOA	Admin domain matches?
#Domains	# domains hosted on the IP
#Queries_IP	# times the IP is accessed
ASN	Autonomous System Number
Org	Organization owning the ASN

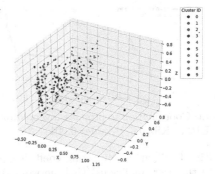

Fig. 3. Malicious URL clusters based on the lexical and hosting features. Each point is a URL, and it is colored according to the cluster it belongs.

computing infrastructure, we believe a key reason why malicious actors utilize such services is to improve attack agility and stay below the radar of malicious domain detection mechanisms in place. This observation is further reinforced with the increased utilization of public cloud computing infrastructure (33.5% of all malicious URLs) sharing hosting IPs with tens of thousands of unrelated domains, which are mostly benign. Such shared IPs are usually not blocked in practice due to the collateral damage.

5   User Co-sharing Practices

As reported previously [16], there are signs of coordination to spread the disinformation content. In this work, one of our assumptions is that the coordination is based on the content being shared (i.e., URLs mentioned in the tweets). One of our objectives is to measure the extent of coordination in URL sharing activities. We construct two bipartite networks to quantify the amount of such coordination effort. The first network connects an author to a low credibility URL mentioned in a tweet, and the second network connects an author to a malicious URL mentioned in a tweet. These networks would capture any suspicious behavior of promoting a particular URL.

We identify 6,860 low credibility information URLs that are shared by 6,600 users in 26,220 Twitter messages. The five (0.08%) most active users cite low credibility URLs in 6,404 (24%) messages. Their most preferred information source is the Turkish outlet of the sputnik media network.

We identify 441 malicious URLs that are shared by 357 users in 571 Twitter messages. While a user might share a malicious URL unwittingly, it is suspicious to note users who share multiple malicious URLs. Specifically, we identify 51 users who share more than 1 malicious URL, and 21 users who share more than 2 malicious URLs.

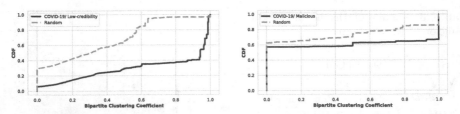

(a) Coordination to promote low credibility URLs (KS=0.6, p-value=0.001)

(b) Coordination to promote malicious URLs (KS=0.27, p-value=$4.45e^{-12}$)

Fig. 4. Bipartite clustering coefficient of a) the user-(low credibility) URL network, b) the user-(malicious) URL network. We also compare the network clustering values with a random bipartite network using the Newman's configuration model [15]. The deviation of the clustering values from the random bipartite network shows potential coordination effort to promote these URLs on Twitter.

Figures 4a and 4b show the distributions of bipartite clustering coefficients [13] for URLs in the two bipartite networks respectively. Clustering coefficient values are higher for URLs when they are shared by a group of users who engage with other URLs together. We compare the similar clustering values which are calculated from the identical random bipartite networks. We construct two random bipartite networks using the Newman's configuration model [15] for the comparison. Given the original user-URL bipartite network, we match the two degree sequences in the users and URLs in the random bipartite network. We noticed a significant deviation of clustering coefficient values for URLs in both bipartite networks compared to URLs in the random bipartite networks (as shown in Figs. 4a and 4b). For example, there are 4,320 (63%) low credibility URLs with a clustering coefficient value greater than 0.8 than those number of URLs (217) in the random bipartite network. In the malicious URL network, there are 121 URLs with a clustering coefficient value greater than 0.8 that are promoted by the same set of users (the expected number of URLs is 36 in the random network). To confirm our observation, we also perform the Kolmogorov-Smirnov (KS) test between the clustering coefficient values from the original and random bipartite networks. KS-statistic values (statistically significant) are 0.6 and 0.27 for the low credibility URL network and malicious URL network, respectively. This suggests the sustained effort of users to amplify both misleading and malicious content. There may be different types of users (e.g., bots, cyborgs, paid activists) who amplify these URLs. That would remain as a future work to identify these types of users.

6   Conclusions

In times of crisis, whether political or health-related, online disinformation is amplified by social media promotion of alternative media outlets [12]. This study adds to the growing body of work [11] that analyzes the misinformation activity during the COVID-19 crisis by studying the sharing of URLs on Twitter between August 17 and September 10, 2020. Our contributions complement previous observations [2, 20] in multiple ways. We discover a strong presence of malicious and low credibility information sources shared on Twitter messages in multiple languages. Not only that URLs from low credibility sources, as classified by independent bodies such as NewsGuard and Media Bias/Fact Check, were present, but many were shown to point to pages with malicious code. A significant portion of these URLs (25%) were in shortened form (compared to under 1% of the non-malicious URLs) and hosted on well-established, reputable content delivery networks in an attempt, we believe, to avoid detection.

We also discovered potential signs of coordination to promote malicious and low credibility URLs on Twitter. Specifically, we discovered unusual clustering of user activity related to the sharing of such URLs. We use a null model and several statistical tests to compare expected behavior with what we suspect to be coordinated behavior as seen in this dataset.

In general, we unmask malicious strategies that exploit Twitter to promote shady objectives in times of crisis. Further work is needed to understand these objectives. For example, bad actors might have chosen this event strategically to maximize the spread of malicious URLs. These actors can deploy the same strategy in the future conversations, thus having content moderation techniques to limit what they can share is important. On the other hand, the low credibility news sources might have reported this event opportunistically in an attempt to promote vaccine hesitancy. People might have engaged with these low quality sources to watch out the information space around COVID-19 vaccines. According to Smith et al. [21], there is a deficit of high quality information sources to seek vaccine information. Bad actors use this information deficit as an advantage to push low quality information. We believe this analysis can be extended in understanding the role of bad actors during similar emotionally charged conversations in the future.

Acknowledgements. This work is partially supported by the DARPA SocialSim Program and the Air Force Research Laboratory under contract FA8650-18-C-7825. The authors would like to thank Grand Challenge, North American Social Network Conference for providing data.

References

1. Nasn 2021: Grand challenge. https://www.insna.org/nasn-2021-grand-challenge
2. Spanish-language vaccine news stories hosting malware disseminated via url shorteners. https://fas.org/disinfoblog/spanish-language-vaccine-news-stories-hosting-malware-disseminated-via-url-shorteners/
3. Alexa Top 1M (2019). https://www.alexa.com/topsites. Accessed 1 Dec 2019
4. Media bias/fact check - search and learn the bias of news media (2020). https://mediabiasfactcheck.com/
5. Us won't rely on UK for Covid vaccine safety tests, says Nancy Pelosi (2020). https://www.theguardian.com/society/2020/oct/09/us-wont-rely-on-uk-for-covid-vaccine-safety-tests-says-nancy-pelosi

6. Coronavirus misinformation tracking center (2021). https://www.newsguardtech.com/coronavirus-misinformation-tracking-center/
7. Covid-19 vaccine astrazeneca roller-coaster ride (2021). https://www.reuters.com/business/healthcare-pharmaceuticals/covid-19-vaccine-astrazeneca-roller-coaster-ride-2021-03-24/
8. Bagherpour, A.: Covid misinformation is killing people (2020). https://www.scientificamerican.com/article/covid-misinformation-is-killing-people1/
9. Cyranoski, D., Mallapaty, S.: Scientists relieved as coronavirus vaccine trial restarts-but question lack of transparency. Nature **585**(7825), 331–332 (2020)
10. Farsight Security: DNS Database. https://www.dnsdb.info/. Accessed: 10 Jan 2021
11. Ferrara, E., Cresci, S., Luceri, L.: Misinformation, manipulation, and abuse on social media in the era of Covid-19. J. Comput. Soc. Sci. **3**(2), 271–277 (2020)
12. Horawalavithana, S., Ng, K.W., Iamnitchi, A.: Twitter is the megaphone of cross-platform messaging on the white helmets. In: Thomson, R., Bisgin, H., Dancy, C., Hyder, A., Hussain, M. (eds.) SBP-BRiMS 2020. LNCS, vol. 12268, pp. 235–244. Springer, Cham (2020). https://doi.org/10.1007/978-3-030-61255-9_23
13. Latapy, M., Magnien, C., Del Vecchio, N.: Basic notions for the analysis of large two-mode networks. Soc. Networks **30**(1), 31–48 (2008)
14. Lui, M., Baldwin, T.: langid.py: an off-the-shelf language identification tool. In: Proceedings of the ACL 2012 System Demonstrations, pp. 25–30 (2012)
15. Newman, M.E.: The structure and function of complex networks. SIAM Rev. **45**(2), 167–256 (2003)
16. Pacheco, D., Hui, P.M., Torres-Lugo, C., Truong, B.T., Flammini, A., Menczer, F.: Uncovering coordinated networks on social media. arXiv:2001.05658 (2020)
17. PNNL, P.N.N.L: Socialsim (2018). https://github.com/pnnl/socialsim
18. Robbins, R., Feuerstein, A., Branswell, H.: Astrazeneca Covid-19 vaccine study is put on hold (2020). https://www.statnews.com/2020/09/08/astrazeneca-covid-19-vaccine-study-put-on-hold-due-to-suspected-adverse-reaction-in-participant-in-the-u-k/
19. Silva, R.D., Nabeel, M., Elvitigala, C., Khalil, I., Yu, T., Keppitiyagama, C.: Compromised or attacker-owned: a large scale classification and study of hosting domains of malicious urls. In: 30th USENIX Security Symposium (USENIX Security 21). USENIX Association (2021). https://www.usenix.org/conference/usenixsecurity21/presentation/desilva
20. Singh, L., Bode, L., Budak, C., Kawintiranon, K., Padden, C., Vraga, E.: Understanding high-and low-quality URL sharing on Covid-19 twitter streams. J. Comput. Soc. Sci. **3**(2), 343–366 (2020)
21. Smith, R., Cubbon, S., Wardle, C.: Under the surface: Covid-19 vaccine narratives, misinformation and data deficits on social media (2020). https://firstdraftnews.org/long-form-article/under-the-surface-covid-19-vaccine-narratives-misinformation-and-data-deficits-on-social-media/
22. VirusTotal, Subsidiary of Google: VirusTotal - Free Online Virus, Malware and URL Scanner. https://www.virustotal.com/. Accessed 14 Jan 2021

How Political is the Spread of COVID-19 in the United States?
An Analysis Using Transportation and Weather Data

Karan Vombatkere, Hanjia Lyu$^{(\boxtimes)}$, and Jiebo Luo

University of Rochester, Rochester, NY 14627, USA
kvombatk@u.rochester.edu, hlyu5@ur.rochester.edu, jluo@cs.rochester.edu

Abstract. We investigate the difference in the spread of COVID-19 between the states won by Donald Trump (Red) and the states won by Hillary Clinton (Blue) in the 2016 presidential election, by mining transportation patterns of US residents from March 2020 to July 2020. To ensure a fair comparison, we first use a K-means clustering method to group the 50 states into five clusters according to their population, area and population density. We then characterize daily transportation patterns of the residents of different states using the mean percentage of residents traveling and the number of trips per person. For each state, we study the correlations between travel patterns and infection rate for a 2-month period before and after the official states reopening dates. We observe that during the lock-down, Red and Blue states both displayed strong positive correlations between their travel patterns and infection rates. However, after states reopened we find that Red states had higher travel-infection correlations than Blue states in all five state clusters. We find that the residents of both Red and Blue states displayed similar travel patterns during the period post the reopening of states, leading us to conclude that, on average, the residents in Red states might be mobilizing less safely than the residents in Blue states. Furthermore, we use temperature data to attempt to explain the difference in the way residents travel and practice safety measures.

Keywords: COVID-19 · Infection analysis · Transportation pattern

1 Introduction

There have been many analyses done recently on investigating the spread of COVID-19 in relation to the state policies, political ideology regarding travel guidelines, mask-usage and state-specific reopening rules. Both news media[1] and academic researchers [10] have intended to investigate the relationship between the difference of infection growth and the politicization across residents' behaviours and psychological factors. In the 2016 presidential election, Hillary Clinton won 21 (Blue) states and Donald Trump won 30 (Red) states.

---

[1] https://www.cnn.com/2020/07/08/politics/what-matters-july-8/index.html.

© Springer Nature Switzerland AG 2021
R. Thomson et al. (Eds.): SBP-BRiMS 2021, LNCS 12720, pp. 13–22, 2021.
https://doi.org/10.1007/978-3-030-80387-2_2

We intend to study the spread of COVID-19 in these two sets of states by mining transportation patterns and attempting to correlate them with the daily infection rate. The graph of daily confirmed infections in Trump vs. Clinton states was publicized by CNN[2] in July 2020 and caught our attention. The goal of the analysis in this paper is to use transportation patterns along with weather data to provide insight into why COVID-19 cases might be growing at a higher rate in the Red states than the Blue states, after states started to officially reopen in May 2020.

Our contributions are summarized as follows:

– We use transportation data to analyze the difference in the spread of COVID-19 between the states won by Donald Trump (Red) and the states won by Hilary Clinton (Blue) in the 2016 presidential election.
– We study the correlations between travel patterns and infection rates before and after the states' reopening dates.
– We discover that there are differences in the correlations before and after the states reopened, and temperature might have affected how people traveled.

2   Related Work

Data from two MTurk studies with US respondents revealed an ideological divide in adherence to social distancing guidelines during the COVID-19 pandemic [10]. Political ideology represents shared beliefs, opinions, and values held by an identifiable group or constituency [4,6], and it endeavors to describe and interpret the world and envision the world as it should be [5]. Many studies have been done to investigate the relationship between travel restrictions and the spread of COVID-19. [1] used a global metapopulation disease transmission model to project the impact of travel limitations on the national and international spread of the pandemic. [8] found that an unconstrained mobility would have significantly accelerated the spreading of COVID-19. In our investigation, we use publicly available transportation data provided by the US Department of Transportation to study the transportation patterns of state residents. We use the John Hopkins COVID-19 reporting data to study the infection rate for each state.

3   Data Collection

We use two primary data sources for the analysis conducted in this paper. We also augment our analysis with data from a couple of other sources, which are mentioned below.

---

[2] https://www.cnn.com/2020/07/08/politics/what-matters-july-8/index.html.

3.1   Department of Transportation Data

This data set [11] is publicly available and can be accessed through the Socrata API[3]. The travel statistics are produced from an anonymized national panel of mobile device data from multiple sources, and a weighting procedure expands the sample of millions of mobile devices, so the results are representative of the entire population in a nation, state, or county. We use the state-level transportation data to compute the following travel feature vectors, which are then normalized by population, i.e., per person basis. Trips under 25 miles and 10 miles are chosen to examine the local mobility of residents, to better capture characteristics of community transmission.

- Mean percentage of the total population that traveled on a given day, $m_1(t)$
- Mean number of trips taken, per person, $n(t)$
- Number of under 10 mile trips taken, per person, $n_{10}(t)$
- Number of under 25 mile trips taken, per person, $n_{25}(t)$

3.2   JHU CSSE COVID-19 Dataset

This is the data repository [2] for the 2019 Novel Coronavirus Visual Dashboard operated by the Johns Hopkins University Center for Systems Science and Engineering (JHU CSSE[4]). We used the time series reported data to obtain the daily confirmed cases data for each state for the period from March to July 2020. We use this dataset to calculate the daily exponential infection growth rate, $r(t)$ for each state.

3.3   Other Sources

We leverage the US Census 2010 data[5] to obtain Area and Population data for each state. We also use data from the BallotPedia[6] to get the official states reopening dates used in our analysis.

4   Methodology

4.1   K-Means Clustering

We use the K-means clustering algorithm to cluster all US states based on their population, area and population density. The rationale behind using these features to cluster is to create clusters of similar states based on only the area and population, in an attempt to create a more even and fair comparison baseline for travel within the Red and Blue states.

---

[3] https://data.bts.gov/resource/w96p-f2qv.json.
[4] https://github.com/CSSEGISandData/COVID-19.
[5] https://www.census.gov/programs-surveys/decennial-census/data.html.
[6] Data available at: https://ballotpedia.org.

4.2 Calculating Infection Rate

We use a simple logistic model for the infection growth rate [9]. The cumulative number of confirmed cases, $C(t)$ at time t can be approximated by

$$\frac{d}{dt}C(t) = rC(t)(1 - \frac{C(t)}{K})$$ (1)

where r is the exponential infection growth rate of interest to us, and $K = \lim_{t\to\infty} C(t)$.

We see that a solution to this model for the number of confirmed cases $C(t)$ is proportional to e^{rt}, so in the discrete case where t is measured in days we can calculate the daily infection growth rate by taking the ratio of the daily difference in cumulative confirmed cases data. For a given day t, the number of new cases $N(t)$ for that day can be calculated by

$$N(t) = C(t) - C(t-1)$$ (2)

The daily exponential growth rate, $r(t)$ can then be calculated for each day by taking the ratio of successive $N(t)$, i.e.

$$r(t) = \frac{N(t)}{N(t-1)}$$ (3)

This method is used to calculate the daily infection growth rate, $r(t)$ for each state from the cumulative confirmed COVID-19 case JHU dataset from March 2020 through July 2020. The $r(t)$ vector for each state was then used as the input to the correlation analysis conducted with the transportation pattern data.

4.3 Correlation Analysis

We use the Pearson correlation coefficient as a measure of the similarity between the time series data for our analysis. We set a p-value threshold of 0.05 to ensure a 95% confidence interval for all correlations reported. It is well accepted that there is a leader-follower relationship between travel and infection, since the infection growth caused by travel would only manifest itself in confirmed cases data after a delay of at least 2–3 weeks based on the incubation period of COVID-19 [7]. Thus, in all correlation analysis we use a lag factor of 21 days to account for this incubation period and the time needed for an individual to obtain a confirmed test result. Based on empirical tests with the data with different lag factors we are able to confirm this pattern.

Based on the graph of daily confirmed cases publicized by CNN, we observe the daily cases for Red states start to increase in early June 2020. Since almost all states opened officially in late April - May 2020, we investigate travel patterns before and after these reopening dates, to attempt to explain the increase in spread of COVID-19 in Red states. This rationale defines our methodology for correlation analysis, and is outlined below. For each state, within each state cluster, we (a) calculate the correlations between daily travel feature vectors $(m_1(t), n(t), n_{10}(t), n_{25}(t))$ and the infection growth rate vector $r(t)$ from March 01, 2020 to the state's official reopening date; (b) calculate the correlations between daily travel feature vectors $(m_1(t), n(t), n_{10}(t), n_{25}(t))$ and the infection growth rate vector $r(t)$ from the state's official reopening date to July 15, 2020. Note that a lag factor of 21 days is applied to the infection growth rate data. We then examine the difference between the correlation coefficients calculated by step (a) and (b) for Red and Blue states on a cluster-wise basis. The time period for both step (a) and (b) is approximately 60 days, since early May 2020 serves as the mean time during which US states officially reopened.

5   Analysis Results

5.1   Clustering US States

The K-means clustering algorithm is used to cluster the states into five clusters based on their area, population and population density. Note that Washington DC and Alaska are identified as outliers and are not included in the original clustering model, but are added back into our analysis by using the resulting model to assign them to the nearest cluster. The choice of $k = 5$ optimal clusters is empirically determined by using the Silhouette Score and the Elbow Method together to find an appropriate grouping of states, and this clustering is visualized on a map in Fig. 1.

Fig. 1. K-Means clustering of US states into $k = 5$ clusters.

5.2   Correlation Analysis

Correlation analysis is done between the mined transportation patterns and the COVID-19 infection rate for each state. We examine the difference in correlations between Red and Blue states within each cluster for an approximate two month period before and after each state's reopening date.

March 1, 2020 to States Reopening Dates. It is observed that during the quarantine period from March 1, 2020 to the states reopening dates, Red and Blue states both displayed strong positive correlations between their travel feature vectors: $m_1(t)$, $n(t)$, $n_{10}(t)$, $n_{25}(t)$ and infection growth rate vectors, $r(t)$[7]. The bar plot in Fig. 2a visualizes this relationship for the mean value of the correlation within each of the five state clusters. The heat-map in Fig. 2b shows the consistency of the mean correlations calculated between the travel feature vectors $m_1(t)$, $n(t)$, $n_{10}(t)$, $n_{25}(t)$ and the infection growth rate $r(t)$. It becomes readily apparent that travel patterns in both Red and Blue states exhibit a strong correlation with the infection growth rate $r(t)$ across all five state clusters.

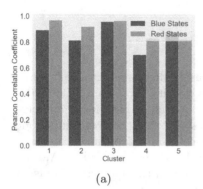

(a)

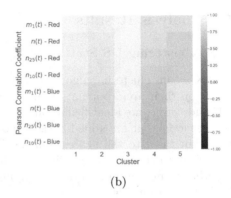

(b)

Fig. 2. (a) Cluster-wise correlations between $m_1(t)$ and $r(t)$ from March 1, 2020 to states reopening dates. (b) Heat-map of correlations between travel feature vectors and $r(t)$ from March 1, 2020 to states reopening dates.

States Reopening Dates to July 15, 2020. After states were officially reopened, correlation analysis between the travel feature vectors $m_1(t)$, $n(t)$, $n_{10}(t)$, $n_{25}(t)$ and the infection growth rate $r(t)$ yielded a very different result for Red and Blue States. These correlations are computed for the time period between the states' official reopening dates to July 15, 2020. We observe that, while travel patterns for Red States continued to display weaker positive correlations with the infection growth rate $r(t)$, there is either no correlation or negative correlations for Blue States within the same clusters. The bar plot in Fig. 3a captures this relationship for the mean value of the correlation within each of the five state clusters. A similar relationship is observed for the correlations of $n(t)$, $n_{10}(t)$, $n_{25}(t)$ and $r(t)$ in this time period. The heat-map in Fig. 3b shows the mean correlations calculated between the travel feature vectors $m_1(t)$, $n(t)$, $n_{10}(t)$, $n_{25}(t)$ and the infection growth rate $r(t)$ for the time period

---

[7] The correlation coefficient of WY is the only one that is not statistically significant ($p > 0.05$).

between the states' official reopening dates to July 15, 2020. From the heat-map, we observe that there is a distinct difference between the correlations coefficients of the Red states and the correlation coefficients of the Blue states, for all four travel feature vectors and across all five clusters. We do see that for cluster 5, this difference is less significant. This could perhaps be attributed to the slower spread of infection in the Midwestern states due to low population density and hence lower community transmission rates.

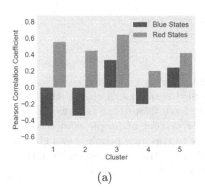 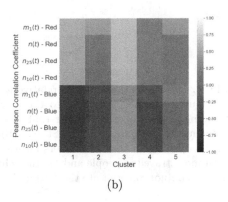

         (a)   (b)

Fig. 3. (a) Cluster-wise correlations between $m_1(t)$ and $r(t)$ from states reopening dates to July 15, 2020. (b) Heat-map of correlations between travel feature vectors and $r(t)$ from states reopening dates to July 15, 2020.

5.3 Further Analysis for Travel Patterns After States' Reopening Dates

To better understand the difference of the correlation coefficients during the periods before and after the states reopening dates, we separate the states into four groups according to the growth pattern of the number of newly confirmed cases and the travel pattern. We apply the K-means algorithm to the relative growth rates of the newly confirmed cases of all the states[8]. Figure 4a shows the patterns of the relative growth rates of newly confirmed cases for both clusters. Green lines represent the states of cluster 1 and red lines represent the states of cluster 2. The dark lines represent the mean relative growth rates of each cluster. The difference of the relative growth patterns of the two clusters is clear where the relative growth rates of the newly confirmed cases of cluster 1 was trending down between May 30 and June 20, while that of the cluster 2 was trending up. Due to the difference of the pattern of relative growth rates, we separate the states into two groups: one contains the states that had a decreasing trend and the other displayed an increasing trend. Similarly, by applying K-means to the percent of people who traveled during reopen, we separate the states into

---

[8] The highest Silhouette Score (0.522) is achieved when the number of clusters is set to two.

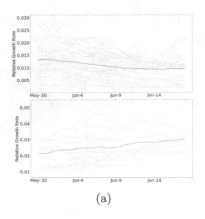
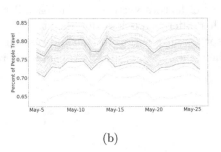

(a)   (b)

Fig. 4. (a) The patterns of the relative growth rate of newly confirmed cases for the two clusters formed according to the newly confirmed cases, i.e. a cluster that was trending up and another cluster that was trending down. Note that here we do not differentiate the Red and Blue States. (b) The daily percent of people who traveled post reopening for the two clusters formed according to traveling people, i.e. a cluster with more traveling people and another cluster with fewer traveling people. Note that here we do not differentiate the Red and Blue States. (Color figure online)

two groups[9]. Figure 4b shows the patterns of the percent of people who traveled during reopening (roughly from May 5 to May 25).

The lines for both clusters are periodic which is expected since people might travel differently on weekdays and weekends. We separate all the states into two groups based on the difference of the travel pattern.

Using this clustering, we split the states into four groups according to the combination of the pattern of relative growth rate and the travel pattern. Figure 5a shows the splitting results. Take the bottom left section as an example: this section represents the states that had a decreasing relative growth rate and a lower percent of people who traveled post reopening. The red states are the states won by Trump in the 2016 presidential election and the blue states are the states won by Clinton. Except for the top right section where the states' relative growth rate were trending up and had a higher percent of people who traveled, the other three sections contain both Blue States and Red States. There are Blue and Red States in the top left and the bottom right sections providing an explanation for the difference of the correlation coefficients during the periods before and after the start of reopening. One the one hand, there are Red States that had a higher percent of people who traveled during reopen and meanwhile they had a decreasing relative growth rate. On the other hand, Blue States such as California, Oregon and Nevada had a lower percent of people who traveled, but they still suffered an increasing relative growth rate.

To obtain a better insight into the patterns of the relative growth rate, we compare the states of the first row and the second row and find that most states in

---

[9] Two clusters yield the highest Silhouette Score (0.569).

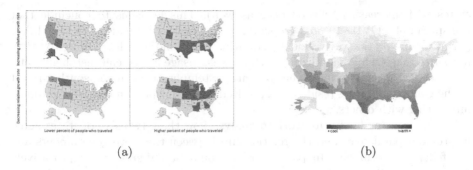

Fig. 5. (a) Combination of the pattern of relative growth rate and the travel pattern. (b) Average temperature by state in May, 2020.

the first row are in the southern part of the country and the majority states in the second row are in the northern part of the country. This leads to a hypothesis that the difference of the relative growth rate could be related to the temperature. Figure 5b shows the average temperature in May 2020[10]. The temperature map roughly corresponds to the patterns of the relative growth rate where in the south, states were warmer and they displayed an increasing relative growth rate, while in the north, states were cooler and they had a decreasing relative growth rate. Previous study has also shown that there is no evidence supporting that case counts of COVID-19 could decline when the weather becomes warmer [12]. We think the relative growth rate increases in the warmer states because the higher temperature could change how people behave and interact outdoors - people in the warmer states might stay outdoors for a longer period of time increasing the chance of interacting with others. In addition, people might also find it less comfortable to wear a mask in warmer weather which potentially reduces the value in curtailing community transmission and the burden of the pandemic [3].

6   Conclusions and Future Work

Through our analysis of the mined transportation data and the daily confirmed cases COVID-19 infection data, we gain a better understanding of the spread of COVID-19 in the US states along the political lines. We observe that during the quarantine period, when stay-at-home orders were imposed nationally, both the Red and Blue states exhibit a similar level of correlation between their travel patterns and infection growth rates. However post-reopening, we observe that the Red states continue to correlate at a significantly higher level than the Blue states. Upon investigating post-reopening travel and infection data further, we find that both sets of states display similar travel patterns (there are Blue

---

[10] https://www.climate.gov/maps-data/data-snapshots/averagetemp-monthly-cmb-2020-05-00?theme=Temperature.

states with increasing travel trends as well). This leads us to conclude that the spread of COVID-19 post-reopening might be less dependent on travel, but perhaps more on the way residents mobilize. This could be related more to safety measures taken while mobilizing, such as mask-usage and social distancing, etc. Another factor explored was that of warmer temperatures in the southern part of the country allowing residents in the Red states to spend more time outdoors interacting with others.

There is scope for future work to analyze how these two sets of states acted in previous pandemics and the relationship between the state governments and the federal government. In particular, it would be valuable to quantitatively measure the adherence to the safety measures when people travel in different states. This could be studied in terms of mask-usage, indoor versus outdoor interactions and safety regulations imposed by different states. It would also be valuable to investigate the spread of COVID-19 at the county level, to understand community transmission and socio-economic factors that could help explain the increase in growth rate for Red states.

References

1. Chinazzi, M., et al.: The effect of travel restrictions on the spread of the 2019 novel coronavirus (Covid-19) outbreak. Science **368**(6489), 395–400 (2020)
2. Dong, E., Du, H., Gardner, L.: An interactive web-based dashboard to track Covid-19 in real time. Lancet Inf. Dis. **20**(5), 533–534. https://doi.org/10.1016/S1473-3099(20)30120-1
3. Eikenberry, S.E., et al.: To mask or not to mask: modeling the potential for face mask use by the general public to curtail the Covid-19 pandemic. Infect. Dis. Model. **5**, 293–308 (2020)
4. Freeden, M.: Reassessing Political Ideologies: The Durability of Dissent. Routledge, Milton Park (2004)
5. Jost, J.T., Federico, C.M., Napier, J.L.: Political ideology: Its structure, functions, and elective affinities. Ann. Rev. Psychol. **60**, 307–337 (2009)
6. Knight, K.: Transformations of the concept of ideology in the twentieth century. Am. Polit. Sci. Rev. 619–626 (2006)
7. Lauer, S.A., et al.: The incubation period of coronavirus disease 2019 (Covid-19) from publicly reported confirmed cases: estimation and application. Ann. Intern. Med. **172**(9), 577–582 (2020)
8. Linka, K., Peirlinck, M., Sahli Costabal, F., Kuhl, E.: Outbreak dynamics of Covid-19 in Europe and the effect of travel restrictions. Comput. Methods Biomech. Biomed. Eng. **23**(11), 1–8 (2020)
9. Ma, J.: Estimating epidemic exponential growth rate and basic reproduction number. Infect. Dis. Model. **5**, 129–141 (2020)
10. Rothgerber, H., et al.: Politicizing the Covid-19 pandemic: Ideological differences in adherence to social distancing (2020)
11. United States Department of Transportation, Research and Innovative Technology Administration: Daily travel during the Covid-19 public health emergency. Bureau of Transportation Statistics (2020)
12. Xie, J., Zhu, Y.: Association between ambient temperature and Covid-19 infection in 122 cities from China. Sci. Total Environ. **724**, 138201 (2020)

Applying an Epidemiological Model to Evaluate the Propagation of Misinformation and Legitimate COVID-19-Related Information on Twitter

Maryam Maleki$^{(\boxtimes)}$ ⓘ, Mohammad Arani ⓘ, Erik Buchholz ⓘ, Esther Mead ⓘ, and Nitin Agarwal ⓘ

COSMOS Research Center, UA – Little Rock, Little Rock, AR 72204, USA
{mmaleki,mxarani,ejbuchholz,elmead,nxagarwal}@ualr.edu

Abstract. Evaluating the propagation of ideas on social media platforms such as Twitter allows researchers to gain an understanding of the characteristics of online communication patterns. The expansion and popularity of Twitter has increased the instances of the propagation of rumors and misinformation all over the world. To distinguish the dynamics of the spread of misinformation and legitimate hashtags related to COVID-19, in this paper, we utilized the SEIZ (Susceptible, Exposed, Infected, Skeptic) epidemiological model. We evaluated the trend of the propagation of misinformation and legitimate hashtags from three different campaigns: lockdown, face mask, and vaccine. Our findings showed that the propagation of misinformation and legitimate hashtags can be modeled by the SEIZ model. Leveraging a mathematical model can lead to an increased understanding of the trends of the propagation of misinformation and legitimate information on Twitter and ultimately help to provide methods to prevent the propagation of misinformation while promoting the spread of legitimate information.

Keywords: COVID-19 · Misinformation · SEIZ model · Mathematical modeling · Twitter

1 Introduction

Due to the growing popularity of online social media, Twitter has become a source of information and idea-sharing for many users all over the world. However, this popularity, along with accessibility, low-barrier for posting, and its crowd-sourced nature, Twitter is continually battling the propagation of misinformation, especially during the global COVID-19 pandemic. Spreading misinformation about healthcare subjects can be more dangerous and can pose a serious threat to people's lives [1]. Multiple types of false information and hoaxes have been propagated on different online social media platforms about COVID-19 topics, such as its causes, consequences, spreading, prevention, medicine, treatment, and vaccination. The preponderance of COVID-19 misinformation

© Springer Nature Switzerland AG 2021
R. Thomson et al. (Eds.): SBP-BRiMS 2021, LNCS 12720, pp. 23–34, 2021.
https://doi.org/10.1007/978-3-030-80387-2_3

is so prolific that it is increasingly referred to as an "infodemic"[1]. Tasnim et al. (2020) showed that misleading information propagates more broadly and more quickly than reliable information on social media, causing harm to the authenticity balance of the information ecosystem. Another study demonstrated that over 25% of the most viewed COVID-19-related YouTube videos included false information and accounted for 62 million worldwide views [2]. A poll by Ofcom[2] in the UK showed that 46% of the people in Britain were exposed to misleading information about COVID-19. Almost 66% reported viewing COVID-19 misinformation on a daily basis, which is a significant issue because repeated exposure can cause people to believe misleading information [3]. This matter can be a major public health concern because being exposed to a lot of information can cause "media fatigue", resulting in the reduction of healthy behaviors that protect from infection [4]. This study is motivated by the need to understand the influence that the propagation of misinformation on social media has on the minds of information consumers, specifically relative to public health issues. To the best of our knowledge, there has been no prior study that has applied the SEIZ epidemiological model to the propagation of COVID-19 misinformation hashtags and compared the trend with the corresponding legitimate hashtags. The main goal of this study is to find a mathematical model that represents the propagation of COVID-19 misinformation and corresponding legitimate information on Twitter during the entire year of 2020.

The remainder of this paper is organized as follows. Section 2 presents the related work that has been done regarding epidemiological modeling for the spread of news, rumors, and misinformation. In Sect. 3, the methodology used in this paper is described, as well as our data collection process. We also briefly discuss the basics of two of the traditional epidemic models (SIS, SIR). These models are then compared to the SEIZ model, which is used in this study. Section 4 discusses the overarching themes and impact of our research. Finally, Sect. 5 concludes the paper with ideas for future work.

2 Literature Review

Applying a mathematical model to evaluate the spread of information on an Online Social Network (OSN) can provide us with the opportunity to acquire effective information toward its propagation. As a result, we can set the stage for developing useful approaches and policies to control this propagation [5]. Bettencourt et al. (2005) contend that quantitative measures derived from the modeling of epidemics and the spread of infections "generalize naturally to the spread of ideas and provide a simple means of quantifying sociological and behavioral patterns." Using the SIR, SEI (Susceptible, Exposed, Infected), and SEIZ models, the authors found values for parameters that they believe generally characterize the spread of ideas. These parameters include the adoption rate, trajectory, and the basic reproductive number (R_0) [6]. Xiong et al. (2012) proposed an online information diffusion model containing four possible states: susceptible, contacted, infected, and refractory (SCIR). The refractory state refers to nodes that cannot

---

[1] https://www.who.int/news/item/23-09-2020-managing-the-covid-19-infodemic-promoting-healthy-behaviours-and-mitigating-the-harm-from-misinformation-and-disinformation.

[2] https://www.ofcom.org.uk/about-ofcom/latest/features-and-news/half-of-uk-adults-exposed-to-false-claims-about-coronavirus.

be infected due to immunity [7]. Rodrigues and Fonseca (2016) demonstrated that the SIR model can be applied to viral marketing techniques. They altered several parameters in their data simulations, finding that greater infectivity rates increase message sharing, but greater recovery rates decrease message sharing [8]. Over the years, different epidemiological models taken from the SIR model were applied to evaluate the propagation of information and rumors in a population [9]. Zhao et al. (2012) proposed a new rumor spreading model they termed SIHR (Susceptible-Infected-Hibernator-Removed). The authors state that the SIHR model is an extension of the traditional SIR model in that it accounts for a direct link from ignorants (similar to Susceptible) to stiflers and adds a new type of user compartment called Hibernators [10]. In another study, Jin et al. (2013) used an epidemiological model to evaluate the propagation of news and rumors on Twitter. The authors applied the SEIZ model to evaluate the diffusion trends of four news items and four rumors on Twitter [5]. Isea et al. (2017) used a version of the SEIZ model to evaluate the spread of a rumor for a specific case in Venezuela. In their model, the rumor propagates between two various scenarios Z_1 and Z_2 that do not share information with each other [11]. Rystrøm (2020) used the SEIZ model to study the propagation of some concepts during the Danish election on 2019 [12].

Gavric and Bagdasaryan (2019) proposed a way of expanding the SI model, suggesting the addition of a fuzzy sharing coefficient. This coefficient is based on an influence power number, which involves evaluating several of a user's characteristics that cause other users to believe a tweet that they decide to share [13]. Holme and Rocha (2019) simulated inaccurate reporting (e.g., misinformation) in networks by randomly altering a fraction of node IDs and contact times in their datasets. They ran SIR models on the original data, then the altered data, to determine disease outbreak and extinction time. They found that the misinformation affected outbreak size and time to extinction in a similar magnitude [14]. In another study, Maleki et al. (2021) leveraged the SEIZ epidemiological model to study the propagation of misinformation on Twitter. They evaluated the misinformation regarding a non-existent blackout in Washington, D.C. during the Black Lives Matter movement in March 2020. Their results showed that the SEIZ model can accurately describe the spread of a specific misinformation campaign on social media [9].

Although previous studies have applied different epidemiological models to the spread of the news, rumors, concepts and misinformation, our work stands out in that it applies the SEIZ model to robustly identify, compare, and contrast the spreading parameters of three different COVID-19-related misinformation campaigns as well as the spreading parameters of the corresponding legitimate information for each of the three misinformation campaigns. Additionally, by concentrating on the "Infected" compartment, this work further contributes to the continued evaluation of the strengths and weaknesses of the SEIZ model.

3   Methodology

This section describes our data collection process. We also briefly discuss some of the fundamental epidemiological models. We end this section with a focused description of the SEIZ model, which was ultimately used as the model for our datasets.

Data Collection and Processing. We used Twitter premium APIs[3] to collect tweets related to COVID-19 for the entirety of 2020. We collected data for different misinformation and legitimate hashtags that could best cover a broad range of topics related to COVID-19. These topics included the following: lockdown, face mask, and vaccine. Because both legitimate and misinformation hashtags were collected for each topic, there were a total of six datasets. The list of hashtags and the number of tweets for every hashtag is shown in Table 1.

Modeling. To evaluate the propagation of legitimate and misinformation hashtags on Twitter, we used an epidemiological model, which divides the population into various compartments. We first explain two different basic epidemiological models, SIS and SIR, then elaborate on the SEIZ model, which is used in this study.

Table 1. Tweet counts for different COVID-19 hashtags

Campaign	Hashtag	Type	# of tweets	Error	Run time (s)
Lockdown	#Lockdownskill	Misinformation	11630	0.115	54.02
	#Lockdownswork	Legitimate	444	0.315	84.90
Face mask	#Nofacemask	Misinformation	740	0.052	131.35
	#Wearafacemask	Legitimate	2200	0.038	102.13
Vaccine	#Novaccineforme	Misinformation	12225	0.140	106.13
	#Vaccinesavelives	Legitimate	305	0.122	43.96

SIS and SIR Model. The SIS model is a basic epidemiological model that divides the population into two compartments: Susceptible (S) and Infected (I). To adapt this model to the idea of the spread of the information on Twitter, we used a new definition for these compartments. A user is Infected if they post a tweet using the specific hashtag (either legitimate or misinformation), and Susceptible if they have not yet posted tweets using the mentioned hashtags [5]. SIR is another epidemiological model with the addition of a Recovered (R) compartment, which means users have not subsequently posted tweets with the specific hashtag within a certain time frame.

SEIZ Model. One important limitation of the basic epidemiological models (SIS and SIR) is that when a Susceptible individual is contacted by an Infected one, there is just one possible action, which is that the individual will become Infected. However, this assumption does not apply properly to the propagation of information, specifically on social media, and particularly on Twitter. Users may have different mindsets when they are exposed to information. In addition, some users can be skeptical of the correctness of the information to which they are exposed and decide to never show any reaction to

---

[3] https://developer.twitter.com/en/products/twitter-api/premium-apis.

tweets with these specific hashtags. These possibilities can happen in reality but are not covered by the basic epidemiological models, such as SIS and SIR. Consequently, we were motivated to apply the SEIZ model to our data, utilizing a stronger model which is more applicable to the propagation of information on Twitter. In the context of analyzing the propagation of information on Twitter, the different compartments of the SEIZ model (Fig. 1) are outlined below.

- Infected (I) relates to users who have used the specific hashtag in their tweets.
- Susceptible (S) represents users who are the followers of the Infected individuals.
- Exposed (E) represents the users who have been Exposed to the tweets containing a specific hashtag and had a delay of time before posting a tweet using the specific hashtag.
- Skeptic (Z) refers to individuals who have encountered the information via a tweet but decide not to tweet and use the hashtag. We assume no a-priori beliefs for skeptics and are simply adopting published terminology [5].

The following system of Ordinary Differential Equations (ODE) represents the SEIZ model [5]. The parameters for the ODEs are defined in Table 2.

$$\frac{dS}{dt} = - \beta S \frac{I}{N} - bS \frac{Z}{N} \tag{1}$$

$$\frac{dE}{dt} = (1-p) \beta S \frac{I}{N} + (1-l)bS \frac{Z}{N} - \rho E \frac{I}{N} - \varepsilon E \tag{2}$$

$$\frac{dI}{dt} = p \beta S \frac{I}{N} + \rho E \frac{I}{N} + \varepsilon E \tag{3}$$

$$\frac{dZ}{dt} = lbS \frac{Z}{N} \tag{4}$$

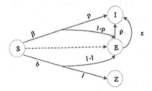

Fig. 1. SEIZ model

Table 2. Parameters of the SEIZ model

Parameter	Definition
B	Contact rate between S and I
B	Contact rate between S and Z
P	Contact rate between E and I
P	Probability of S to I given contact with I
1-p	Probability of S to E given contact with I
E	Transition rate of E to I (Incubation rate)
L	Probability of S to Z given contact with Z
1-l	Probability of S to E given contact with Z

4 Analysis and Results

This section presents the research findings in two parts. First, a preliminary analysis evaluates the frequency of hashtags over time in three different campaigns: lockdown, face mask, and vaccine. Second, the SEIZ model was applied to fit our different datasets to the Infected (I) compartment of the model. Before applying the epidemiological model, we compared the spread of misinformation and legitimate hashtags in three different campaigns: lockdown, face mask, and vaccine (Figs. 4, 5, 6). The figures follow the same format, displaying: (a) the cumulative sum of tweets for misinformation campaigns, (b) the cumulative sum of tweets for legitimate campaigns, (c) the spread of tweets during different months of 2020 for both types of campaigns, and (d) the cumulative sum of tweets for both types of campaigns. The figures include blue lines for legitimate hashtags and red lines for misinformation hashtags.

Lockdown Campaign
The misinformation hashtag #Lockdownskill initially spread in April 2020, demonstrated slow growth until the middle of October 2020, and then sharply increased (Fig. 2a). By contrast the legitimate hashtag #Lockdownswork initially spread in March 2020, a month earlier. It demonstrated slow growth until November, wherein it sharply increased (Fig. 2b). When comparing monthly new tweets, legitimate hashtags demonstrated two small spikes in April and November, but misinformation hashtags demonstrated multiple spikes: one spike in May, many small spikes several months later, and several large spikes in mid-October, mid-November, and December (Fig. 2c). Additionally, the increase in legitimate hashtags in November corresponds to a drop in misinformation hashtags. When viewed on a wider scale, the misinformation campaign was more successful, demonstrating more tweets and a faster rate of spreading. In fact, the legitimate hashtag was almost spread at a constant rate (Fig. 2d).

Vaccine Campaign
While the legitimate hashtag #Vaccinesavelives started spreading in January (Fig. 3b), the misinformation hashtag #Novaccinforme started spreading in late March (Fig. 3a).

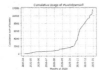

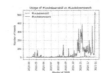

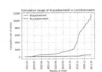

Fig. 2a. Cumulative sum of tweets for #Lockdownskill

Fig. 2b. Cumulative sum of tweets for #Lockdownswork

Fig. 2c. Number of tweets for misinformation and legitimate hashtags in lockdown campaign

Fig. 2d. Cumulative sum of the tweets for misinformation and legitimate hashtags in lockdown campaign

As if in response, use of the legitimate hashtag sharply increased in March, but slowly increased in later months (Fig. 3b). The misinformation hashtag sharply increased afterward, between April and June, then slowly increased in later months (Fig. 3a). When comparing monthly new tweets, the misinformation hashtag demonstrated sharp increases in May, July, and December (Fig. 3c). Because the volume of tweets for the misinformation hashtag dwarfed the legitimate hashtag, the growth in the use of the legitimate hashtag appears negligible (Figs. 3c and 3d). In fact, the final cumulative tweet count for the misinformation hashtag was 12,225 compared to 305 for the legitimate hashtag.

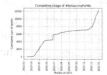

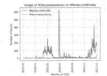

Fig. 3a. Cumulative sum of tweets for #Novaccineforme

Fig. 3b. Cumulative sum of tweets for #Vaccinesavelives

Fig. 3c. Number of tweets for misinformation and legitimate hashtags in vaccine campaign

Fig. 3d. Cumulative sum of the tweets for misinformation and legitimate hashtags in vaccine campaign

Face Mask Campaign

Both hashtags started spreading near to each other in the year (only 10 days apart): February 2nd for #Nofacemask and January 23rd for #Wearafacemask (Table 1). Likewise, they demonstrated sharp increases between April and August (Figs. 4a and 4b). However, when evaluating new monthly tweets, the largest spikes do not overlap. For instance, use of the misinformation hashtag spike twice in July, but use of the legitimate hashtag spike in August, October, and December (Fig. 4c). When viewed on a wider scale, both hashtags display an s-shape. In addition, the volume of legitimate tweets exceeds misinformation tweets (Fig. 4d), which contrasts with the lockdown and vaccine campaigns, where misinformation tweets exceeded legitimate tweets. One possible explanation is that people may be more likely to believe in face masks and to encourage others to wear face masks to prevent spreading the virus.

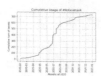
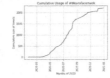
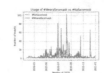
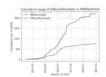

Fig. 4a. Cumulative sum of tweets for #Nofacemask

Fig. 4b. Cumulative sum of tweets for #Wearafacemask

Fig. 4c. Number of tweets for misinformation and legitimate hashtags in face mask campaign

Fig. 4d. Cumulative sum of the tweets for misinformation and legitimate hashtags in face mask campaign

Comparison of All Campaigns

When placing misinformation hashtags on the same scale, three patterns emerge: (1) slow and slight s-shape (#Nofacemask), (2) slow s-shape with sudden spike (#Lockdownskill), and (3) fast s-shape with two spikes, which means that after one s-shape curve is created, another s-shape curve started (#Novaccineforme). Additionally, the Twitter misinformation discourse about vaccine and lockdown appear to be far more popular than the misinformation discourse about face masks (Fig. 5a). When placing legitimate hashtags on the same scale, three patterns emerge: (1) slow and slight s-shape (#Wearafacemask), (2) two s-shaped curves that are still increasing (#Vaccinesavelives), and (3) one slow and one fast s-shaped curve (#Lockdownswork). Also, the twitter legitimate discourse for face mask is far more popular than the legitimate discourse about lockdown and vaccine (Fig. 5b). When all hashtags are placed on the same scale, the use of the misinformation hashtags for lockdown and vaccine clearly set the stage to minimize the use trends of the other hashtags. Also, the use of these hashtags is still increasing and has not yet reached a stable state, while the use of other hashtags is stable in comparison (Fig. 5c).

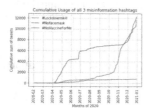

Fig. 5a. Cumulative sum of the tweets for three misinformation campaign

Fig. 5b. Cumulative sum of the tweets for three legitimate campaign

Fig. 5c. Cumulative sum of the tweets for six campaign (three misinformation and three legitimate)

Fitting Datasets to Infected (I) Compartment of the SEIZ Model

We fit the number of Infected people (those users who used the hashtag of interest in each experiment) in each 24-h time interval as the Infected (I) compartment in the SEIZ model by using Matlab. Model fit results for hashtags were graphed in Fig. 6a, 6b, 6c, 6d, 6e, 6f. We used the lsqnonlin[4] function, which is a nonlinear least square curve fitting function in Matlab and utilizes a trust-region reflective algorithm, to fit our model to the different datasets. To solve the ODEs, we used ode45, which is a built-in function of Matlab. Results were obtained from a laptop with Intel Core i5 CPU and 12 GB of RAM coded on Matlab Language 2019b. For every dataset there are a set of optimal parameters which can minimize the error between the real number of tweets (e.g., users of hashtags) and the estimated number of users in the Infected compartment, $|I(t) - tweets(t)|$. Using these parameters can help us to calculate the change of every compartment in the ODE's (dS/dt, dE/dt, dI/dt, dZ/dt) in the specific time interval (24 h in this case study) in real time. Consequently, we can use this mathematical model to predict the future trend of every compartment, and specifically the I compartment, which in this case study are people who spread the specific hashtag.

While the end point for tweets was the same for hashtags, the starting times were different (Table 1). The "error" column in Table 1 displays the difference between the actual number of tweets for every hashtag and the Infected compartment predicted by the SEIZ model, reflecting the relative error in 2-norm [5].

$$\frac{||I(t) \; - \; tweets \, (t)||_2}{||tweets(t)||_2} \tag{5}$$

When comparing misinformation to legitimate hashtags, comparable error values were obtained for #Nofacemask and #Wearafacemask (0.052 vs 0.038), as well as #Novaccineforme and #Vaccinesavelives (0.140 vs. 0.122). In addition, graphs of these hashtags demonstrated similar trends: S-shaped for both #Nofacemask and #Wearaface-mask, and curved power functions for #Novaccineforme and #Vaccinesavelives. So, concerns about face masks and vaccines could be modeled effectively in similar ways, suggesting that their discourse is similar in infection and decay. By contrast, #Lock-downskill and #Lockdownswork did not display similar error rates (0.1158 vs. 0.315) or linear trends. Consequently, the SEIZ model may not be appropriate for modeling the legitimate hashtag #Lockdownswork.

---

[4] https://www.mathworks.com/help/optim/ug/lsqnonlin.html.

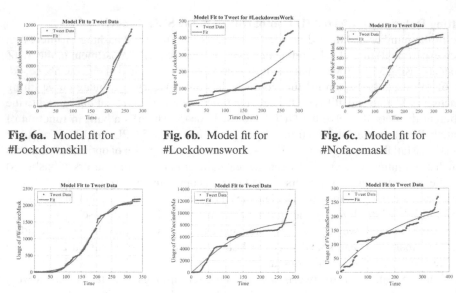

Fig. 6a. Model fit for #Lockdownskill

Fig. 6b. Model fit for #Lockdownswork

Fig. 6c. Model fit for #Nofacemask

Fig. 6d. Model fit for #Wearafacemask

Fig. 6e. Model fit for #Novaccineforme

Fig. 6f. Model fit for #Vaccinesavelives

To summarize, the key findings of this study are:

- The quantity of tweets spreading misinformation hashtags about the topic categories of lockdown and vaccine outnumber the tweets spreading legitimate hashtags for these topic categories. However, tweets spreading legitimate hashtags about the topic category of face masks outnumber the tweets spreading misinformation hashtags.
- For all but one category (face mask), the misinformation campaigns were more successful and dominant than the legitimate information campaigns in terms of spreading the hashtags.
- The SEIZ model can efficiently fit the spreaders of some misinformation and legitimate hashtags more than the others. For example, in the cases where the datasets are more s-shaped (e.g., #Nofacemask and #Wearafacemask), the error between the actual number of tweets and the fitted I compartment is relatively low compared to others (0.052 and 0.038, respectively).

5   Conclusions and Future Work

In this study, we demonstrated how the propagation of misinformation and legitimate hashtags about COVID-19 on Twitter can be modeled by applying the SEIZ model. Such findings present both a strength and limitation to the SEIZ model: while it can be effectively applied to many misinformation and legitimate hashtags, it is not perfect to some of them. Future research will involve applying other epidemiological models to more datasets. Using the mathematical model to evaluate the propagation of misinformation and legitimate hashtags on social media, especially Twitter, can provide the opportunity to study and predict trends of propagation. This evaluation can help policy makers develop suitable strategies for controlling and preventing the propagation of misinformation, as well as increasing the propagation of legitimate information on social media. In the future, we plan to fit the datasets to the other compartments of the SEIZ model, such as the Skeptic (Z) compartment. As a result, we hope to find a way to transfer more users of misinformation hashtags from the Susceptible compartment to the Skeptics compartment which refers to users who decide not to tweet and use the misinformation hashtag. Alternatively, we may be able to find a good strategy to guide people into spreading legitimate information, effectively leading Susceptible to the Infected compartment. Future research also includes applying the SEIZ model and other epidemiological models to different hashtags from different domains, such as politics, healthcare, and religion; we can then compare the accuracy of those different models on datasets. Lastly, we can apply the SEIZ model to datasets collected from other social media platforms, such as Reddit, Facebook, YouTube, WhatsApp, and Instagram.

Acknowledgements. This research is funded in part by the U.S. National Science Foundation (OIA-1946391, OIA-1920920, IIS-1636933, ACI-1429160, and IIS-1110868), U.S. Office of Naval Research (N00014-10-1-0091, N00014-14-1-0489, N00014-15-P-1187, N00014-16-1-2016, N00014-16-1-2412, N00014-17-1-2675, N00014-17-1-2605, N68335-19-C-0359, N00014-19-1-2336, N68335-20-C-0540, N00014-21-1-2121), U.S. Air Force Research Lab, U.S. Army Research Office (W911NF-20-1-0262, W911NF-16-1-0189), U.S. Defense Advanced Research Projects Agency (W31P4Q-17-C-0059), Arkansas Research Alliance, the Jerry L. Maulden/Entergy Endowment at the University of Arkansas at Little Rock, and the Australian Department of Defense Strategic Policy Grants Program (SPGP) (award number: 2020-106-094). Any opinions, findings, and conclusions or recommendations expressed in this material are those of the authors and do not necessarily reflect the views of the funding organizations. The researchers gratefully acknowledge the support.

References

1. Van Bavel, J.J., et al.: Using social and behavioural science to support COVID-19 pandemic response. Nat. Hum. Behav. **4**(5), 460–471 (2020)
2. Li, H.O.-Y., Bailey, A., Huynh, D., Chan, J.: YouTube as a source of information on COVID-19: a pandemic of misinformation? BMJ Glob. Heal. **5**(5), e002604 (2020)
3. Pennycook, G., Cannon, T.D., Rand, D.G.: Prior exposure increases perceived accuracy of fake news. J. Exp. Psychol. Gen. **147**(12), 1865 (2018)
4. Tasnim, S., Hossain, M.M., Mazumder, H.: Impact of rumors and misinformation on COVID-19 in social media. J. Prev. Med. Public Heal. **53**(3), 171–174 (2020)
5. Jin, F., Dougherty, E., Saraf, P., Cao, Y., Ramakrishnan, N.: Epidemiological modeling of news and rumors on twitter. In: Proceedings of the 7th Workshop on Social Network Mining and Analysis, pp. 1–9 (2013)
6. Bettencourt, L.M.A., Cintrón-Arias, A., Kaiser, D.I., Castillo-Chávez, C.: The power of a good idea: quantitative modeling of the spread of ideas from epidemiological models. Phys. A Stat. Mech. Appl. **364**, 513–536 (2006)
7. Xiong, F., Liu, Y., Zhang, Z., Zhu, J., Zhang, Y.: An information diffusion model based on retweeting mechanism for online social media. Phys. Lett. A **376**(30–31), 2103–2108 (2012)

8. Rodrigues, H.S., Fonseca, M.J.: Can information be spread as a virus? Viral marketing as epidemiological model. Math. Methods Appl. Sci. **39**(16), 4780–4786 (2016)
9. Maleki, M., Mead, E., Arani, M., Agarwal, N.: Using an Epidemiological Model to Study the Spread of Misinformation during the Black Lives Matter Movement (2021). arXiv Preprint arXiv:2103.12191
10. Zhao, L., Wang, J., Chen, Y., Wang, Q., Cheng, J., Cui, H.: SIHR rumor spreading model in social networks. Phys. A Stat. Mech. Appl. **391**(7), 2444–2453 (2012)
11. Isea, R., Lonngren, K.E.: A new variant of the SEIZ model to describe the spreading of a rumor. Int. J. Data Sci. Anal. **3**(4), 28–33 (2017)
12. Rystrøm, J.H.: SEIZ Matters. J. Lang. Work. Studentertidsskrift **5**(1), 78–91 (2020)
13. Gavric, D., Bagdasaryan, A.: A fuzzy model for combating misinformation in social network twitter. J. Phys. Conf. Ser. **1391**(1), 12050 (2019)
14. Holme, P., Rocha, L.E.C.: Impact of misinformation in temporal network epidemiology. Netw. Sci. **7**(1), 52–69 (2019)

Optimization of Mitigation Strategies During Epidemics Using Offline Reinforcement Learning

Alina Vereshchaka[✉] and Nitin Kulkarni

Department of Computer Science and Engineering,
State University of New York at Buffalo, Buffalo, USA
{avereshc,nitinvis}@buffalo.edu

Abstract. Emerging infectious diseases affect a large number of people throughout the world. Preventing the spread of viruses and mitigating their adverse societal and economic effects are major challenges facing all institutions in society. When reacting to events occurring in real-time, approaches based on human decision-making systems usually encounter difficulties in sorting out the most efficient mitigation strategies. In this paper, we present the framework for a real-time data-driven decision support tool for policymakers. Our framework is based on a reinforcement learning algorithm meant to optimize governmental responses to the state of the epidemic at each time-step. This framework adapts to changes in epidemic-spread given the advances in disease treatment methods and public health interventions. The mitigation strategy is adjusted based on the government's priorities in a specific region. Our model is validated based on the COVID-19 data collected from New York state, USA.

Keywords: Optimization · Reinforcement learning · Markov decision process · Epidemic model · Sequential decision making · Mitigation regulations · SIR model · Infection disease · COVID-19

1 Introduction

The spread of emerging infectious diseases is increasing in both prevalence and scale. Preventing the spread of viruses and mitigating adverse effects are some of the major challenges facing all institutions in society. Policy-making authorities face high-stake decisions regarding strategies for controlling the spread. The potential harm posed by the spread of emerging infectious diseases has been illustrated in recent history by incidents involving the Spanish influenza, H3N2, H1N1, Ebola, Zika and SARS-Cov-2 (COVID-19)[1,2].

Rapidly evolving epidemics can result in the entrance of a high influx of patients into the healthcare system, stressing clinical and human resources [1].

---

[1] https://www.who.int/emergencies/diseases/en/.
[2] https://www.cdc.gov/flu/pandemic-resources/.

© Springer Nature Switzerland AG 2021
R. Thomson et al. (Eds.): SBP-BRiMS 2021, LNCS 12720, pp. 35–45, 2021.
https://doi.org/10.1007/978-3-030-80387-2_4

Without a timely response, an epidemic can lead to an exponential growth in cases [2]. When reacting to events occurring in real time, approaches based on human decision-making systems usually encounter difficulties sorting out the most efficient mitigation strategy. One of the main reasons for this difficulty is the lack of information about the disease characteristics, the large number of decision factors involved as well as the stochastic nature of the infection spread.

In this paper, we address the challenge of introducing a mitigation strategy in the midst of an epidemic, as the strategy must attempt to lower the spread of an infection while simultaneously minimizing economical loss and maximizing personal freedom and happiness. This real-time framework is based on a reinforcement learning algorithm meant to optimize governmental responses to the state of the epidemic at each time-step, the framework is thus able to adapt to the rapidly changing characteristics of the epidemic spread. This work presents a novel mathematical modeling framework that integrates both outbreak dynamics and outbreak control into a decision-support tool for mitigating the health and social effects of infectious disease outbreaks.

To represent the epidemiological model, we use a modified version of a SIR model, one of the more commonly used compartmental models [3]. This model calculates the amount of people affected by an infectious disease over a time in a closed population. SIR-based models were applied to simulate epidemic processes [4–6] and to model influenza [7]. They were also applied to other domains including that of social media in an attempt to explain the spread of hate-speech on YouTube [8]. Our framework extended a SIR model to consider an additional transition state and the stochastic nature of the disease spread. Reinforcement learning can be applied to solve sequential decision-making problems in dynamic systems with specified objective functions. It has previously been successfully applied to optimize the lockdown policies [9] and to solve complex domains, including that of robotics [10], Atari games, Go, Chess and Shogi [11], and dynamic resource allocation [12].

In this paper, we formalize the problem using the Markov decision process (MDP) framework and apply an offline reinforcement learning algorithm to optimize it. We aim to help policymakers minimize the overall loss associated with the objectives. The model we develop can be applied to the study of optimal policy making strategies that apply different priority levels. The contributions of the paper are summarized as follows: (1) We formulate the SIR-model as the reinforcement learning framework with particular consideration of the stochastic nature of the disease spread; (2) We extend the SIR-model with an additional state to represent the propagation of the infectious disease; (3) We propose the automated method to extract the domain knowledge; (4) We demonstrate empirical performance of reinforcement learning algorithms based on the COVID-19 data from New York, USA.

2   Background

2.1   SIR-Based Model for Epidemic Simulation

An epidemic model is a compartment model that divides a population into different groups with underlined transition dynamics shared between them [13]. In the SIR model the total population of interest (M) is divided into three compartments: individuals who have not been infected (susceptible) X^S, individuals who can infect others (infected) X^I and individuals who have been removed from the infection system (removed) X^R. All individuals in the population of interest belong to one group at a time and assumes similar properties of that group. The deterministic SIR model (Fig. 1a) describes how these groups evolve at each time-step t using the following differential equations: $X_{t+1}^S = X_t^S - \beta(a_t)X_t^S X_t^I/M$, $X_{t+1}^I = X_t^I + \beta(a_t)X_t^S X_t^I/M - \gamma X_{I,t}^I$, $X_{t+1}^R = X_t^R + \gamma X_t^I$, where $\beta(a_t)$ is the infection rate that depends on the control action a_t and γ is the removal rate.

To capture the stochastic nature of the spread of infection we use a Generalised SIR (GSIR) [14] model that adds noise to each of the population group counts:

$$
\begin{aligned}
X_{t+1}^S &= X_t^S - e_t^S, \quad e_t^S \sim \text{Poisson}(\beta(a_t)X_t^S X_t^I/M) \\
X_{t+1}^R &= X_t^R + e_t^R, \quad e_t^R \sim \text{Binomial}(X_t^I, \gamma) \\
X_{t+1}^I &= M - X_{t+1}^S - X_{t+1}^R
\end{aligned}
\tag{1}
$$

2.2   Markov Decision Process

We model the problem in the framework of a Markov decision problem (MDP) which can be formalized by a tuple $(\mathcal{S}, \mathcal{A}, r, p, \gamma)$, where $\mathcal{S}$ denotes the set of states that the agent can move into $s \in \mathcal{S}$ and $\mathcal{A}$ refers to the set of actions that the agent can take $a \in \mathcal{A}$. The reward function $r : \mathcal{S} \times \mathcal{A} \to \mathbb{R}$ determines the immediate reward. The transition probability $p : S \times A \times S$ characterizes the stochastic evolution of states in time $P(s_{t+1}|s_t, a_t)$. The constant $\gamma \in [0, 1)$ is a scalar discount factor. At each time step t, the agent takes action a. The agent learns a policy $\pi : \mathcal{S} \to \mathcal{A}$ that maps states to actions.

The RL objective, $J(\pi)$ can be written as an expectation under the trajectory distribution:

$$
J(\pi) = \mathbb{E}_{\tau \sim p_\pi(\tau)}\left[\sum_{t=0}^{T} \gamma^t r(s_t, a_t)\right]
\tag{2}
$$

3   Methodology

In this section, we first propose an extension to the compartment model in order to make it applicable to recent novel infectious diseases. Following this, we formulate the regulation optimizations as a reinforcement learning framework, and finally we detail our algorithm for solving the problem.

3.1   Extended Model for Epidemic Simulation (SI2R)

We extend the GSIR model (described in Sect. 2.1) to consider a group of individuals who recovered from the disease (X_t^{Rec}). Transition dynamics between the groups contained within the proposed Susceptible-Infected-Recovered-Removed (SI2R) model are presented in Fig. 1b. Transition flow of individuals between the four compartments in consideration of their stochastic nature can be summarized as follows:

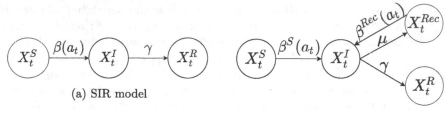

(a) SIR model

(b) Extended SIR model (SI2R)

Fig. 1. (a) Block diagram of the SIR model representing the transition dynamic from the group of individuals who have not been infected (X_t^S) to the group that has been removed from the infection system (X_t^R). (b) Block diagram of the extended SIR model that represents the transition dynamic from the group of individuals who have not been infected (X_t^S) to the group of individuals who have recovered (X_t^{Rec}) or have been removed from the infection system (X_t^R).

$$
\begin{aligned}
X_{t+1}^S &= X_t^S - e_t^S, \quad e_t^S \sim \text{Poisson}(\beta^S(a_t)X_t^S X_t^I / M) \\
X_{t+1}^{Rec} &= X_t^{Rec} + c_t - d_t, \quad c_t \sim \text{Binomial}(X_t^I, \mu), d_t \sim \text{Poisson}(\beta^{Rec}(a_t)X_t^{Rec} X_t^I / M) \\
X_{t+1}^R &= X_t^R + e_t^R, \quad e_t^R \sim \text{Binomial}(X_t^I, \gamma) \\
X_{t+1}^I &= M - X_{t+1}^S - X_{t+1}^{Rec} - X_{t+1}^R
\end{aligned}
\tag{3}
$$

where $\beta^S(a_t)$ and $\beta^{Rec}(a_t)$ are the infection rates that depend on the control action a_t for individuals who have not been infected and individuals who have recovered, respectively; μ is the recovery rate and γ is the removal rate, such that $(\mu + \gamma) \leq 1$.

We consider that the recovered population likely has acquired an immunity [15], thus the probabilities of getting infected are lower compared to the susceptible members of the population, such that $\beta^S(a_t) > \beta^{Rec}(a_t)$.

3.2   Reinforcement Learning for Optimization of Regulations

We use SI2R model proposed in Sect. 3.1 as a compartmental model to mathematically describe the spread of infectious disease. The independent variable is time t, measured in days.

Initial: To initiate the SI2R model simulator, it is necessary to pass initial values of susceptible people (X_0^S), the number of infected people (X_0^I), number of recovered people (X_0^{Rec}) and number of people removed (X_0^R). We also need to extract the domain knowledge including μ, γ, β_j^S and β_j^{Rec} for $j \in \{0, \ldots, A\}$, where A is the total number of actions available in the environment.

Observations: At each time step, the agent receives information about the number of infected individuals X_t^I, the economic and social rate E_t, the previous action a_{t-1}, the penalty associated with a frequent change of action $\phi(a)$. The state at timestep t is $s_t = \{X_t^I, E_t, a_{t-1}, \phi(a)\}$.

Actions: Actions represent the policymaker's response to the current state. Our framework considers two possible actions $\mathcal{A}$: {no_lockdown, lockdown}. During the 'Lockdown' it is assumed that public health interventions, reduction in contact rates and other measures will result in lower infection rates. During 'No lockdown' the infection rate β is assumed to be higher.

Rewards: We design our reward function to motivate the agent to minimize the number of infected people, while keeping the economy operating and maintaining the freedom and happiness of citizens. We also penalize our agent for frequent action changes assuming there is a significant cost associated with frequent change in regulations. Policymakers define the priorities by $\alpha = [\alpha_0, \alpha_1, \alpha_2]$. For certain regions some of the parameters can be omitted by setting $\alpha_i = 0$. We aim to maximize the following objective function:

$$r_t = \alpha_0 \frac{E_t}{E_0} - \alpha_1 \frac{X_t^I}{M} - \alpha_2 \phi(a) \tag{4}$$

where E_t represents the level of the economy and the overall freedom and happiness of the citizens with the initial value of $E_0 = 100$. This parameter is updated in accordance with the action.

Our framework is represented as a flowchart in Fig. 2

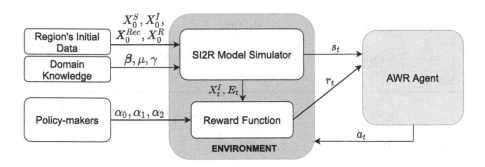

Fig. 2. Mitigation response decision making process. To initiate the framework, it is necessary to pass initial values of susceptible people (X_0^S), the number of infected people (X_0^I), recovered people X_0^{Rec} and removed individuals (X_0^R). It is also necessary to extract the domain knowledge μ, γ and $\beta_j, j \in \{0, \ldots, A\}$, where A is the total number of actions available in the environment

3.3   Offline Reinforcement Learning

Offline reinforcement learning methods extract policies with the maximum possible utility out of the available data, thereby allowing for automation of a wide range of decision-making domains. In our project we use an advantage-weighted regression (AWR) algorithm [16].

Offline RL methods aim to effectively utilize previously collected data, with limited need for additional online data collection. This allows for an effective utilisation of large datasets as powerful decision-making engines. In offline RL, the agent is provided with a static dataset of transitions $\mathcal{D} = \{(s_t, a_t, s_{t+1}, r_t)\}$ and it aims to learn the optimal policy using only this dataset. Here we can regard $\mathcal{D}$ as the training set for policy.

Our methodology can be summarised in Algorithm 1.

Algorithm 1: Epidemic Mitigation Strategy using AWR

Input   : Regional initial data $X_0^I, X_0^S, X_0^R, X_0^{Rec}$; model coefficients μ, β, γ;
           level of priorities for each component of the reward function (α)
Output: policy at time step t
Initialize random policy π_1 with weights θ and replay buffer $\mathcal{D} \leftarrow \emptyset$
for *iteration $k = 1, ..., k_{max}$* **do**
    for *episode $= 1,... episode_{max}$* **do**
        for *$t = 1, ..., t_{max}$* **do**
            From the current policy $\pi_k(s_t; \theta)$ choose action a_t;
            Execute action a_t and estimate the state s_{t+1} following Eq. 3;
            Estimate the reward (r_t) following Eq. 4;
            Add trajectory $\mathcal{T} (s_t, a_t, r_t, s_{t+1})$ to $\mathcal{D}$.
        end
    end
    Compute Monte Carlo returns $R_{s,a}^{\mathcal{D}} = \sum_{t=0}^{T} \gamma^t r_t$;
    Compute the state values $V(s)$;
    $V_k^D \leftarrow \arg\min_V \mathbb{E}_{s,a\sim\mathcal{D}} \left[||\mathcal{R}_{s,a}^{\mathcal{D}} - V(s)||^2 \right]$;
    $\pi_{k+1} \leftarrow \arg\max_\pi \mathbb{E}_{s,a\sim D} \left[\log \pi(a|s) \exp(\frac{1}{\beta}(\mathcal{R}_{s,a}^{\mathcal{D}} - V^{\mathcal{D}}(s))) \right]$.
end

---

4   Experiments

For our experiments we applied our framework to the SARS-CoV-2 (COVID-19) data collected from New York (NY), USA. We demonstrate that our proposed methodology enables optimization under various sets of priorities. We compare the results with DQN-based RL algorithms and NY, USA data from October 1, 2020 till March 12, 2021.

4.1   Dataset Description

We collected daily observations from New York State Government's recorded COVID-19 data for the period of March 12, 2020 till March 30, 2021[3]. Based on the observed cases, we extracted the domain knowledge for the region (Table 1), including the daily infection rate (β) under various regulation constrains, the recovery rate (μ) and the removal rate (γ), all of which we then used with our model to model a real-world simulation of the pandemic. The recovery rate (μ) and removal rate (γ) are dynamic, representing the changes associated with the advances in disease treatment methods and public health interventions.

Table 1. Model parameters based on the COVID-19 NY, USA data

Infection rate				Recovery rate,	Removal rate,
$\beta^S(a_t^l)$	$\beta^S(a_t^n)$	$\beta^{Rec}(a_t^l)$	$\beta^{Rec}(a_t^n)$	μ	γ
0.003238	0.031756	0.0003238	0.0031756	$\mathcal{N}(0.658, 0.364)$	$\mathcal{N}(0.022, 0.009)$

4.2   Results

We run the experiments using the NVIDIA Tesla V100 Tensor Core 16GB GPU. Deep Q-Network (DQN) [17] and Double Deep Q-Network (DDQN) [18] were used to provide benchmark comparisons for our AWR agents. DQN is a value based approximation algorithm based on Q-learning, and DDQN aims to reduce over-estimations by decomposing the max operation in the target into action selection and action evaluation.

To initiate our framework, we considered the following values for all agents in our experiments. The compartment X_0^S took the initial value of 19,453,561 based on the NY state population in 2019 according to CENSUS[4]. The compartments X_0^I, X_0^{Rec} and X_0^R took the initial values of 65,935, 395,584 and 33,266 respectively based on the data reported on October 1, 2020 in NY, USA (see Footnote 3).

We considered a time-frame of 180 days covering the period between October 1, 2020 and March 30, 2021. The comparison between different agents is presented in Fig. 3. Based on the results, we can note that AWR agent outperforms other models resulting in the lowest number of infected and removed individuals by the end of the selected period.

---

[3] https://covid19tracker.health.ny.gov/.
[4] https://www.census.gov/quickfacts/NY.

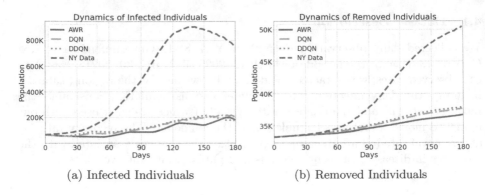

(a) Infected Individuals   (b) Removed Individuals

Fig. 3. Comparison of population dynamics for different balanced agents.

Our framework provides optimal strategies under various sets of priorities. In our experiments, we considered three different agents that prioritize health, economy or maintaining a balanced strategy (results are in Fig. 4).

The AWR Balanced agent (Fig. 4a) learns the strategy to enter lockdown for an extended time period initially to flatten the curve and prevent the exponential growth of the infected individuals. Once the economic and social rate drops to ≈ 60% it introduces the lifting of lock-down in order to keep the economy operating at around 60%. The AWR Health agent (Fig. 4b) learns the strategy to introduce a lockdown for a larger time period as compared to the AWR Balanced agent with the priority of reducing the number of infections albeit at a cost to the economy. The AWR Economy (Fig. 4c) agent learns the strategy to enter an initial lockdown period that is shorter than the other agents as it prioritizes the economy and it keeps the economic and social rate above 60%.

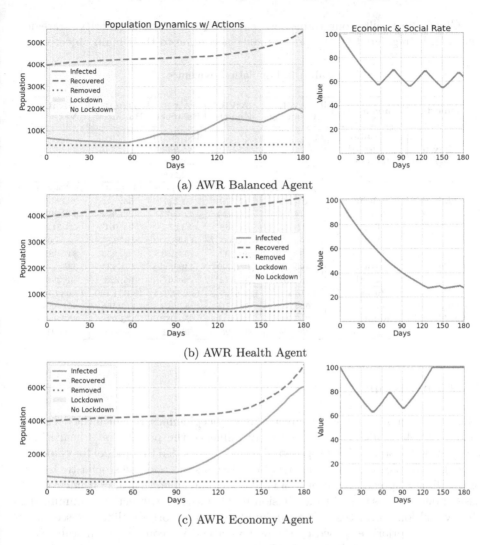

(a) AWR Balanced Agent

(b) AWR Health Agent

(c) AWR Economy Agent

Fig. 4. AWR agents with different priorities.

Results for AWR agents with different priorities, DDQN balanced and real values are summarized in Table 2. The AWR Balanced agent is able to reduce the number of infected people while keeping the economy operating at around 60% of the initial value. The AWR Health agent significantly reduces the number of infected people and towards the end keeps the economy operating at around 30% of the initial value. The AWR Economy agent prioritizes the economy so the economy and social rate does not drop below 60% of the initial value and makes a recovery towards the end of the selected period.

Given the results, our framework can adjust to various priority-strategies that can be regulated by the policy-maker according to their main objectives.

Table 2. Population dynamics

Timestep	Compartment	AWR Economy	AWR Health	AWR Balanced	DDQN Balanced	Real Values
Day 0 10/01/2020	Healthy			18,958,776		
	Infected			65,935		
	Recovered			395,584		
	Death			33,266		
Day 60 11/29/2020	Healthy	18,928,126	18,948,670	18,944,698	18,903,450	18,774,450
	Infected	67,231	47,179	51,117	85,998	202,590
	Recovered	424,287	423,844	423,873	429,952	442,009
	Death	33,917	33,868	33,873	34,161	34,512
Day 120 01/28/2021	Healthy	18,775,763	18,939,977	18,843,977	18,802,552	18,034,076
	Infected	196,779	46,222	134,181	162,955	841,087
	Recovered	445,227	432,690	440,036	451,831	535,198
	Death	35,792	34,672	35,367	36,223	43,200
Day 180 03/30/2021	Healthy	18,072,137	18,888,848	18,684,711	18,664,160	17,684,706
	Infected	604,818	58,976	182,101	173,485	851,888
	Recovered	737,782	470,611	550,087	578,038	867,946
	Death	38,824	35,126	36,662	37,878	49,021

5　Conclusion

This paper is motivated by the adverse effects of epidemics, in which the decisions made by policymakers can significantly influence the population's health and their economic status. We proposed SI2R as an extended SIR model for epidemic propagation to consider an additional transition state and the stochastic nature of the disease spread. We applied an offline reinforcement learning framework to assist policymakers in real-time decision-making with the objective of minimizing the overall long-term costs. We validated our framework on different scenarios based on the priorities posed by the policymakers. We compared our framework against the DQN and DDQN algorithms. The proposed approach shows great performance based on the COVID-19 collected data from NY, USA.

A potential direction for improvement would be to extend the cost function to include more objectives and constraints. We can extend the set of actions to reflect a higher range of strictness-measurement available during the epidemic. This can help the agent to learn more optimal strategies and improve the results.

References

1. World Health Organization, et al.: Pandemic influenza preparedness and response: a WHO guidance document. World Health Organization (2009)
2. Li, Q., et al.: Early transmission dynamics in Wuhan, China, of novel coronavirus-infected pneumonia. New Engl. J. Med. **38**(13), 1199–1207 (2020)

3. Cooper, I., Mondal, A., Antonopoulos, C.G.: A SIR model assumption for the spread of Covid-19 in different communities. Chaos Solitons Fractals **139**, 110057 (2020)
4. Ahmad, R., Xu, K.S.: Continuous-time simulation of epidemic processes on dynamic interaction networks. In: Thomson, R., Bisgin, H., Dancy, C., Hyder, A. (eds.) SBP-BRiMS 2019. LNCS, vol. 11549, pp. 143–152. Springer, Cham (2019). https://doi.org/10.1007/978-3-030-21741-9_15
5. Acemoglu, D., Chernozhukov, V., Werning, I., Whinston, M.D.: Optimal targeted lockdowns in a multi-group SIR model, vol. 27102, National Bureau of Economic Research (2020)
6. Bjørnstad, O.N., Finkenstädt, B.F., Grenfell, B.T.: Dynamics of measles epidemics: estimating scaling of transmission rates using a time series sir model. Ecol. Monogr. **72**(2), 169–184 (2002)
7. Smith, M.C., Broniatowski, D.A.: Modeling influenza by modulating flu awareness. In: Xu, K.S., Reitter, D., Lee, D., Osgood, N. (eds.) SBP-BRiMS 2016. LNCS, vol. 9708, pp. 262–271. Springer, Cham (2016). https://doi.org/10.1007/978-3-319-39931-7_25
8. Obadimu, A., Mead, E., Maleki, M., Agarwal, N.: Developing an epidemiological model to study spread of toxicity on YouTube. In: Thomson, R., Bisgin, H., Dancy, C., Hyder, A., Hussain, M. (eds.) SBP-BRiMS 2020. LNCS, vol. 12268, pp. 266–276. Springer, Cham (2020). https://doi.org/10.1007/978-3-030-61255-9_26
9. Khadilkar, H., Ganu, T., Seetharam, D.P.: Optimising lockdown policies for epidemic control using reinforcement learning. Trans. Ind. Nat. Acad. Eng. **5**(2), 129–132 (2020)
10. Kahn, G., Abbeel, P., Levine, S.L.: Learning to navigate from disengagements. IEEE Rob. Autom. Lett
11. Schrittwieser, J., et al.: Mastering atari, go, chess and shogi by planning with a learned model. Nature **588**(7839), 604–609 (2020)
12. Vereshchaka, A., Dong, W.: Dynamic resource allocation during natural disasters using multi-agent environment. In: Thomson, R., Bisgin, H., Dancy, C., Hyder, A. (eds.) SBP-BRiMS 2019. LNCS, vol. 11549, pp. 123–132. Springer, Cham (2019). https://doi.org/10.1007/978-3-030-21741-9_13
13. Keeling, M., Danon, L.: Mathematical modelling of infectious diseases. Br. Med. Bull. **92**(1), 33–42 (2009)
14. Wan, R., Zhang, X., Song, R.: Multi-objective reinforcement learning for infectious disease control with application to Covid-19 spread. arXiv preprint arXiv:2009.04607 (2020)
15. Dan, J.M., et al.: Immunological memory to SARS-Cov-2 assessed for up to 8 months after infection. Science **371**, eabf4063 (2021)
16. Peng, X.B., Kumar, A., Zhang, G., Levine, S.: Advantage-weighted regression: simple and scalable off-policy reinforcement learning. arXiv preprint arXiv:1910.00177 (2019)
17. Mnih, V., et al.: Human-level control through deep reinforcement learning. Nature **518**(7540), 529–533 (2015)
18. van Hasselt, H., Guez, A., Silver, D.: Deep reinforcement learning with double q-learning. In: Proceedings of the AAAI Conference on Artificial Intelligence, vol. 30, pp. 2094–2100 (2016)

Mining Online Social Media to Drive Psychologically Valid Agent Models of Regional Covid-19 Mask Wearing

Peter Pirolli[1]([⊠]) [ORCID], Kathleen M. Carley[2] [ORCID], Adam Dalton[1] [ORCID],
Bonnie J. Dorr[1] [ORCID], Christian Lebiere[2] [ORCID], Michael K. Martin[2] [ORCID],
Brodie Mather[1] [ORCID], Konstantinos Mitsopoulos[2] [ORCID], Mark Orr[3] [ORCID],
and Tomek Strzalkowski[1] [ORCID]

[1] Institute for Human and Machine Cognition, 40 South Alcaniz Street,
Pensacola, FL 32502, USA
ppirolli@ihmc.org
[2] Carnegie Mellon University, 5000 Forbes Avenue, Pittsburgh, PA 15213, USA
[3] University of Virginia, Charlottesville, VA 22904, USA

Abstract. Understanding how humans respond to an ongoing pandemic and interventions is crucial to monitoring and forecasting the dynamics of viral transmission. Heterogeneous response over time and geographical regions may depend on the individual beliefs and information consumption patterns of populations. To address the need for more precise and accurate epidemiological models we are researching Psychologically Valid Agent models of human responses to epidemic information and non-pharmaceutical interventions during the COVID-19 global pandemic with input drivers induced from sources including online media that provide indicators of pandemic awareness, beliefs, and attitudes.

Keywords: COVID-19 · ACT-R · Twitter · Stance detection

1 Introduction

Understanding how humans respond to an ongoing pandemic and how they respond to interventions is crucial to monitoring and forecasting the dynamics of viral transmission and resulting infected cases and deaths. Important factors such as $R(t)$ (effective reproduction number) have large heterogeneity over time and geographical regions that may depend on the individual beliefs and information consumption patterns of their populations [5]. To address the need for more precise and accurate epidemiological models we are researching novel computational models of human responses to epidemic information and non-pharmaceutical interventions (NPIs) during the COVID-19 pandemic.

Our focus [11] is on modeling beliefs, attitudes, intentions and behavior that are assumed to influence the transmission of infectious disease (e.g., COVID-19). This is because human behavior is a key determinant of viral transmission, and behavior change is crucial during the early stages of pandemics when the only weapons are NPIs

© Springer Nature Switzerland AG 2021
R. Thomson et al. (Eds.): SBP-BRiMS 2021, LNCS 12720, pp. 46–56, 2021.
https://doi.org/10.1007/978-3-030-80387-2_5

[7, 16]. Our approach uses Psychologically Valid Agents (PVAs) implemented in the ACT-R architecture [1], with input drivers induced from heterogeneous sources including online media such as Twitter that provide indicators of pandemic awareness, beliefs, and attitudes [11]. We model awareness-driven behavior changes [15] because humans appear to change their behavior in response to awareness of epidemics, often before government interventions take effect [3]. PVAs are used to model behavior change, such as social distancing and mask wearing in response to information dynamics in the physical and digital environments. We go beyond analysis of online narratives to actually predict their impact on the attitudes and behavior of our citizens, given the heterogeneity of mindsets and information consumption across regions.

We describe specific datasets we are currently using in our modeling mask-wearing behavior in 4 U.S. states. One set of data provides *inputs* to our analytic and modeling pipeline. A second set of data provides empirical measurements of *attitudes and behavioral responses* to the pandemic and NPIs that can serve to test and refine our models and predictions. We present two analyses of Tweets relevant to mask-wearing. Each analysis also results in inputs to the PVA models. The first Twitter analysis concerns pro- vs con-mask-wearing using hashtags. A second Twitter analysis provides a more refined analysis of cognitive content using Natural Language Processing (NLP) techniques. Finally, we present PVA models of mask-wearing attitudes and behaviors and compare model behavior to mask-wearing behaviors in 4 states over the past year.

2 Data

2.1 Mask-Wearing Data

For the study presented in this paper, we use survey data provided by the Covid States Project, which is a 50-state COVID-19 project launched in March 2020 by a multi-university group of researchers.[1] Data collection methods are described in detail in Lazer et al. [9]. The data were collected since April 2020 by PureSpectrum via an online, nonprobability sample, with state-level representative quotas for race/ethnicity, age, and gender. Survey results are conducted in waves approximately every three weeks and aggregated at the state level.

2.2 CMU CASOS/IDeaS COVID-19 Twitter Data

A superset of Tweets based on COVID related keywords in multiple languages were selected for the period of February 2020 to the beginning of December 2020 from a large corpus collected using the Twitter API. Machine learning algorithms were then used to label these tweets by location country, and then again within country by state. The data was selected for four regions are the US states: Pennsylvania, New York, California, and Florida. For only the tweets from the four regions, a more focused subset about mask wearing and social distancing was extracted using additional

---

[1] https://covidstates.org.

keywords in hashtags or text. This subset had either: (a) 'mask', 'shield', 'stay home', 'isolat*', 'stay at home', 'cough' or (b) ('face' or 'mouth') AND 'cover' in the tweets. The mean number of Tweets per day in all four states was 18,257 (median = 14,392; SD = 12,693). Machine learning algorithms were then used to add additional labels including: is it a bot, is it a news agency, and is it a government actor. Finally, using a second location classifier these tweets were then cross-identified by city, where feasible. For each of the states, there were tweets in the four major cities. In addition, there are a few tweets from smaller cities. This data was then de-identified while retaining labels.

We then conducted a high-level stance analysis. For this we identified a set of pro-facemask and anti-facemask hashtags. Then using a label propagation algorithm [8] available in ORA-PRO [2] we labeled each tweet and so each tweeter as being either Contra −1 (anti-facemask) or Pro +1 (pro-facemask), along with the confidence in this label. Tweets not containing hashtags are by definition neutral. We then defined those tweets with stance x confidence > .2 as being Pro, and those with a stance x confidence < −.2 as being Con. For each individual, stance is determined as the average of the stance × confidence across the individual's tweets.

2.3 Twitter Coronavirus (COVID-19) Geo-Tagged Tweets Dataset

We also used a dataset of tweets with geographic location information gathered by a third party and hosted on the IEEE dataport website [13]. This dataset monitors Twitter's data feed and extracts the IDs of tweets that contain any of the 90+ keywords that the curators determined to be relevant to COVID-19.

3   Pro/Con Tweets/Retweets in Four States

Figure 1 shows the volume of Pro vs Con tweets in the CMU Casos/IDeaS dataset for four major cities within each of the four states of interest for February – to beginning of December 2020. Figure 1 shows that Pro tweets vastly outnumber Con tweets. Although there appear to be autocorrelations amongst the time series data, one can also observe that Pro tweet dynamics are not entirely correlated with Con tweet dynamics and there appears to be some location-specific spikes. Closer inspection reveals that the location-specific spikes are often correlated with a local event. For instance, the early spike in the Sacramento data aligns with the police shooting of Stephon Clark in Sacramento, the resulting marches, and tweets about "no masks" at the marches. There are regional differences and it is possible to discriminate activity at the city level.

Questions arise about the factors that drive the volume of Pro/Con tweets regarding masks, over time, by region. Differences in volume across states are largely due to difference in population size in urban settings. It is possible that the variation in regional volumes is driven by what the rest of the US states are tweeting, by case counts in the state, and/or announcements or enactment, state and local. An autoregression analysis addressing this question is presented in Table 1. All states' tweet volumes follow the national volume. California's policy had a significant effect on CA volume. Cases counts in Florida had significant impact on volume. These results suggest that we need to model heterogeneity in what information users attend to by region, possibly due to social and cultural differences.

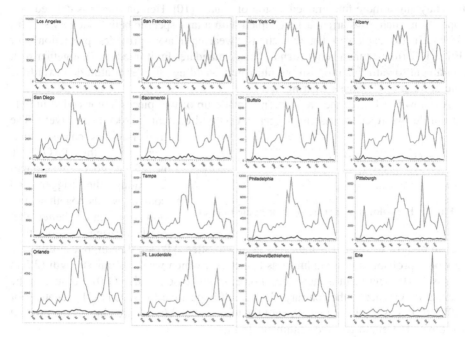

Fig. 1. Covid-related tweet volumes per day for the four major cities in each of the four states (Red = Con; Blue = Pro). (Color figure online)

The Twitter data can be used to instantiate PVA models at the state and city level. However, we cannot assume that this data is necessarily reflecting the authentic chatter of people in these cities. Next steps include separating out the bot, news, and human

Table 1. Autoregression analysis of factors driving volume

State	Regression Pr(> F)	Total R^2 for Autoregression	Parameters: Pr > abs(t)		
			Other states	Case count	Policies
CA	< .0001	0.5851	< .0001	0.3963	**0.0008**
FL	< .0001	0.6210	< .0001	**<.0001**	0.7777
NY	< .0001	0.6732	< .0001	0.4000	0.5350
PA	< .0001	0.4000	< .0001	0.8958	0.6322

activity to provide insight into the drivers of human attitudes as expressed in Twitter.

While stance detection via hashtags provides a high-level indicator of general attitudes it does not provide fine-grained notions, e.g., *belief*, needed to predict action at an individual level. In the next section we describe a finer-grained stance analysis that, taken together with hashtag-based stance detection, enables a richer framework for inducing attitudes.

4 Stance Detection

We have been developing a Natural Language Processing (NLP) pipeline for detecting and analyzing a more fine-grained notion of *stance* [10]. Here, a stance is defined as a proposition, such as wear(a mask), coupled with a task-specific belief category such as PROTECT (to capture a belief such as "wearing masks provides protection") or RESTRICT (to capture a belief such as "mask mandates limit freedom."), together with a sentiment toward this proposition, ultimately yielding an overall attitude toward the topic at hand (e.g., "mask wearing").

As shown in Fig. 2, the proposition is made up of two pieces, color-coded in blue and purple; we refer to these as trigger (wear) and content (mask), respectively. The trigger-content pair is associated with the PROTECT belief category (in orange) indicating the trigger and content words together support a potential belief that mask wearing is protective.

Fig. 2. Breakdown of a stance (Color figure online)

The strength of this belief is indicated in green–often 3.0 as a default–but in this case the word "probably" is used almost as a "hedge", which weakens this strength (2.5 in this case). The sentiment is indicated in blue, in this case positive (1.0), because in the absence of evidence to the contrary, the default is positive for PROTECT, negative for RESTRICT. Lastly, the attitude toward the topic at hand (mask wearing) is the product of the belief strength and the sentiment, in this case 2.50.

We developed stance detection to rapidly discover contrasting views in conversations where different groups champion their own beliefs while attacking others who hold opposing beliefs, causing extreme polarization. For example, messages composed that directly attack anti-mask groups convey beliefs and belief strengths that masks do provide protection, that social distancing does not restrict freedom, and that it's important to show solidarity. This leads to a collective positive attitude toward masks.

The two main constraints applied in this work are: (1) belief categories are semi-automatically generated based on the central the notions of function (e.g., a mask is worn) and purpose (e.g., a mask protects) based on generative lexicon theory [12]; (2) the mapping from natural language input to belief categories relies on a new notion of one sense per domain/content [10], an extension of one sense per discourse in the word sense disambiguation literature [17]. In our work, we adapt this notion to constrain words like "wear" to the PROTECT category given its appearance in the context of the domain-specific content word mask. We note that belief categories are specialized forms of LOSE (e.g., restrict) and GAIN (e.g., protect), as described below, and generally form a small set, i.e., less than 10 per domain.

Our prior work on stance detection [10] compares stance-induced attitudes to state-of-the-art (SoA) sentiment analysis [4], yielding two results: (a) higher accuracy for attitudes produced by stance detection over SoA; and (b) accommodation of more than one output (or perspective) for a given input. Higher accuracy derives from the ability to assign attitude to proposition-level elements in contrast to SoA's single sentence-level assignment. Accommodation of more than one output (perspective) is a byproduct of proposition-level assignments combined with detection of multiple underlying beliefs: "The governor requires mask wearing [but I prefer this for safety reasons]." By contrast, SoA sentiment detection assigns one sentiment overall, whether it is a tweet or a very long sentence, and without regard to (potentially) multiple underlying beliefs.

The comparison above was applied to a representative set of notional tweets related to mask wearing [10]. We focus here on two examples:

- I don't like not wearing masks.
  - Stance Rep: <PROTECT[wear[masks]], 2.75, 1.0>
  - Stance attitude: 2.75; AllenNLP sentiment: −3.0
- The governor requires mask wearing.
  - Stance Rep: <RESTRICT[require[mask]], 3.00, −1.0>
  - Stance Attitude: −3.0; AllenNLP sentiment: −2.0
  - Stance Rep: <PROTECT[wear[mask]], 2.50, 1.0>
  - Stance attitude: 2.5; AllenNLP sentiment: −2.0.

The first is an illustration of higher accuracy: "I don't like not wearing masks" yields a positive stance attitude of 2.75, whereas AllenNLP sentiment yields −3.0, as the double negation is not taken into account. Stance detection is able to determine that the attitude is not negative. Also, "hedge terms" (don't like not) are being used, which softens the belief strength from 3.0 to 2.75, and the product of belief strength and sentiment yields an overall attitude of 2.75. In general, negation (and other related terms like "certainly", "probably", "unlikely", etc.) can be quite complex and cannot be resolved at the sentence level which is what SoA sentiment analysis does, yielding a very negative sentiment of −3.0. Stance detection breaks down the input into smaller chunks (trigger-content pairs), and then applies linguistic constraints at multiple, recursive levels, and this increases accuracy overall.

The second example is an illustration of multi-stance assignment, detection. We interpret the neutral nature of this statement to mean that multiple perspectives are potentially available (e.g., masks are restrictive, masks are protective). We do not artificially constrain stance detection to just one output in computing an attitude value toward masks. Instead, we allow more than one stance to be active, leaving open the possibility that additional context may constrain the final choice of stance output. Specifically, "The governor required mask wearing", is interpreted from one point of view that mask wearing is restrictive, so the stance attitude is negative (−3.0); the other point of view is that masks are protective, so the stance attitude is positive (2.5). Trigger-content pairs are crucial to recognizing the existence of more than one underlying belief, and thus more than one possible stance. By contrast, SoA detection assigns only one value to the statement, without any additional context, in this case a negative sentiment (−2.0); the positive interpretation is ignored.

Our work is further augmented in the context of the Covid-19 domain, through the adaptation of a small set of ontologically organized belief categories, e.g., RESTRICT and PROTECT, derived from broader, domain-independent, ontological categories: LOSE and GAIN. These two ontological categories are expected to carry across domains and tasks, most notably ones we have explored at various levels of readiness, as shown in Fig. 3. These high-level categories are a starting point for semi-automatic

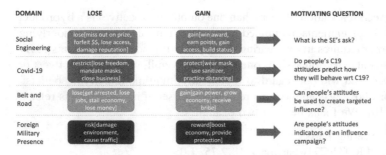

DOMAIN	LOSE	GAIN	MOTIVATING QUESTION
Social Engineering	lose[miss out on prize, forfeit $$, lose access, damage reputation]	gain[win award, earn points, gain access, build status]	What is the SE's ask?
Covid-19	restrict[lose freedom, mandate masks, close business]	protect[wear mask, use sanitizer, practice distancing]	Do people's C19 attitudes predict how they will behave wrt C19?
Belt and Road	lose[get arrested, lose jobs, stall economy, lose money]	gain[gain power, grow economy, receive bribe]	Can people's attitudes be used to create targeted influence?
Foreign Military Presence	risk[damage environment, cause traffic]	reward[boost economy, provide protection]	Are people's attitudes indicators of an influence campaign?

Fig. 3. Ontological belief categories occur across domains

generation of belief categories at opposite ends of the spectrum, for topics that are generally viewed as polarizing (e.g., mask wearing or vaccinations). Choices for specific belief categories are further constrained by domain according to the two constraints described above, within four different domains: social engineering, Covid-19, Belt and Road Initiative, and foreign military presence. Each domain has specific terminology under the heading of LOSE and GAIN and is associated with a different motivating question, as shown on the right-hand side of the figure.

Stance detection has provided a means for analysis of attitudes within different domain-specific datasets, e.g., the IEEE Geotagged dataset for the Covid domain [13]. The specialization of LOSE/GAIN for this domain has led to the identification of two belief categories for the majority of stances detected for this dataset: PROTECT (for GAIN) and RESTRICT (for LOSE). Specifically, PROTECT and RESTRICT make up 76% of the stance outputs, and the remaining 24% are split across 6 additional belief categories (e.g., SPREAD_ILLNESS). This is an indication that the generality of GAIN and LOSE, in terms of data coverage, is quite high, and that PROTECT and RESTRICT are indeed representative of the types of beliefs encountered in the Covid domain.

Future work for stance detection validation will include an extrinsic evaluation through the use of surveys to compare stance detection to reported beliefs in order to determine if detected beliefs are predictive of people's behavior. Moreover, stance detection will be applied to additional domains, thus enabling additional comparison to SoA sentiment analysis, for validation across broader narratives. Stance output provides attitudes, beliefs, etc. to agent simulations which can then be used to model a population. Through this we are able to evaluate stance detection and its effectiveness at taking the pulse of a population with regard to specific topics (i.e., mask wearing). The modeling system is presented in detail below.

5   Psychologically Valid Agents (PVAs)

PVAs simulate people, at multiple levels, with a range of attitudes, beliefs, and credibility assessments that determine their intentions and decisions in response to NPIs. Populations of PVAs will ultimately be embedded in an agent-based modeling (ABM) framework for epidemiological predictions. These models can be tested in multiple ways. For instance, we can use the PVAs to generate and test predictions about differences in epidemiological outcomes arising from the natural experiments across US regions that differ in public health interventions and public mindsets. This can be achieved by embedding PVAs in existing ABM epidemiological models. We can also test forecasts about observable behavior that mediates or moderates viral transmission, such as mobility data or continuous polls of mask wearing behavior as well as attitudes.

The cognitive models of mask wearing behavior discussed here were developed using the ACT-R cognitive architecture [1]. Our models are implemented using the ACT-UP framework [14], which provides flexibility in representing the decisions of a population, community or individual. In this simple simulation, the single model represents the entire population, with the set of instances representing the overall belief system, the action representing the percentage of population engaging in that behavior (mask wearing, in our case), and the outcomes (as well as the situation) representing the system variables resulting from the behavior (e.g., the population level of infection). An individual model would be substantially similar, but involves individual beliefs, the probability of individual action, and individual outcomes. The cognitive architecture provides a computational implementation of a unified theory of cognition that specifies representations and mechanisms for cognitive functions such as perception and attention, memory, decision-making, and action selection. Within the ACT-R architecture we rely on instance-based learning (IBL). IBL [6] specifies that decisions are based on memories of communications and experiences, mediated by cognitive mechanisms.

We have built upon existing ACT-R models to formulate a first approximation model of the dynamics of attitudes and behavior in response to experience and information sources regarding NPIs. These models are generally consistent with traditional psychological theories of decision making, behavior change, and mathematical models of attitude change. There are many advantages that result from using ACT-R to provide a computational formulation of these disparate theories. Amongst these advantages are the integration of multiple factors into a single predictive theory, the modeling dynamics of attitudes and behavior in response to the "dosing" effects of NPI messages, and a foundation for modeling effects of (in)coherence of messaging and sources on credibility.

5.1 Dynamics of Attitudes and Infection

We explore the PVA modeling of the dynamics of attitudes and infection. We assume that normative (socially influenced) attitudes reflect inputs from the PVA's local and global information network, social/political affiliations and polarization dynamics.

As discussed above, intentions (intended actions) are formalized as ACT-R chunks—for example, *wear-mask* vs *not-wear-mask* intended

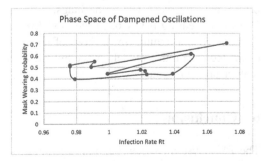

Fig. 4. Dynamics of attitude and infection as predicted by ACT-R PVAs

actions. This representation of competing actions fits the IBL approach and is con- sistent with many microeconomic or decision-theoretic analysis of discrete choice behavior. The underlying base-level activation of the competing alternative action chunks drives action choice and determines the probability of action choice.

Our PVAs capture a signature temporal phenomenon observed in virtually all regions: the damped oscillation pattern of the effective transmission rate (R_t). That is, there is typically a rapid decline from $R_t > 1$ as people react to the initial spread of the virus, followed by an oscillation around $R_t = 1$. The PVAs effectively provide a psychologically-based approximation of a Proportional-Integral-Derivative (PID) con- trol system, which can inform control-theory approaches to on-off imposition of NPIs.

The ACT-R PVA dynamics produce similar predictions of damped oscillation of infection rates as observed in the phase space diagram in Fig. 4. These arise because of feedback delays in information propagation and aggregation as well as responses to external controls (e.g., mandates, weather) and competing forces ("hope" vs "fear"). High mask wearing compliance reduces initial infection rates. Lower infection rates lead to a relaxation of mark wearing. Lower mask wearing leads to rises in infection rate. Rising infection rates results in increased mask wearing. The PVAs include representations for situation/context features that capture situation dynamics in the form of a time-integrated version of R_t reflecting aggregation of information in memory.

5.2   Match to Data

We have developed PVAs that inte- grate inputs from the CASOS Pro/Con analysis and Stance analysis to make predictions that can be matched to mask wearing survey data. Figure 5 presents preliminary validation against mask wearing data from four large states (CA, FL, NY, PA) in the CovidStates dataset. In the PVA model, memory strengths of norms result from rehearsal reflecting the ratio of pro- and con-mask Twitter hashtags.

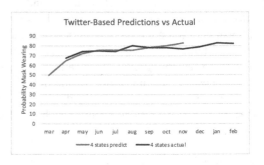

Fig. 5. Mask wearing aggregate validation.

The mask-wearing behavior of the model reflects forgetting and learning mechanisms in memory that exhibit power laws of recency and frequency. Only one parameter is estimated to fit the model, reflecting an estimate of "social cohesion" controlling winner-take-all dynamics.

6   Conclusion

The PVA system is now beginning to input Twitter and NLP analyses and integrate three levels of modeling. Pro- and anti-mask wearing behavior is driven by prevalence of social norms in a context-free process. Attitude-driven reactions are context-sensitive to perceptions of pandemic expansion or contraction ($\sim R_t$), and driven by affective reactions such as "hope" and "fear." Decision-making under risk and uncertainty is driven by a combination of risk messaging and direct experience, and reflects sampling dynamics, e.g., risk aversion and confirmation bias. The three levels of reactive, affective and reflective decision making are combined in a single IBL model.

Acknowledgements. Authors 2-10 are in alphabetical order. Work on this material is supported by the National Science Foundation Grant No. 2033390 and the Office of the Director of National Intelligence (ODNI), Intelligence Advanced Research Projects Activity (IARPA), via 2020-20092500001.

References

1. Anderson, J.R., Bothell, D., Byrne, M.D., Douglass, S., Lebiere, C., Qin, Y.: An integrated theory of the mind. Psychol. Rev. **111**(4), 1036–1060 (2004)
2. Carley, K.M.: ORA: a toolkit for dynamic network analysis and visualization. In: Alhajj, R., Rokne, J. (eds.) Encyclopedia of Social Network Analysis and Mining, pp. 1–10. Springer , New York (2017). https://doi.org/10.1007/978-1-4614-7163-9_309-1
3. Christakis, N.A.: Apollo's Arrow. Hachette Book Group, New York (2020)
4. Gardner, M., et al.: AllenNLP: a deep semantic natural language processing platform. arXiv: 1803.07640 (2017)
5. Gollwitzer, A., et al.: Partisan differences in physical distancing are linked to health outcomes during the COVID-19 pandemic. Nat. Hum. Behav. **4**(11), 1186–1197 (2020)
6. Gonzalez, C., Lerch, J.F., Lebiere, C.: Instance-based learning in dynamic decision making. Cogn. Sci. **27**, 591–635 (2003)
7. Harvey, A.G., Armstrong, C.C., Callaway, C.A., Gumport, N.B., Gasperetti, C.E.: COVID-19 prevention via the science of habit formation. Curr. Dir. Psychol. Sci (2021). https://doi.org/10.1177/0963721421992028
8. Kumar, S., Carley, K.M.: Tree LSTMs with convolution units to predict stance and rumor veracity in social media conversations. In: Proceedings of the 57th Conference of the Association for Computational Linguistics, pp. 5047–5058 (2019)
9. Lazer, D., et al.: The COVID states project #26: trajectory of COVID-19-related behaviors (2021)
10. Mather, B., Dorr, B., Rambow, O., Strzalkowski, T.: A general framework for domain-specialization of stance detection. In: The International FLAIRS Conference Proceedings (2021)
11. Pirolli, P., Bhatia, A., Mitsopoulos, K., Lebiere, C., Orr, M.: Cognitive modeling for computational epidemiology. In: 2020 International Conference on Social Computing, Behavioral-Cultural Modeling & Prediction and Behavior Representation in Modeling and Simulation (SPB-BRIMS 2020). Springer, Washington, DC (2020)
12. Pustejovsky, J.: The Generative Lexicon. The MIT Press, Cambridge (1995)
13. Rabindra, L.: Coronavirus (COVID-19) geo-tagged tweets dataset
14. Reitter, D., Lebiere, C.: Accountable modeling in ACT-UP, a sclable, rapid-prototyping ACT-R implementation. In: Salvucci, D.D., Gunzelmann, G. (eds.) Proceedings of the 10th International Conference on Cognitive Modeling, Drexel, Philadelphia, PA, pp. 199–204 (2010)

15. Weitz, J.S., Park, S.W., Eksin, C., Dushoff, J.: Awareness-driven behavior changes can shift the shape of epidemics away from peaks and toward plateaus, shoulders, and oscillations. Proc. Natl. Acad. Sci. **117**(51), 32764 (2020)

16 West, R., Michie, S., Rubin, G.J., Amlôt, R.: Applying principles of behaviour change to reduce SARS-CoV-2 transmission. Nat. Hum. Behav. **4**(5), 451–459 (2020)

17. Yarowsky, D.: Unsupervised word sense disambiguation rivaling supervised methods. In: Proceedings of the ACL, Cambridge, MA, pp. 189–196 (1995)

Fine-Grained Analysis of the Use of Neutral and Controversial Terms for COVID-19 on Social Media

Long Chen[✉], Hanjia Lyu, Tongyu Yang, Yu Wang, and Jiebo Luo

University of Rochester, Rochester, NY 14627, USA
{lchen62,tyang20}@u.rochester.edu, hlyu5@ur.rochester.edu,
w.y@alum.urmc.rochester.edu, jluo@cs.rochester.edu

Abstract. During the COVID-19 pandemic, 'Chinese Virus' emerged as a controversial term for coronavirus. To some, it may seem like a neutral term referring to the physical origin of the virus. To many others, however, the term is in fact attaching ethnicity to the virus. While both arguments have justifications of their own, quantitative analysis of how these terms are being used in real life is lacking. In this paper, we attempt to fill this gap with fine-grained analysis. We find that tweets with a controversial term can be easily distinguished from those with neutral terms using state-of-the-art classifiers, that they cover different substantive topics, and that they possess different linguistic features reflecting sentiment and psychological states. All evidence suggests that the real life use of the controversial terms is distinctively different from that of the neutral terms.

Keywords: COVID-19 · Social media · Twitter · Controversial term · Linguistic analysis · Classification

1 Introduction

Since late 2019, the COVID-19 pandemic has rapidly impacted over 200 countries, areas, and territories. As of September 4, according to the World Health Organization (WHO), 26,121,999 COVID-19 cases were confirmed worldwide, with 864,618 confirmed deaths[1]. This disease has tremendous impacts on people's daily lives worldwide.

In light of the deteriorating pandemic situation in the United States where high positive case counts and death counts were observed, discussions of the pandemic on social media have drastically increased since March 2020 [7]. Within these discussions, an overwhelming trend is the use of controversial terms targeting Asians and, specifically, the Chinese population, insinuating that the virus originated in China. On March 16, the former President of the United States,

---

[1] https://www.who.int/docs/default-source/coronaviruse/situation-reports/wou-4-se
ptember-2020-approved.pdf?sfvrsn=91215c78_2.

© Springer Nature Switzerland AG 2021
R. Thomson et al. (Eds.): SBP-BRiMS 2021, LNCS 12720, pp. 57–67, 2021.
https://doi.org/10.1007/978-3-030-80387-2_6

Donald Trump, posted on Twitter calling COVID-19 the "Chinese Virus". Around March 18, media coverage of the term "Chinese Flu" also took off. Although most public figures who used the controversial terms claimed them to be non-discriminative, such terms have stimulated racism and discrimination against Asian-Americans in the US, as reported by New York Times[2], the Washington Post[3], the Guardian[4], and other main stream news media. A recent work was done with social media data to characterize users who used controversial or non-controversial terms associated with COVID-19 and found the associations between demographics, user-level features, political following status, and geo-location attributes with the use of controversial terms [7].

In contrast to the above previous study that focuses on analyzing the *metadata* of the related tweets, we ask fine-grained questions: are there differences in terms of language, topic, and sentiment between tweets with controversial and those with neutral terms? In answering this question, we make the following contributions:

1. We use state-of-the-art transformer-based models to validate that tweets with controversial terms are linguistically different from those with neutral terms, with a high F1 score of 0.95.
2. We then use LDA models [1] to learn topic distribution and show that controversial terms are more concentrated in topics on racism and on the Chinese government.
3. We analyze LIWC features [11] and demonstrate that controversial terms differ from neutral terms in time orientation, sentiment, and personal concerns.

2   Related Work

Our work builds on previous works on text mining using data from social media during influential events. Studies have been conducted using topic modeling, a process of identifying topics in a collection of documents. The commonly used model, Latent Dirichlet Allocation (LDA), provides a way to automatically detect hidden topics in a given number [1]. Previous research has been conducted on inferring topics on social media. Chen et al. [3] found LDA-generated topics from e-cigarette related posts on Reddit to identify potential associations between e-cigarette uses and various self-reporting health symptoms.

A large number of studies were performed with Linguistic Inquiry and Word Count (LIWC), an API[5] for linguistic analysis of documents. Tumasjan et al. [13] used LIWC to capture the political sentiment and predict elections with Twitter. The API was also used by Zhang et al. [17] to provide insights into the sentiment of the descriptions of crowdfunding campaigns.

---

[2] https://www.nytimes.com/2020/03/23/us/chinese-coronavirus-racist-attacks.html.
[3] https://www.washingtonpost.com/nation/2020/03/20/coronavirus-trump-chinese-virus/.
[4] https://www.theguardian.com/world/2020/mar/24/coronavirus-us-asian-american s-racism.
[5] https://www.liwc.wpengine.com/.

Previous studies have attempted to make textual classification on social media data. Mouthami et al. [9] implemented a classification model that approximately classifies the sentiment using Bag of words in Support Vector Machine (SVM) algorithm. Huang et al. In addition, a number of other studies performed textual classifications for various purposes using social media data, including the use of pre-trained language models (e.g. BERT) [2,6,10,16].

3 Data and Methodology

In this section, we describe data collection, pre-processing and methods to analyze data.

3.1 Data Collection and Pre-processing

We adopt the Twitter dataset used by the previous study in [7], with the controversial dataset (CD) and the non-controversial dataset (ND) from March 23 to April 5. The controversial keywords consist of 'Chinese virus" and '#ChineseVirus", whereas non-controversial keywords include 'corona", 'covid-19", 'covid19", 'coronavirus", '#Corona" and '#Covid 19". We remove any tweet that contains both controversial keywords and non-controversial keywords in the best effort to separate 'controversial tweets' from non-controversial ones. In total, 2,607,753 tweets for CD and 69,627,062 tweets for ND are collected. We then randomly sample datasets into 3 sizes (2M, 500K, and 100K), respectively, with perfect balance between CD and ND. For preprocessing, URLs, emails, and newlines are removed, as they are not informative for language analysis.

3.2 Classification

We design a classification task to see if the context of tweets can provide sufficient clues to differentiate CD and ND tweets, *in the absence of the keywords*[6] *in question*. Using 'COVID-19', the official non-controversial term, as our anchor, we test whether 'Chinese Virus' is equivalent to 'COVID 19' in its real-world usage in the sense that they are interchangeable. To operationalize this idea, we first mask out all appearances of our streaming keywords for CD and ND. Our idea of masking comes from the original BERT paper where masked language modeling is used to train the BERT model from scratch [4]. We mask out the neural terms and controversial terms of COVID-19. We illustrate our masking with the following sample tweet:

'*The Chinese virus, originated in Wuhan, has killed thousands of people.*"
  We replace all streaming keywords with a token '[MASK]", as follows, respectively for the two tweet:
  '*The [MASK], originated in Wuhan, has killed thousands of people.*"

---

[6] The streaming keywords for data streaming purposes.

Next, state-of-the-art textual classification models, including BERT [4] and XLNet [14], are fine-tuned. We also include the Bi-LSTM model as a baseline.

Classification Datasets. Since we want to compare the significance of corpus size on classification performance, all three sizes of datasets are used for the models. First, we merge CD and ND datasets for all sizes of corpus respectively to form perfectly balanced datasets (50:50 ratio). Next, a number of words and phrases need to be masked to prevent data leakage. Therefore, the aforementioned streaming keywords are removed from CD and ND, respectively. We also remove hashtags from the dataset, as hashtags contain concise but focused meaning and can potentially be the trigger words for classifications. The datasets are then converted into model-compatible formats. A 90-10 split is made on datasets of different sizes to form training and development sets. We make a universal testing set by splitting the big dataset into a 80-10-10 distribution for training, development and testing set, respectively, so that the evaluation metrics are comparable between different datasets.

3.3   Latent Dirichlet Allocation

We use Latent Dirichlet Allocation to extract topics from the tweets in CD and ND. To see the difference in discussed topics between CD and ND, we merge the CD and ND dataset to generate one LDA model. We utilize our medium dataset (500K) from both CD and ND, resulting in a 1-million dataset. Optimal number of topics ($num\_topics = 8$) is found with the objective of maximizing the coherence score of $C_v = 0.378$[7]. Since the objective of topic modeling is to find what people talk about when using controversial or non-controversial terms, we mask all the appearances of the aforementioned streaming keywords by deleting them out of the dataset. We then perform a comparative analysis to find topics that have significantly more tweets from CD or ND, and then discover differences and similarities among topics.

3.4   LIWC2015

Linguistic Inquiry and Word Count (LIWC2015) is used to extract the sentiment of the tweets of CD and ND. LIWC2015 is a dictionary-based linguistic analysis tool that can count the percentage of words that reflect different emotions, thinking styles, social concerns, and capture people's psychological states[8]. We focus on 4 summary linguistic variables and 12 more detailed variables that reflect psychological states, cognition, drives, time orientation, and personal concerns of the Twitter users of both groups. We follow the similar methodology used by Yu et al. [15] by concatenating all tweets posted by the users of CD and ND,

---

[7] C_v is a performance measure based on a sliding window, one-set segmentation of the top words and an indirect confirmation measure that uses normalized pointwise mutual information (NPMI) and the cosine similarity.

[8] https://liwc.wpengine.com/how-it-works/.

respectively. One text sample is composed of all the tweets from the aforementioned sampled dataset of CD, and the other is composed of all the tweets from that of ND. We then apply LIWC2015 to analyze these two text samples. In the end, there are 16 linguistic variables for the tweets of both groups.

4 Empirical Results

4.1 Classification

Textual classification for predicting whether a tweet uses controversial terms associated with COVID-19 is used to test whether neutral terms and their controversial counterparts are linguistically interchangeable. We assume that a low classification accuracy can be interpreted as easy interchangeability of the two groups of terms, supporting the view that controversial terms such as 'Chinese Virus' is simply 'COVID-19' plus its origin, vice versa. Classification results are reported in Table 1. For robustness, we use three models with different discriminating powers and three sizes of dataset. We use F1 score evaluation metric.

Based on our experiments, the highest F1 score we are able to achieve is 0.9521 by XLNet with 500K training samples. Substantively, this high accuracy indicates a low interchangeability between the two groups of terms and supports the view that 'Chinese Virus' is not a straightforward substitute for 'COVID-19'.

Table 1. F1 scores of the classification models with different dataset sizes. No attempt is made with XLNet on the 2M dataset due to limited computing power.

	100K	500K	2M
Bi-LSTM	0.6723	0.6831	0.7050
BERT-Base, Cased	0.8734	0.9136	0.9302
XLNet-Base, Cased	**0.9499**	**0.9521**	–

4.2 LDA

The eight topics generated by the LDA model are reported in Table 2, with the top 20 topic words in each topic. We first discover the distribution of documents associated with each topic by averaging the weight of documents for each topic, as shown in Fig. 1. Proportion z-test is then performed to find significant difference between CD and ND for each topic. We identify that topic 2, 3, and 4 are discussed significantly more in CD, while topic 5, 6, and 7 are discussed more significantly in ND. No significant difference is discovered for topic 1 and 8 between the two datasets.

Next, we manually assign each topic a topic name to generalize what would most likely be discussed under the topic by looking at topic words that either explicitly refer to a person or an entity that is closely associated with the pandemic, or contain significant emotion or opinion. We then analyze the topics

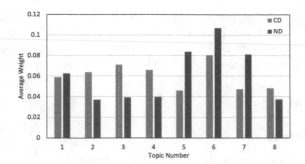

Fig. 1. Average weights of documents of CD and ND for topics. Topics 2, 3 and 4 have significantly higher weights of CD, while Topics 5, 6 and 7 have higher weights of ND. No difference is discovered for Topics 1 and 8.

with significant discussion from only one dataset but not both, and compare such topics to evaluate different topics in the discussion of CD and ND tweets.

We first consider topics that are CD dominant. Topic 2 (Lie and Racism) is CD dominant and contains very strong opinion words, such as 'lie" and 'racist". It also has 'government", 'let", and 'kill" as keywords, indicating a likely discussion about how government's decisions in response to the virus jeopardized people's health. As another CD dominant topic, topic 4 (Chinese Government and the Virus Outbreak) contains more specific keywords, including 'chinese", 'government", 'must" and 'responsible", which indicate how the Chinese government must be held responsible for the spread of the pandemic. It also contains more opinionated keywords such as "propaganda" and 'cover", suggesting misinformation shared by the Chinese government and, to some degree, misendeavor by the government to 'cover up" the situation in the early stage of the pandemic. The other CD dominant topic, topic 3 (Doctors Fighting the Virus) is the most neutral of the three, with focuses on 'doctor", 'hospital", and 'combat" against COVID-19.

On the other hand, the ND dominant topics tend to be more factual. In topic 5 (health workers), most keywords are about doctors and health workers trying to give 'medical", 'support", or 'help" to the patients. In topic 6[9] (stay home), few meaningful keywords associated with COVID-19 are found, except 'Trump" and 'home". In topic 7 (Test, Cases and Death), a large number of keywords are about testing and positive cases (e.g. 'test", 'case", 'report", 'new", 'positive", 'confirm", and 'total"), along with 'death".

One finding is that all three CD dominant topics contain the topic word 'chinese" in the top keywords, even though we have removed all keywords/phrases related to 'Chinese virus" in the documents for LDA. This suggests that the discussions in CD are closely related to China or the Chinese people/government.

---

[9] The titling of this topic is rather difficult, as most keywords do not contain significant meaning associated with COVID-19. We finally choose 'Stay Home' as the topic because a keyword 'home' is in the topic.

Table 2. Topics generated by the LDA model with the top 20 topic words. Topics in blue have significantly more CD tweets, while those in orange have significantly more ND tweets.

Topic title	Topic words
1. Trump and Economy	virus, make, trump, chinese, say, bill, pandemic, people, crisis, use, pay, economy, vote, stop, country, video, president, take, day, grow
2. Lie and Racism	virus, call, people, stop, pandemic, kill, infect, let, originate, together, pig, chinese, lie, racist, must, take, government, spread, need, help
3. Doctors Fighting the Virus	virus, chinese, world, fight, doctor, spread, hospital, say, try, trump, call, new, know, still, clear, let, time, save, combat, treatment
4. Chinese Government and Virus Outbreak	virus, world, spread, chinese, must, outbreak, much, humanity, government, people, logical, responsible, know, propaganda, tweet, drag, strip, cover, news
5. Health Workers	help, need, get, thank, test, work, health, well, support, medical, good, worker, speak, state, fight, stay, home, strong, positive, night
6. Stay Home	virus, say, go, come, know, call, get, take, people, time, keep, trump, think, see, home, make, life, work, outside, meet
7. Test, Cases and Death	virus, case, death, test, people, report, die, new, positive, day, number, say, week, break, world, confirm, country, total, state, last
8. Anecdotes and Reports	say, man, send, hospital, right, pass, go, chinese, take, handle, medium, situation, donate, today, give, agree, month, deadly, cases, rate

In addition, two of the three CD dominant topics have keyword 'government", while none in ND dominant topics except one keyword 'Trump" in topic 6. Such difference indicates that the discussion of ND is more factual, while that of CD tends to be more political.

These discrepancies in the topic modeling result contradict the claim of 'only referring to the geo-locational origin of the pandemic" by some public figures who employed the use of 'Chinese virus" when referring to COVID-19.[10] Nevertheless, such words have provoked, to a certain degree, racist or xenophobic opinions and hate speeches towards China or people of Chinese ethnicity on social media.

4.3 LIWC Sentiment Features

Figure 2a shows four summary linguistic variables for CD and ND. We observe that the clout scores for CD and ND are similar. A high clout score suggests

---

[10] https://www.youtube.com/watch?v=E2CYqiJI2pE.

that the author is speaking from the perspective of high expertise [11]. At the same time, analytical thinking, authentic and emotional tones scores for ND are higher than those for CD. The analytical thinking score reflects the degree of hierarchical thinking. A higher value indicates a more logical and formal thinking [11]. A higher authentic score suggests that the content of the text is more honest, personal, and disclosing [11]. The emotional tone scores for CD and ND are both lower than 50, indicating that the overall emotions for CD and ND are negative. This is consistent with our expectation. However, the emotional tone score for ND is higher than that for CD, indicating that the Twitter users in ND are expressing relatively more positive emotion.

Table 3. Scores of 'i", 'we", 'she/he", 'they", and present orientation.

Variables	CD	ND
i	0.96	1.04
we	1.25	1.00
she/he	0.69	0.70
they	1.05	0.71
present orientation	9.37	9.22

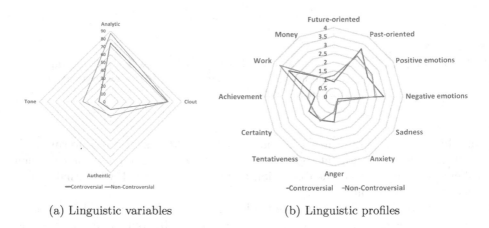

(a) Linguistic variables   (b) Linguistic profiles

Fig. 2. Linguistic variables and profiles for the tweets of CD/ND.

Figure 2b shows 12 more detailed linguistic variables of the tweets of CD and ND. The scores of 'future-oriented" and 'past-oriented" reflect the temporal focus of the attention of the Twitter users by analyzing the verb tense used in the tweets [12]. The tweets of ND are more future-oriented, while those of CD are more past-oriented. To better understand this difference, we conduct a similar analysis as in a previous study by Gunsch et al. [5]. We extract 5 more linguistic variables, including four pronoun scores and a one-time orientation score. The scores of 'i", 'we", 'she/he", 'they", and present-orientation are shown in Table 3. The tweets of CD show more other-references ('they"), whereas more self-references ('i", 'we") are present in the tweets of ND. The scores of 'she/he" of CD and ND are close. The score of present orientation of CD is higher than that of ND. From this observation (similar to the findings of Gunsch et al. [5]), we can infer that the tweets of CD focus on the past and present actions of the others, and the tweets of ND focus on the future acts of themselves. Research shows that LIWC can identify the emotion in language use [12]. From the aforementioned discussion, the tweets of both CD and ND are expressing negative emotion, and the emotion expressed by the Twitter users of ND is relatively more positive. This is consistent with the positive emotion score and negative emotion score.

However, there are nuanced differences across the sadness, anxiety, and anger scores. When referring to COVID-19, the tweets of ND express more sadness and anxiety than those of CD do. More anger is expressed through the tweets of CD. The certainty and tentativeness scores reveal the extent to which the event the author is going through may have been established or is still being formed [12]. A higher percentage of words like 'always" or 'never" results in a higher score for certainty, and a higher percentage of words like 'maybe" or 'perhaps" leads to a higher score for tentativeness [11]. We observe a higher tentative score and a higher certainty score for the tweets of CD, while these two scores for the tweets of ND are both lower. We have an interesting hypothesis for this subtle difference. Since 1986, Pennebaker et al. [11] have been collecting text samples from a variety of studies, including blogs, expressive writing, novels, natural speech, New York Times, and Twitter to get a sense of the degree to which language varies across settings. Of all the studies, the tentative and certainty scores for the text of the New York Times are the lowest. However, these two scores for expressive writing, blog, and natural speech are relatively higher. This observation leads to our hypothesis that the tweets of CD are more like blogs, expressive writing, or natural speeches that focus on expressing ideas, whereas the tweets of ND are more like newspaper articles that focus on describing facts.

As for the score of 'achievement", McClelland et al. [8] found that the stories people told in response to drawings of people could provide important clues to their needs for achievement. We hypothesize that the higher value of the 'achievement" score for the tweets of ND reflects the need of these Twitter users to succeed in fighting against COVID-19. As for personal concerns, the scores of 'work" and 'money" of ND are both higher than those of CD, which shows that the Twitter users of ND focus more on the work and money issue (e.g. working from home, unemployment). According to the reports of the U.S. Department of Labor, the advance seasonally adjusted insured unemployment rate reached a historical high of 8.2% for the week ending April 4.

5   Conclusion

We have presented a fine-grained study of the use of controversial and non-controversial terms associated with COVID-19 on Twitter. Instead of them being a mere description of the geographical origin of the virus, we find that controversial terms differ from neutral terms in three key aspects. First, at a high level, tweets with controversial terms can be distinguished with high accuracy from those with neutral terms even after proper masking. Second, at the topic level, tweets with controversial terms discuss substantially different topics such as racism and the Chinese government. Lastly, tweets with controversial terms possess quite different LIWC features, including time orientation, sentiment, certainty, tentativeness and personal concerns. Researchers interested in COVID 19, hate speech and racism should find our work useful.

References

1. Blei, D.M., Ng, A.Y., Jordan, M.I.: Latent dirichlet allocation. J. Mach. Learn. Res. **3**, 993–1022 (2003)
2. Chatzakou, D., Vakali, A.: Harvesting opinions and emotions from social media textual resources. IEEE Internet Comput. **19**(4), 46–50 (2015)
3. Chen, L., et al.: A social media study on the associations of flavored electronic cigarettes with health symptoms: observational study. J. Med. Internet Res. **22**(6), e17496 (2020)
4. Devlin, J., Chang, M.W., Lee, K., Toutanova, K.: BERT: pre-training of deep bidirectional transformers for language understanding (2018). arXiv:1810.04805
5. Gunsch, M.A., Brownlow, S., Haynes, S.E., Mabe, Z.: Differential forms linguistic content of various of political advertising. J. Broadcasting Electron. Media **44**(1), 27–42 (2000)
6. Lukasik, M., Srijith, P., Vu, D., Bontcheva, K., Zubiaga, A., Cohn, T.: Hawkes processes for continuous time sequence classification: an application to rumour stance classification in Twitter. In: Proceedings of the 54th Annual Meeting of the Association for Computational Linguistics (Volume 2: Short Papers), pp. 393–398 (2016)
7. Lyu, H., Chen, L., Wang, Y., Luo, J.: Sense and Sensibility: Characterizing social media users regarding the use of controversial terms for covid-19. IEEE Trans. Big Data 1 (2020)
8. McClelland, D.C.: Inhibited power motivation and high blood pressure in men. J. Abnormal Psychol. **88**(2), 182 (1979)
9. Mouthami, K., Devi, K.N., Bhaskaran, V.M.: Sentiment analysis and classification based on textual reviews. In: 2013 International Conference on Information Communication and Embedded Systems (ICICES), pp. 271–276. IEEE (2013)
10. Müller, M., Salathé, M., Kummervold, P.E.: COVID-Twitter-BERT: A natural language processing model to analyse covid-19 content on Twitter. arXiv preprint arXiv:2005.07503 (2020)
11. Pennebaker, J.W., Boyd, R.L., Jordan, K., Blackburn, K.: The development and psychometric properties of LIWC2015. Technical report (2015)
12. Tausczik, Y.R., Pennebaker, J.W.: The psychological meaning of words: LIWC and computerized text analysis methods. J. Lang. Soc. Psychol. **29**(1), 24–54 (2010)
13. Tumasjan, A., Sprenger, T.O., Sandner, P.G., Welpe, I.M.: Predicting elections with Twitter: What 140 characters reveal about political sentiment. In: Fourth International AAAI Conference on Weblogs and Social Media (2010)

14. Yang, Z., Dai, Z., Yang, Y., Carbonell, J., Salakhutdinov, R.R., Le, Q.V.: XLNet: generalized autoregressive pretraining for language understanding. In: Advances in Neural Information Processing Systems, pp. 5754–5764 (2019)
15. Yu, B., Kaufmann, S., Diermeier, D.: Exploring the characteristics of opinion expressions for political opinion classification. In: Proceedings of the 2008 International Conference on Digital Government Research. pp. 82–91. Digital Government Society of North America (2008)
16. Zhang, D., Li, S., Wang, H., Zhou, G.: User classification with multiple textual perspectives. In: Proceedings of COLING 2016, the 26th International Conference on Computational Linguistics: Technical Papers, pp. 2112–2121 (2016)
17. Zhang, X., Lyu, H., Luo, J.: What contributes to a crowdfunding campaign's success? Evidence and analyses from GoFundMe data. arXiv preprint arXiv:2001.05446 (2020)

Methodologies

Assessing Bias in YouTube's Video Recommendation Algorithm in a Cross-lingual and Cross-topical Context

Baris Kirdemir[1]([⊠]), Joseph Kready[1], Esther Mead[1], Muhammad Nihal Hussain[1], Nitin Agarwal[1], and Donald Adjeroh[2]

[1] COSMOS Research Center, UA - Little Rock, Little Rock, AR, USA
{bkirdemir,jkready,elmead,mnhussain,nxagarwal}@ualr.edu
[2] Department of Computer Science and Electrical Engineering, West Virginia University, Morgantown, WV, USA
donald.adjeroh@mail.wvu.edu

Abstract. Recently, a growing collection of interdisciplinary literature has been suggesting that algorithmic bias in recommender systems may cause severe consequences for sociotechnical systems. Audits of multiple social media platforms documented the implicit bias in content, products, information, or connections recommended to users, sometimes based on gender, sociocultural background, or political affiliation. Within the given context, YouTube's video recommendation algorithm has been subject to a popular public debate in terms of its broad societal impact and alleged contribution to the spread of harmful content. However, the current literature still lacks a comprehensive understanding of, and broad agreement on, whether the video recommendations on YouTube are biased in favor of a small set of items in generalizable conditions. In addition, the characterization of potential recommendation bias is a non-trivial problem given the black-box characteristics of the system. This study addresses the given pair of problems by adopting a graphical probabilistic approach and assessing the structural properties of video recommendation networks. We adopt a stochastic approach and examine PageRank distributions over a diverse set of recommendation graphs (256,725 videos, 803,210 recommendations) built upon 8 distinct datasets categorized by languages, topics, and system entry-points. As a result, we find structural and systemic bias in video recommendations, although the level and behavior of biased recommendations vary between experiments. This study sets the stage for further research in creating comprehensive evaluation methodologies that would address the severe effects of bias in terms of fairness, diversity, coverage, exposure, security, and utility.

Keywords: Recommender Bias · YouTube Algorithm · PageRank · Power law

1 Introduction

In recent years, several academic and journalistic investigations focused on the societal impact of YouTube's search and recommendation algorithms. Among the subjects of

© Springer Nature Switzerland AG 2021
R. Thomson et al. (Eds.): SBP-BRiMS 2021, LNCS 12720, pp. 71–80, 2021.
https://doi.org/10.1007/978-3-030-80387-2_7

the ongoing debate are the potential roles played by YouTube's recommendation, personalization, and ranking algorithms in the spread and influence of harmful content as well as formation of echo-chambers and polarization as a cumulative function of how the algorithm impacts the consumption patterns on the platform [5, 8, 12, 16–19]. However, the majority of the records regarding the alleged effects originate from journalistic accounts and anecdotal evidence [17, 19], while the number and scope of the systematic scholarly studies gradually increase with significant lagging. Therefore, currently, studies lack an established understanding of, and agreement on, the exact role of YouTube's recommendation algorithm in the mentioned set of problems, nor their extent.

The current literature still lacks comprehensive and systematic evaluations of whether the video recommendation system on the platform tends to be biased in favor of a small number of videos in realistic scenarios and generalizable ways. In addition, characterization of any detected algorithmic bias remains non-trivial given the variety of use-case scenarios and platform architectures. In this study, we approach the given pair of problems from a graphical probabilistic perspective, focusing on the node-centric probabilistic distributions and network topologies of recommendations. Also, we design a set of experiments that correspond to content and real-world scenarios in multiple languages, topics and themes. We utilize PageRank computations in the context of recommendation networks.

Implicit bias in recommender systems may have significant social implications. The most recent examples of systematically documented cases range from gender bias in Facebook's job ads [9] to Amazon's book recommendations that favor misinformation about COVID-19 vaccines [15]. Referring to the public and academic debates mentioned above, exploration and characterization of emergent algorithmic biases in YouTube's video recommendation system using the given graph analysis approach would provide a significant contribution to the literature of algorithmic bias, fairness, transparency, and accountability. In addition, characterization and evaluation of potential biases in YouTube's recommendation systems relate to misinformation, disinformation, information operations, and other important pressing issues threatening the well-being and security of individuals, social groups, and both governments and non-governmental entities. Thus, by evaluating the structural tendencies of bias in YouTube recommendation graphs, we aim to contribute to the given field with regards to implicit and structural biases, feedback loops, and complex networks across social media platforms. This study aims to extend the assessment of potential, structural, inherent, and emergent video recommendation bias by focusing on the diversification of data collection methodology and experiment settings and enabling future work that would characterize the dynamics and evolution of bias vis-à-vis fairness, coverage, exposure, diversity, security, and utility.

The paper is organized as follows: Sect. 2 briefly outlines the recent literature on the impact and behavior of YouTube's recommendation algorithm, as well as the use of network analysis methods to examine recommender systems. Section 3 describes our data collection methodology and experimental design. Section 4 documents the results of our experiments, while the final section briefly discusses our findings and sets the stage for further studies on the evaluation of bias in recommender systems.

2 Related Work

In recent years, several studies examined whether YouTube's recommendation algorithm leads its users to harmful content, or whether it contributes to the formation of echo-chambers, polarization, or radicalization [5, 8, 12, 16, 18]. Ribeiro et al. [16] suggested that users are exposed to increasingly extreme content if they had previously viewed other videos associated with radicalization. In contrast, Ledwich and Zaitsev [12] suggested that the YouTube algorithm does not promote radical or far-right content. Instead, they claim that the algorithm consistently leads users towards moderately left or mainstream channels. Hussein et al. [8] analyzed non-political misinformation topics and claimed that the algorithm is associated with the filter-bubble formation and spread of specific misinformation topics like chemtrails, flat earth, and moon landing.

Faddoul et al. [5] measured the amount of conspiracy videos in YouTube's video recommendations in a longitudinal setting, by developing a classifier to detect conspiracy theory videos on the platform, showing a reduction in conspiratorial recommendations over time after YouTube's announcement to limit radical content [20]. The authors note, however, that this finding alone does not prove that the "problem of radicalization on YouTube" has diminished. Roth et. al. [18] focused on the formation of "filter bubbles" or the "confinement" effect that allegedly exists on YouTube. They followed a research protocol that is based on a platform-level and node-centric design, similar to our approach. They used a random walk model on a "recursively crawled" recommendation network and measured the entropy as a representation of diversity to study the dynamics of confinement. The researchers found that recommendations tended to lead to topical confinement, which seemed "to be organized around sets of videos that garner the highest audience" [18].

Recent literature suggests that the use of network analysis approaches may enable examining latent characteristics and impact of YouTube's recommendation algorithm, especially from the perspective of its users. Typically, graphs are used to represent videos, channels, pre-defined content categories, or inferred classes of videos as nodes, while recommendations constitute edges. Both in "node-centric" [18] and "user-centric" (personalization) [8] research settings, the basic structure and topology of the recommendation graph are used in the operationalization of concepts such as bias, fairness, confinement, filter bubbles, echo-chambers, polarization, and radicalization, often being captured through multiple hops of graph crawling or in longitudinal terms. Le Merrer and Trédan [11] suggested "detection" of recommendation biases that are explicitly induced by the "service provider" through network analysis of "successive" recommendations. They highlighted an application of their evaluation methodology on YouTube's video recommendations. Specifically relevant to our work is their view that graph topologies show different features in biased and unbiased settings. Their experiments revealed that the distribution of path lengths in the graph becomes "more compact" when explicitly "biased edges" are introduced, resembling well-known features of small-world networks [11].

In this study, we assert that a stochastic analysis of the video recommendation networks on YouTube leads to a better understanding and characterization of the bias in recommender systems that may have significant influence on the dissemination, platforms, consumers, and providers of information recommended. As a core pillar of our approach, we also contend that to better represent and operationalize real-world scenarios, our data collection methodology should include a careful diversification of the initial data points (seeds), based on different scenarios involving multiple languages, themes, topics, and platform entry-points. The following section describes the datasets we used in our experiments.

3 Datasets and Data Collection Methodology

To capture and explore the recommendation graphs on YouTube, we start with building a diverse set of criteria to crawl multiple datasets of recommended videos. We created distinct lists of seed videos that were categorized in terms of main topics, video languages, and platforms where the lists were acquired. Overall, we collected eight different datasets, with the number of recommendations ranging from 70,200 to 126,548, and a total of 803,210 recommendations. We then built the directed recommendation graphs on collected datasets, in which "parent" videos link to recommended videos or channels. Parent videos are the source content being consumed, while recommended videos are presented to the user on the same page with the parent.

Our approach is based on the assumption that we should first capture the general behavioral patterns of the recommendation algorithm as they tend to apply to a diverse set of conditions across the platform. We share the perspective that personalization plays a role in the evolution of recommendations to a specific logged-in user, and that the platform also utilizes the user's web traces through cookies. Nevertheless, a system level, structural analysis that pays special attention to the node-specific probabilistic computations is required to understand the inherent properties of the examined behavior. Therefore, we posit that we can fairly examine recommendation algorithm bias by excluding the user personalization aspect of the YouTube platform.

Data collection was conducted by using the YouTube Tracker tool [13] and the data collection method described by Kready et. al. [10]. The process was initiated in two ways: 1) a keyword search, and 2) a list of seed videos. We then followed "related videos" for each initial set to gather details about additional videos recommended by the YouTube algorithm. These videos would typically appear on YouTube under the "Up Next" heading or in the list of the recommended videos. We determined that two consecutive hops of crawling would enable a sizable amount of data for each experiment, while also allowing us to diversify our collection.

The first distinction in the categorization of the seed list is based on the topics of the videos: a) a generalized category that includes videos about the COVID-19 pandemic, and b) a set of highly engaged political topics. This distinction enables us to observe an additional phenomenon and its impact in terms of potential implicit biases or the distribution of recommended videos in general. We also included two additional political topics which attracted high-level user engagement, indicated by views counts, number of likes, and comments. Our alternative datasets relate to Canadian politics and elections, and geopolitical events involving Turkey, Russia, and NATO.

We also diversified the sources of seed videos. Four of the eight seed lists are the direct results of a search query on YouTube, while the other four were compiled from other platforms. Our collections of seeds from the YouTube Data API search endpoint contain 50 videos for each, which were later used for crawling the related videos through multiple hops. For the alternative seed list related to Canadian politics, we used a dataset including instances of disinformation during the federal election in Canada [6]. For seed videos in English about the COVID-19 pandemic, we utilized a large dataset of misinformation during the pandemic from the COSMOS COVID-19 misinformation repository [4]. The repository contains a continually curated collection of COVID-19 misinformation items that are actively being spread via multiple platforms and websites. In addition, we combined the information from the repository with a sample of highly engaged video URLs collected from Twitter. Similarly, for videos in Turkish, we sampled two lists of video URLs collected from Twitter.

The language and topic diversity in our 8 datasets enabled us to conduct a comparative analysis of recommendation graphs to identify similarities and differences among them. This way, we were able to study English versus non-English networks, and political versus health misinformation networks.

4 Experiments and Findings

The first step of the analysis involved building the recommendation graphs for each dataset we acquired, by drawing edges between "parent videos" and "recommended videos". To explore and evaluate potential biases in video recommendations, we utilized a probabilistic approach focusing on the nodes potentially gaining influence over the rest of the graph as a result of the recommendations. We extracted the PageRank [14] scores acquired by each node in the recommendation graph and explored the PageRank distributions of each recommendation graph. The PageRank algorithm enabled us to demonstrate the probabilities of being visited for each video in a relatively realistic random walk scenario. In addition, the PageRank metric allows us to use a "damping factor" value which defines a probability for the random walker to jump to a random node instead of strictly following the graph connections during the entire walk. Furthermore, as we indicated earlier, we compared graphs we built in terms of language-based, topical, and seed source categories to see potential variances between PageRank probability distributions and curves of biases that favor small fractions of video items.

4.1 Bias Detection

In a graph $G(V, E)$ where V is the videos denoted by vertices and E is edges from "parent video" to "recommended video", O_v denotes the out-degree or number of "recommended videos" for $v \in V$ and I_v denotes the in-degree or number of "parent videos" for $v \in V$ in the graph. Let $p \in [0,1]$ be the "damping factor" ($d = 0.85$) representing the probability for the random walker to jump to a random node instead of strictly following the graph connections (or in this case YouTube's recommendations) during the entire walk, and N is the number of nodes in the graph. We aim to study the potential influence of alternative damping factors in further iterations of this work. The PageRank for each $v \in V$ is computed by:

$$PageRank_i = \frac{p}{N} + (1 - p) * \sum_{j \in V} \frac{PageRank_j}{O_j}$$

The number of total recommendations for each dataset vary. This variance partially emanates from the size of the lists that contained the seed video IDs, as five of the eight lists (including the results of the request on YouTube API's search endpoint) contain 50 videos, while we had one list with 35 seeds, one list with 30 seeds, and another list with 40 seeds. Nevertheless, we also observed limited variance in the number of rows (total number of recommendations) even when data crawls were initiated with the same number of seeds.

Table 1 provides the description of the 8 datasets and associated experiment results. For each item, it gives details on the size of the seed set, language of the videos, and description of the dataset. It also lists the number of nodes, edges, average degree, and maximum likelihood estimate of alpha values, as described in the remainder of this section.

4.2 Bias Variance

Further, we plot the complementary cumulative distribution functions (CCDF) for each recommendation graph. Briefly, CCDF shows the probability of a variable (PageRank in our case) being above a particular value and it is previously documented as an effective technique to explore scale-free characteristics of real-world networks. To explore the variance between different recommendation graphs and their structural features based on PageRank or in-degree distributions, we compare how the given CCDF distributions fit the power-law based on the maximum likelihood estimation as developed by Clauset et al. [3]. We use a Python implementation [1] of the log-likelihood functions that enable comparative computations suitable for our purpose.

4.3 Findings

This study of potential biases of the YouTube recommendation algorithm is based on eight different experiments with a variety of distinct seed lists categorized along their topics, languages, and entry points to the YouTube ecosystem. As a result, we collected distinct datasets and explored the structural properties as well as node-centric features of the recommendation graphs. In all experiments, the bias of the recommendation algorithm in favor of a small fraction of videos seems to emerge as a basic, scale-free characteristic that is evident on a topological level. In particular, cumulative probability distributions of PageRank values demonstrates that a few videos turn out to be far more likely to be visited by a user following the recommended items with some randomness included. Our experiments also show that the shape, skewness, and proportion of the bias varies between different use-case scenarios. Although our data collection methodology with lingual, topical, and system entry categorization enable initial observations, the variance of bias should be a subject of further investigations.

The distribution of PageRank values, as shown in Fig. 1, followed a long tail distribution indicating that a small fraction of nodes received a large number of connections. The log-log representation of the same confirms the highly skewed nature of the distribution. Figure 1 shows the distribution of PageRank scores for one of the recommendation graphs (ID = 2). We observed similar distributions in other experiments with the remaining seven datasets. Thus, initial results of the PageRank computations demonstrate emerging biases in the recommendation graphs we inspected.

Table 1. Properties of YouTube datasets and recommendation graphs

Dataset ID	Seed size	Lang.	Description of seed videos	Rows (total recom.)	Nodes	Edges	Avg. InDegree	Maximum likelihood estimate of α^1
1	50	English	A sample of YouTube video IDs collected from COSMOS repository [4] and Twitter by querying "COVID 19 hoax" and "COVID 19 weapon"	119,094	50,491	92,850	1.8389	2.498
2	50	Mixed	YouTube search results for the query "covid hoax"	126,548	36,627	77,895	2.1267	2.757
3	50	Turkish	YouTube search results for the query "koronavirus biyolojik"	123,986	31,108	73,902	2.3757	2.155
4	35	Turkish	A sample of YouTube videos IDs collected from Twitter and Facebook wrt. COVID-19	77,286	21,521	58,310	2.7094	2.135
5	30	English	YouTube videos containing misinformation wrt. elections in Canada (from Galeano et al. [11])	70,200	31,032	61,618	1.9856	2.261
6	50	English	YouTube search results for the query "Canada election Trudeau"	71,290	31,466	62,805	1.9960	2.252
7	50	Turkish	YouTube search results for the query "Turkiye Rusya NATO"	115,723	24,561	73,521	2,9934	2.100
8	40	Turkish	A sample of YT. video IDs collected from Twitter and Facebook about "Turkey, Russia, and NATO"	99,083	29,919	87,577	2,9271	2.028

Maximum Likelihood Estimation was computed on each distribution by using the method proposed by Clauset et al. [3].

We observe power law fits and CCDF distributions having some variance, especially in terms of the exponent and how much and how well the distributions fit the power law. For example, Graph 7 seems to fit a likely power law almost perfectly, while the Graph 8 curve has only a partial fit that stops approximately after the mid-point on the plot. Additionally, the CCDF slopes of Graph 5 and Graph 6 diverge from the power law lines at smaller PageRank values than the other graphs.

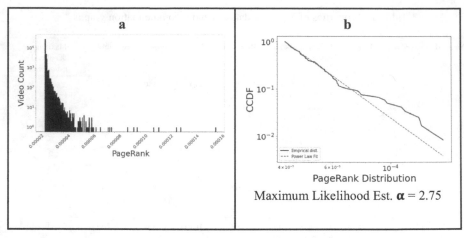

Fig. 1. (**a**) The Distribution of PageRank values (y axis log scale) and (**b**) corresponding CCDF Plot for the recommendation Graph 2. The CCDF curve is fitted on a Power Law line using Maximum Likelihood Estimation method proposed by Clauset et al. [3]

We confirm our previous assertion that one of the core structural features of each graph is an emergent bias that favors a small fraction of recommended videos. In addition, CCDF plots show power law characteristics, although the exponent value and the portion that fits the likely power law line vary between different graphs. Therefore, despite a remarkable confirmation of emergent bias as a common property, characterization of it seems to depend on various other factors that should be considered in further studies. The slopes for the log-log graph showed significant variation across the datasets. This indicates transitions in terms of how distributions of cumulative PageRank probabilities change within a single graph. This variance highlights emergent biases on YouTube, potentially depending on the characteristics of the dataset. As we specified in the description of our data collection approach, such characteristics may include the source of seeds and entry points, language, or primary topics of the content.

5   Conclusions and Future Work

This study aimed to discover, explore, and understand the potential, emergent, and inherent characteristics of bias in YouTube's video recommendations. Despite the black-box characteristics and closed inner-workings of the algorithm itself, our approach enabled identifying the existence of recommendation bias in the studied system. By using probabilistic distributions and the PageRank algorithm to operationalize our stochastic approach, we were able to demonstrate the resulting influence of a small number of videos over the entire network.

We relied on a diversified data collection methodology and an increased quantity of experiments conducted in accordance with realistic scenarios and expected sources of behavioral variance in the studied system. The resulting indicators of such variance between different recommendation networks point to further investigations in future work. Nevertheless, at the current stage, the study has a few limitations that can be addressed in future work. For instance, in our experiments, we did not consider the potential impact and correlation of channel activity, engagement metrics, or content production rates vis-à-vis PageRank distributions in recommendation graphs.

In particular, in the subsequent phases of this effort, we aim to produce models that can aid both the prediction and understanding of the behavioral patterns that lead to the documented bias and its variations. As mentioned in the earlier sections, in the future work we aim to characterize the variance of the detected structural bias, examine the influence of alternative damping factors and limitations such as the potential presence of linking farms with regards to the PageRank algorithm, and further systematize our bias evaluation methodology in relation to the exposure, coverage, diversity, security, utility, fairness, and transparency of recommended information.

Acknowledgements. This research is funded in part by the U.S. National Science Foundation (OIA-1946391, OIA-1920920, IIS-1636933, ACI-1429160, and IIS-1110868), U.S. Office of Naval Research (N00014-10-1-0091, N00014-14-1-0489, N00014-15-P-1187, N00014-16-1-2016, N00014-16-1-2412, N00014-17-1-2675, N00014-17-1-2605, N68335-19-C-0359, N00014-19-1-2336, N68335-20-C-0540, N00014-21-1-2121), U.S. Air Force Research Lab, U.S. Army Research Office (W911NF-20-1-0262, W911NF-16-1-0189), U.S. Defense Advanced Research Projects Agency (W31P4Q-17-C-0059), Arkansas Research Alliance, the Jerry L. Maulden/Entergy Endowment at the University of Arkansas at Little Rock, and the Australian Department of Defense Strategic Policy Grants Program (SPGP) (award number: 2020-106-094). Any opinions, findings, and conclusions or recommendations expressed in this material are those of the authors and do not necessarily reflect the views of the funding organizations. The researchers gratefully acknowledge the support.

References

1. Alstott, J., Bullmore, E., Plenz, D.: Power-law: a Python package for analysis of heavy-tailed distributions. PLoS ONE **9**(1), e85777 (2014)
2. Buntain, C., Bonneau, R., Nagler, J., Tucker, J.A.: YouTube Recommendations and Effects on Sharing Across Online Social Platforms (2020). arXiv preprint arXiv:2003.00970
3. Clauset, A., Shalizi, C.R., Newman, M.E.: Power law distributions in empirical data. SIAM Rev. **51**(4), 661–703 (2009)
4. COVID-19 Misinformation Repository. COSMOS Research Center (2020). https://cosmos.ualr.edu/covid-19
5. Faddoul, M., Chaslot, G., Farid, H.: A Longitudinal Analysis of YouTube's Promotion of Conspiracy Videos (2020). arXiv preprint arXiv:2003.03318
6. Galeano, K., Galeano, L., Mead, E., Spann, B., Kready, J., Agarwal, N.: The role of youtube during the 2019 Canadian federal election: a multi-method analysis of online discourse and information actors. J. Future Conflict. **2**, 1–22 (2020)
7. Google Developers. YouTube Data API, Google (2020). https://developers.google.com/youtube/v3
8. Hussein, E., Juneja, P., Mitra, T.: Measuring misinformation in video search platforms: an audit study on youtube. In: Proceedings of the ACM on Human-Computer Interaction. (CSCW1), vol. 4, pp.1–27 (2020)

9. Imana, B., Korolova, A., Heidemann, J.: Auditing for discrimination in algorithms delivering job ads. In: Proceedings of the Web Conference 2021 (WWW 2021), Ljubljana, Slovenia. ACM (2021). https://doi.org/10.1145/3442381.3450077
10. Kready, J., Shimray, S.A., Hussain, M.N., Agarwal, N.: YouTube data collection using parallel processing. In: 2020 IEEE International Parallel and Distributed Processing Symposium Workshops (IPDPSW), pp. 1119–1122. IEEE (2020)
11. Le Merrer, E., Trédan, G.: The topological face of recommendation. In: Cherifi, C., Cherifi, H., Karsai, M., Musolesi, M. (eds.) COMPLEX NETWORKS 2017 2017. SCI, vol. 689, pp. 897–908. Springer, Cham (2018). https://doi.org/10.1007/978-3-319-72150-7_72
12. Ledwich, M., Zaitsev, A.: Algorithmic extremism: examining YouTube's rabbit hole of radicalization. First Monday (2020)
13. Marcoux, T., Agarwal, N., Obadimu, A., Hussain, M.N.: Understanding information operations using YouTubeTracker. In: 2019 IEEE/ACM International Conference on Advances in Social Networks Analysis and Mining (ASONAM), pp. 709–712 (2019). https://doi.org/10.1145/3341161.3343704
14. Page, L., Brin, S., Motwani, R., Winograd, T.: The PageRank citation ranking: bringing order to the web. Stanford InfoLab (1999)
15. Juneja, P., Mitra, T.: Auditing e-commerce platforms for algorithmically curated vaccine misinformation. In: Proceedings of the 2021 CHI Conference on Human Factors in Computing Systems (CHI 2021). Association for Computing Machinery (2021). https://doi.org/10.1145/3411764.3445250
16. Ribeiro, M.H., Ottoni, R., West, R., Almeida, V.A., Meira Jr, W.: Auditing radicalization pathways on youtube. In: Proceedings of the 2020 Conference on Fairness, Accountability, and Transparency, pp. 131–141 (2020)
17. Roose, K. (2019). The making of a YouTube radical. The New York Times
18. Roth, C., Mazières, A., Menezes, T.: Tubes and bubbles topological confinement of YouTube recommendations. PLoS ONE 15(4), e0231703 (2020)
19. Tufekci, Z.: YouTube, the great radicalizer. In: The New York Times, vol. 10, p. 2018 (2018)
20. Wakabayashi, D.: YouTube Moves to Make Conspiracy Videos Harder to Find. The New York Times (2019). https://www.nytimes.com/2019/01/25/technology/youtube-conspiracy-theory-videos.html

Formal Methods for an Iterated Volunteer's Dilemma

Jacob Dineen$^{(\boxtimes)}$, A. S. M. Ahsan-Ul Haque, and Matthew Bielskas

Department of Computer Science, University of Virginia, Charlottesville, USA
jd5ed@virginia.edu

Abstract. Game theory provides a paradigm through which we can study the evolving communication and phenomena that occur via rational agent interaction [10]. The Volunteer's dilemma is a vastly studied game throughout literature that models agents as cooperative, rather than selfish, entities. In this work, we design a model framework and explore the Volunteer's dilemma with the goals of 1) modeling it as a stochastic concurrent n-player game, 2) constructing properties to verify model correctness and reachability, 3) constructing strategy synthesis graphs to understand how the game is iteratively stepped through most optimally and, 4) analyzing a series of parameters to understand correlations with expected local and global rewards over a finite time horizon.

1 Introduction

One-shot games, i.e. Prisoner's Dilemma, can typically be modeled with a simple payoff matrix. Players in the game choose a strategy and act concurrently and independently of one another. Extensive form games model game theoretic scenarios with sequential mechanisms, in which a subsequent player acts once their predecessor makes known their strategy and state transition. Iterated games, or repeated games, are examples of extensive form games and study longer (possibly infinite) time horizons. Both methods have gleaned valuable insight into behavioral economics and rational choice theory, and fuse many respective fields. Stochastic games are argued to be the most reflective of real-world systems, as they are governed by probabilistic dynamics that many situations incur. These games are typically modeled as being extensive form, and arguably produce more interesting results of long-run behavior. These dynamics have been studied in games involving social welfare (public goods), robot coordination and investing/auction scenarios [3,6,8].

Previous work [1] has analyzed a public good game as a concurrent stochastic game. There, the authors evaluated optimal strategies under a fixed set of parameters. We adopt the finite state methodology, i.e. each agent can choose to share a discrete portion of their initial resources, but study a countering problem in a collective good game. We are interested in expressing the Volunteer's dilemma through Prism Model Checker, which allows for user flexibility of game dynamics and thorough analysis of verification and reachability. To the best of

© Springer Nature Switzerland AG 2021
R. Thomson et al. (Eds.): SBP-BRiMS 2021, LNCS 12720, pp. 81–90, 2021.
https://doi.org/10.1007/978-3-030-80387-2_8

our knowledge, PRISM has not been used to study Volunteer's Dilemma in the form of an iterated game, i.e. a game that repeats and experiences soft resets after each round.

2   Background

2.1   The Volunteer's Dilemma

The Volunteer's dilemma is a game played by multiple agents concurrently that models a situation in which each agent has one of two options: 1) Cooperation: An agent can make a small sacrifice for public good i.e. that benefits everyone, 2) Defect: An agent can wait and freeride, and hope someone else will eventually cooperate. A typical payoff matrix for the Volunteer's dilemma looks like this (Table 1):

Table 1. Payoff matrix

	At least one other cooperates	All others defects
Cooperate	0	0
Defect	1	-10

The agents make the decisions independently of each other. The incentive for an agent to freeride is greater than the incentive to volunteer. However, if no-one volunteers then everyone loses. Conversely, if at least one person volunteers then everyone receives benefit.

The Volunteer's dilemma occurs in various natural scenarios. For example: in a group of meerkats, some act as sentries to let everyone else know if there are any predators nearby. In doing so, those become more vulnerable. It is also important to understand group behaviors like voting behavior in democratic elections. Let's assume an election where a candidate has much more support than all other candidates. The supporters of that candidate have little incentive to go out and vote, since that candidate is predicted to win anyway. However, if all of his supporters think in that way and do not vote, that candidate may end up losing the election.

3   Design Overview

Concurrent stochastic multi-player games (CSGs) are an extension to stochastic games (SGs) popularized in the 1950s. SGs are generalizable to n-player games and present a viable way to model group dynamics, in collaborative or competitive games, where the environment changes given feedback from agents in the system. Beginning from some state $s \in S$, immediate payoff, or reward, is dependent on the actions taken by all agents in the system $v \in V$. Stochastic multi-player games (SMGs) are turn-based and are governed by individual or joint state transitions, where a player chooses from a set of probabilistic transitions to determine the next state [5]. Formally, a CSG can be represented by a tuple not dissimilar from a Markov Decision Process (MDP):

$$G = (N, S, \vec{S}, A, \Delta, \delta, AP, L)$$

where: N is a finite set of players. S is a finite set of states. A is a finite set of actions available to v_i at time t. Δ is an action assignment function. δ is a probabilistic transition function. AP is a set of atomic propositions, and L is a labeling function. In a CSG, similar to how a policy resolves nondeterminism in an MDP, a strategy resolves choice [8]. Our focus is in games where this transition occurs as a product of all agents simultaneously, hence the concurrence. As the environment changes depending on these actions, the choice of a new state is also influenced, and, in turn, expected future payoffs are affected. The design of our CSG is detailed in this section. We use an extension to Probabilistic Symbolic Model Checker (PRISM), PRISM Games [4,5,7], throughout experimentation.

3.1 Game Parameters

1. k: We refer to rounds of the finite length game as episodes. $k \in \{1, 2, 3, ..., k_{max}\}$
2. k_{max}: The maximum number of episodes specified as input into the environment.
3. V: The set of agents within a system. Let, $n = |V|$. In literature, the game is typically played with $n > 2$. Here, we fix $n = 3$. We'll be studying this problem through the lens of a 3-player game.
4. r_{init}: The initial allocation of a resource. Each agent within the system is initialized with $r_{init} = 100$ at each episode. Resources could be generalized to be currency, votes or public goods, etc.
5. c_i: The current resource allocation for agent v_i at round k. c_i is updated throughout gameplay.
6. s_i: The number of shared resources for agent v_i at round k. A player can donate increments ($\{0, 0.5, 1.00\}$) of their procured resource allocation. $s_i \leq c_i$
7. r^{needed}: A specified parameter that dictates the number of resources needed to 'win' a round k in the game. In the traditional game, only a single volunteer is needed. Here, we consider the effects of resource procurement over finite-length runs of the game. E.g., rewards distributed at round k can be used as 'donation' at round $k + 1$. In literature, to reach a winning condition, we generally require donations from strictly less than the total number of agents in the system. This holds here. We require $r_{needed} < 100n$. This parameter is fixed round over round, i.e., it is not dynamically dependent on the values of state variables c_i.

3.2 Action Space

We discretize the allowable actions, i.e. resource donations, to reduce the search space. We present the action space $\{a_0, a_{50}, a_{100}\} \in A$ below (Table 2).

Table 2. Volunteer's dilemma action space

Variable	Name	Definition
a_0	Free ride	A player here chooses to contribute nothing to the pot of r_{needed}. They are hopeful that total group contribution still results in immediate payoff without sacrificing any of their resource allocation
a50	Partial contribution	A player taking this action will contribute $\lfloor (0.5 * c_i) \rfloor$ resources
a100	Total contribution	This action entails contribution in totality. All available resources will be pushed toward r_{needed}. An agent taking this action could be seen as altruistic, as they may perceive the good of the many to outweigh the good of themselves

3.3 Reward Structure

We present a simple reward structure as follows. At the kth round, all agents starting in s_0 are to concurrently choose an action. For a winning condition to be met, the sum of total contributions from all agents $\Sigma_{i=1}^{n} s_i$ must meet or exceed the predefined threshold r^{needed} (Fig. 1).

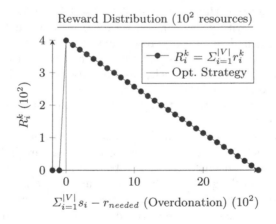

Fig. 1. Reward function given: $r_{needed} = 200, n = |V|, f = 2$. This plot shows donated resources exceeded resources needed and reward (resources) in the 100s of units. When $\Sigma_{i=1}^{n=|V|} s_{i'} < r_{needed}$, that round incurs no reward. When $\Sigma_{i=1}^{|V|} s_{i'} = r_{needed}$, an optimal joint strategy has been found. Because a single agent freerode in this instance, the number of resources at the end of this round exceeds those of when the round began. When $\Sigma_{i=1}^{n=|V|} s_{i'} > r_{needed}$, a winning condition has been met, but resources were expended that didn't need to be. This figure shows the linearly decaying reward function, and current resources at the kth round are found via the update function in Eq. 3.

The immediate reward passed back to each agent subject to a winning condition can be formulated as:

$$r_i^k = \begin{cases} 0 & \left(\Sigma_{i=1}^n s_{i'}^k\right) < r_{needed} \\ \frac{r_{needed} \cdot f}{|V|} & \left(\Sigma_{i=1}^n s_{i'}^k\right) = r_{needed} \\ \frac{(-.014)(\Sigma_{i=1}^n s_{i'}^k - r_{needed}) + r_{needed} \cdot f}{|V|} & \left(\Sigma_{i=1}^n s_{i'}^k\right) > r_{needed} \end{cases} \quad (1)$$

$$R_i^k = \Sigma_{i=1}^{n=|V|} r_i^k \quad (2)$$

$$c_i^{k+1} \leftarrow min\left(rmax, \left(\left\lfloor (c_i^k - s_{i'}^k) + \frac{R_i^k}{|V|} \right\rfloor\right)\right) \quad (3)$$

where the possible resources procured by a single player are constrained by r_{max}. $c_i - s_{i'}$ is the cost incurred by donating resources (initial resources at the start of a round less the donated resources during a round). This can also be thought of as an expenditure. f is a scaling factor to ensure that players who donate aren't penalized more than players that do not donate when a WIN is achieved. $s_{i'}$ is the state that v_i transitions to $c^{k+1} = (s_i, c_i|a_i)$ given their initial round state and the chosen action. Rewards gained at a time-step are re-aggregated in c^{k+1} and are allowable donations in round $k + 1$. In literature, particularly studies involving human psychology, confounding effects may diminish the virtue of altruism in and of itself, as it could be done for ulterior motives [2]. We consider this here by 'punishing' over-donations. If $\Sigma_{i=1}^n s_i > r_{needed}$ the immediate reward for all agents at the kth round R_k decays linearly according to the piecewise function noted above. This can be seen in Fig. 1.

4   Experiments and Results

The size of the state space is generally represented by $|S| = n^{|A|}$ for static games. The winning conditions here are dynamically dependent on the state of the game at a given time-step, There are more possible joint policies that result in WIN/SAT as the game progresses and resources are procured via reward feedback, and, as such, the state space grows exponentially over time. E.g., in this first round of a game, assuming $r_{init} = 1$, there are only $\binom{n}{r_{needed}}$ transitions that induce a 'perfect' WIN, where no decay is met via over-donation. $|S|$ increases as $c_i \rightarrow r_{max}$. With a parameter set of $\{kmax = 4, r_{init} = 100, r_{needed} = 200, n = 3\}$ the size of $|S|$ grows according to $|S| = 1.6978e^{3.0479k}$. We'll constrain k_{max} to be ≤ 4, as extrapolating this to 5 and 6 rounds leads to respectively ≈ 7 mm and ≈ 148 mm possible states.

4.1   Model Correctness

We look at a mostly fixed parameter set: the number of agents $n = 3$, the number of initial resources $e_{init} = 100$, the threshold for resources $r_{max} = 1000$, specified at the local level, and the maximum number of rounds iterated through as $k_{max} = 4$. To ensure that our model is working, we create properties based on a temporal logic, rPATL, which combines PCTL and ATL [9].

With a nonzero probability, we want to ensure that after k rounds, it is *eventually possible* for an agent to have $c_i \geq e_{init}$, which would mean that rewards were accrued during game-play and winning conditions were met. It is not a formal requirement that all agents meet this condition individually, however. If it is not met, the piecewise function in conjunction with the resource update step ensures that $c_i^k < c_i^{k+1}$. If at any point during the game $\Sigma_{i=1}^n c_i < r_{needed}$ it becomes impossible to satisfy this correctness property. Unfortunately, PRISM Games does not support model checking on CTL operators. Ideally, we would want to verify that there exists some state $goal = \Sigma_{i=1}^n c_i > r_{needed}$ across k rounds, such that $E[Fgood]$ evaluates to TRUE. Because this condition is trivially satisfied via the initial state where $n \cdot e_{enit} > r_{needed}$, we look at a case where $2\, r_{needed}$ are required, which can be satisfied only after the first round of the game.

$$goal = \Sigma_{i=1}^n c_i > 2 * r_{\text{needed}}$$
$$<<p1,p2,p3>> \; P^{>=1.0} [F <= k_{\max} + 1 \; \text{``}goal\text{''}] \tag{4}$$

In rPATL, the $<<C>>$ operator specifies a coalition of players [9]. Here, we consider a cooperative game, where players are within a singular coalition aimed at maximizing expected reward. The property in Eq. 4 asserts that there exists a joint strategy, or a collection of policies for each agent, such that the probability of reaching the goal state "goal" within k_{max} steps is at least 1.00. This verifies to FALSE in the first round, and TRUE thereafter to $k_{max} = 4$, suggesting a viable model for our purposes. We can also observe the probabilistic reachability via the PRISM Games GUI for the noted property, detailed in Table III. Intuitively, as the game progresses on, assuming round-wise SAT of the given property, the number of possible states which result in SAT increase. This is due to more resources being injected into the environment, resulting in more possible combinations of donations which result in reward.

Table 3. VGD probabilistic reachability analysis

Round	States	Y	N	M	Y/(Y + N)
1	2 (1 init)	0	2	0	0%
2	55 (1 init)	6	48	1	11.1%
3	1162 (1 init)	141	1009	12	12.3%
4	27065 (1 init)	2724	8766	85	23.7%

4.2 Property Verification

Now that we're sure our model is implemented correctly in PRISM, the next step is to construct properties and verify them so we formulate a reachability analysis for the CSG. Recall "probabilistic reachability" as referred to in the previous subsection and Table 3. For a probability-based property, the direct result is a Boolean that indicates whether the property holds for at least one state in the model. This corresponds to at least one Yes in the aforementioned (Y, N, M) tuple and we emphasize that both outputs are situationally useful for understanding the game. For a property defined by maximizing or minimizing a variable/reward, the direct result is the max/min number while (Y, N, M) has no reason to be recorded. Below we present some property templates we experimented with in PRISM.

$$<<p1, p2, p3>> R\{``r1"\}max =?\{F \ k = k_{max} + 1\} \qquad (5)$$

With our three players in the game, this property returns maximum reward value $r1$ assigned to Player 1 when the game ends *after* k_{max} rounds. Here $r1$ and *done*1 are interchangeable but $r1$ is able to be examined for all $k = 1, ..., k_{max}$.

$$<<p1 : p2, p3>> max =?(R``done1"[Fk = k_{max} + 1]$$
$$+R``done23"[Fk = k_{max} + 1]) \qquad (6)$$

For this property we have Player 1 aligned against Players 2 and 3 for a total of two coalitions. With $done23 = c2 + c3 - 2 \cdot e_{init}$, the returned value is the maximum when these two coalitions are separately trying to maximize reward.

$$<<p1, p2, p3>> P^{\geq 1}[F \ c1 + c2 + c3 < 200] \qquad (7)$$

Here we present the first probability-based property. The direct result obtained is 1 if there must always exist a reachable state where the sum of player resources is below 200. In the PRISM log we can examine (Y, N, M) to see the fraction of states where this inequality holds.

$$<<p1, p2, p3>> Pmax =?[F <= k_{max} + 1c1 < c2] \qquad (8)$$

This property returns the maximum probability that player 2 has more resources than player 1 after k_{max} rounds. This is expected to be 1 since our CSG doesn't impose limitations on how player resources compare *to each other*. Similarly we expect the minimum probability to be zero, and we can obtain a fraction of states that satisfy this from the PRISM log.

4.3 Reward Maximization

We subject the environment to a property involving global reward maximization: $<<p1, p2, p3>> R``done123"max =?[Fk = kmax + 1]$, where the label $R``done123"$ specifies the total resources accrued after round k by all agents in the system. The results can be seen in Fig. 2. Interestingly, we see that when players within the system are instantiated with a lesser initial resource allocation, the maximum possible reward at the end of round 4 is greater than other cases. We believe this relationship to be paradoxical. We can view the slope of the reward plots as an indicator, where $r_{init} < 200$ produces greater rates of change and lesser stabilization as the rounds progress. Because the update step

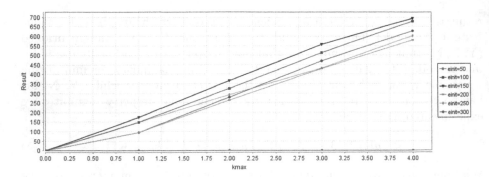

Fig. 2. An iterated run through the system with variable initial resources. The y-axis represents the total, aggregate group reward through time k. The different lines represent varying initial state conditions.

of current resource allocation takes into account expenditures, this leads us to believe that freeriding is a more popular choice of action when initial resources are more scarce. We also note that an optimal strategy is not reached round over round, as the aggregate reward falls below the ceiling of possible reward $300n$. We intend to explore theoretical analysis concerning agent participation in pursuit of cooperative welfare.

5 Limitations and Future Work

Reward Properties: Although it's simple enough to formulate properties involving a max or min over linear combinations of rewards, PRISM doesn't support the usage of probability bounds (max/min) or inequalities ($P >= p$) for such formulas. Luckily in this game all rewards are of the form $c_i - e_{init}$ with each c_i a player resource variable. Therefore this became a non-issue as we realized all reward formulas can be substituted if necessary.

Limitations in Multi-partition Property Analysis: We note that PRISM's support for CSGs is in beta-testing, and additionally the final release may feature limitations to prevent 'obvious' computational intractability. With that in mind, a challenge we faced was the inability to create more than two partitions for properties i.e. maximizing sum of player reward. Of course we are still allowed to feature more than two players in a property. But ultimately we lack the capability to fully analyze this game when each player is in a different coalition, and thus we work around this by extracting as much as we can from 1/2-coalition properties.

Strategy Graphs: Perhaps the most interesting analysis involves the strategy graphs generated for specific properties. Because our state space grows exponentially due to the mechanisms involving game-play, this is exceedingly difficult. For instance, we can look at a strategy graph for one round over three players and the strategy synthesis is easy to conceptualize. As the game progresses, it becomes computational taxing to conduct value iteration with an exploding state space (Fig. 3).

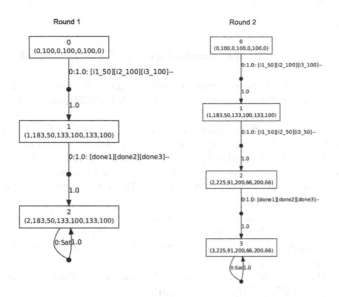

Fig. 3. Strategy Graphs can be used to find an optimal controller given a property. Here, we consider $<<p1 : p2, p3>> R"r1"max =?[Fk = kmax + 1]$ under the specified parameter set noted above. The graphs can be read via [k, c1, s1, c2, s2, c3, s3], where branching is determined by the actions taken concurrently by all agents in the system. Some interesting patterns emerge when looking at global reward maximization against optimal strategies. On the left, results are shown for a single round. From the init state of the game, the optimal strategy is for two players to donate in totality, and for one player to partially donate. On the right, we extend this to round 2. Here, global reward maximization is achieved as a result of full participation via partial contribution. In both cases, no agent within the system freerides.

6 Conclusion

We have presented a viable, working model for studying optimal and sub-optimal behaviors in multi-agent systems under probabilistic dynamics. We have also introduced and verified properties to check the correctness of our holistic approach, as well as having analyzed our reward mechanism under various conditions. Our analysis focused mainly on a single parameter set where a number of variables were fixed. Our model was presented as a concurrent stochastic game in which players were guided to cooperate with one another, but it is entirely possible that agents in a real-world system do not act cooperatively in the presence of such a dilemma. Perhaps, in the case of a democratic voting schema, a coalition of agents gains intrinsic satisfaction from minimizing the collective reward of an opposing coalition. This could be introduced by partitioning coalitions in the form of subgraphs in a graphical dynamic system. There, it would be of interest to explore such games under a combative approach where coalitions would oppose one another.

References

1. Public Good Game. http://prismmodelchecker.org/casestudies/public_good_game. php
2. Askitas, N.: Selfish altruism, fierce cooperation and the predator. J. Biol. Dyn. **12**(1), 471–485 (2018). https://doi.org/10.1080/17513758.2018.1473645, pMID: 29774800
3. Biró, P., Norman, G.: Analysis of stochastic matching markets. Int. J. Game Theory **42**(4), 1021–1040 (2012). https://doi.org/10.1007/s00182-012-0352-8
4. Chen, T., Forejt, V., Kwiatkowska, M., Simaitis, A., Trivedi, A., Ummels, M.: Playing stochastic games precisely. In: Koutny, M., Ulidowski, I. (eds.) CONCUR 2012. LNCS, vol. 7454, pp. 348–363. Springer, Heidelberg (2012). https://doi.org/ 10.1007/978-3-642-32940-1_25
5. Chen, T., Forejt, V., Kwiatkowska, M., Parker, D., Simaitis, A.: PRISM-games: a model checker for stochastic multi-player games. In: Piterman, N., Smolka, S.A. (eds.) TACAS 2013. LNCS, vol. 7795, pp. 185–191. Springer, Heidelberg (2013). https://doi.org/10.1007/978-3-642-36742-7_13
6. Hauser, O.P., Hilbe, C., Chatterjee, K., Nowak, M.A.: Social dilemmas among unequals. Nature **572**(7770), 524–527 (2019)
7. Kwiatkowska, M., Parker, D., Wiltsche, C.: PRISM-Games 2.0: a tool for multi-objective strategy synthesis for stochastic games. In: Chechik, M., Raskin, J.-F. (eds.) TACAS 2016. LNCS, vol. 9636, pp. 560–566. Springer, Heidelberg (2016). https://doi.org/10.1007/978-3-662-49674-9_35
8. Kwiatkowska, M., Norman, G., Parker, D., Santos, G.: Equilibria-based probabilistic model checking for concurrent stochastic games. In: ter Beek, M.H., McIver, A., Oliveira, J.N. (eds.) FM 2019. LNCS, vol. 11800, pp. 298–315. Springer, Cham (2019). https://doi.org/10.1007/978-3-030-30942-8_19
9. Kwiatkowska, M., Parker, D., Wiltsche, C.: PRISM-games: verification and strategy synthesis for stochastic multi-player games with multiple objectives. Int. J. Softw. Tools Technol. Transf. **20**(2), 195–210 (2017). https://doi.org/10.1007/ s10009-017-0476-z
10. Ummels, M.: Stochastic multiplayer games: theory and algorithms. Ph.D. thesis, RWTH Aachen, Germany, January 2010

Privacy Preserving Text Representation Learning Using BERT

Walaa Alnasser[1(✉)], Ghazaleh Beigi[2], and Huan Liu[1]

[1] Arizona State University, Tempe, USA
{walnasse,huanliu}@asu.edu
[2] Google, Sunnyvale, USA
ghazalb@google.com

Abstract. The availability of user generated textual data in different activities online, such as tweets and reviews has been used in many machine learning models. However, the user generated text could be a privacy leakage source for the individuals' private-attributes. In this paper, we study the privacy issues in the user generated text and propose a privacy-preserving text representation learning framework, DP_{BERT}, which learns the textual representation. Our proposed framework uses BERT to extract the sentences embedding to learn the textual representation that (1) is differentially private to protect against identity leakage (e.g., if a target instance in the data or not), (2) protects against leakage of private-attributes information (e.g., age, gender, location), and (3) maintains the high utility of the given text.

Keywords: Privacy · Text representation · Utility · Semantic

1 Introduction

User generated textual data is rich of information that can be used in different tasks such as understanding users' behavior and recommendation systems. From the privacy perspective, user generated textual data can cause a privacy leakage since it contains sensitive information about the individuals. There are several privacy issues related to textual data such as re-identification and private-attributes inference.

Online users who publish textual data may not be aware that their private information can be easily inferred by malicious adversaries. Many research studies have shown that the user generated textual data may reveal private information about the users. The following review from a dataset TrustPilot from Hovy et al. [9] is an example of such textual data:
"...*I should receive my shoes at the end of next week so I waited by the end of the week there was no shoes so my husband called and was told the order never shipped out...*"
It is clear that one can infer the user's gender (female) which is one of the private-attributes that the user may not want to share publicly.

© Springer Nature Switzerland AG 2021
R. Thomson et al. (Eds.): SBP-BRiMS 2021, LNCS 12720, pp. 91–100, 2021.
https://doi.org/10.1007/978-3-030-80387-2_9

Two categories in general of information leakages have been studied: identity disclosure and private-attributes disclosure. Identity disclosure happens when a targeted instance is mapped to an instance in a publicly released dataset while private-attributes leakage happens when the adversary is able to infer some of the sensitive information such as age, gender, and location [2]. To protect user's privacy, various protection techniques have been proposed such as k-anonymity [16] and differential privacy [6] which used to tackle the identity disclosure attack. However, these techniques have shown inefficiency to protect textual generated data for several reasons such as the data being unstructured and contains a huge number of short and informal post [8]. Besides, these techniques do not protect textual information against private-attributes leakage and also may have a negative impact on the utility as they do not take it as a part of the solution.

Our main contribution is proposing a framework, called DP_{BERT}, which learns a privacy preserved text representation that is differentially private to protect against identity leakage (if a target instance is available in the data or not), does not leak private-attributes information (age, gender, location, etc.), and preserves the semantic of the text.

2   Problem Statement

PROBLEM 1. *Given a set of documents X, a set of sensitive attributes P, and a task T, learn a function f that can anonymize the text embedding representation $\tilde{Z}_i$ for each document x_i in X so that, 1) the adversary cannot infer the targeted user's private-attributes P from the privacy-preserving text representation $\tilde{Z}_i$ and 2) the generated private representation $\tilde{Z}_i$ is preserving the utility for a downstream task T. The problem can be mathematically defined as [3]:*

$$\tilde{Z}_i = f(x_i, P, T) \tag{1}$$

3   The Proposed Framework

In this section, we discuss the details of the proposed model framework which is an extension to a novel double privacy preserving text representation learning framework, DBT_{EXT}, [3]. The illustration of the entire model is shown in Fig. 1. Our proposed model framework uses BERT for text representation. BERT [5] is a language representation model which is developed to pre-train deep bidirectional representations from a given text by jointly conditioning on both left and right context in all layers. Then, the differential-privacy-based noise adder adds random noise, e.g., a Laplacian noise, to the original text representation. Furthermore, two discriminators for the semantic meaning D_S and private-attributes D_P are defined to infer the optimal amount of the added noise to preserve the semantic meaning and minimize the leakage of the private-attributes. The final output of the proposed model is the manipulated embedding text $\tilde{Z}$ which is deferentially private, hides the private attributes, and preserves semantic meaning.

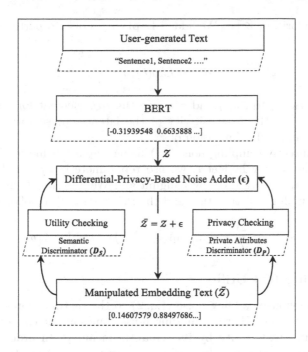

Fig. 1. The framework of DP_{BERT} architecture. It consist of four components, BERT, a differential privacy-based noise adder, a semantic discriminator D_S and a private-attributes discriminator D_P. The manipulated embedding text $\tilde{Z}$ is a noisy representation which is differentially private, hides private information and has semantic meaning

3.1 Sentence Representation Using BERT

Here, we illustrate how to extract the sentence embedding for a given text. Let $X = \{x^1, ..., x^m\}$ be a document that contains m sentences. We use BERT to extract the textual embedding because BERT has shown to be significantly efficient when modeling textual embedding [5,13,17]. In particular, we use SBERT [13] to extract the sentence embedding. SBERT is a modification of BERT network that uses siamese and triplet network structures to derive semantically meaningful sentence embeddings.

3.2 Perturbing Text by Adding Noise

In order to prevent text re-identification which lead to privacy leakage and revealing private information of the individuals, we need to add noise to the textual representation. In this work, we follow [3], where differential privacy technique is used for preserving the privacy of users' data. Differential privacy is one of the traditional privacy-preserving techniques [1]. Output perturbation mechanism is used which is achieving the differential privacy by adding Laplacian noise to the output of an algorithm $\tilde{Z}$ [4]. We added Laplacian noise to perturb the output Z as follows:

$$\tilde{z}_i = z(i) + s(i), s(i) \sim Lap(b), b = \frac{\Delta}{\epsilon}, i = 1, \ldots, d \tag{2}$$

where s the noise vector, $s(i)$ and $z(i)$ are the i-th element for vectors s and z, respectively, Δ is the L_1-sensitivity of the latent representation z, ϵ is the privacy budget and d the dimension of z.

Instead of directly sampling noise $s(i)$ using Laplacian mechanism to learn the value of the privacy budget ϵ, we use reparameterization trick which was introduced first in a work done by Kingma and Welling [11]. It first samples a value r from a uniform distribution and then rewrites $s(i)$ as follows:

$$\mathbf{s}(i) = -\frac{\Delta}{\epsilon} \times \operatorname{sgn}(r) \ln(1 - 2|r|), \quad i = 1, \ldots, d \tag{3}$$

3.3 Preserving Text Utility

As discussed previously, adding noise (Eq. 2) to the textual representation will prevent privacy leakage. However, adding noise comes at the cost of text utility loss. We measure text utility by its semantic meaning. In order to preserve the text's semantic meaning, the optimal amount of noise needs to be added to ensure not too much noise has been added to the textual embedding as it can reduce the utility of the textual information. This addresses the trade-off between preserving privacy and maintaining utility. We train a classifier to learn the amount of added noise with the privacy budget ϵ as:

$$\hat{y} = softmax(\tilde{z}; \theta_{D_S}) \tag{4}$$

where $\hat{y}$ represents the inferred label for the classification and θ_{D_S} are the weights associated with the softmax function.

For the semantic meaning of the text representation, we define a semantic discriminator D_S to assign a correct class label to the perturbed representation as follows:

$$\min_{\theta_{D_S}, \epsilon} \mathcal{L}(\hat{y}, y) = \min_{\theta_{D_S}, \epsilon} \sum_{i=1}^{C} -y(i) \log \hat{y}(i) \tag{5}$$

where $\mathcal{L}$ is the cross entropy loss function, C is the number of classes, y is the ground truth label for the classification task and $y(i)$ represents the i-th element of y.

3.4 Protecting Private Information

As we discussed, adding noise to the textual representation will prevent adversaries from inferring the user information. The other side of preserving text representation privacy is to ensure that the sensitive information of the individuals is not captured in the text representation. In our proposed model, we follow the idea of adversarial learning by training a private-attributes discriminator D_P that identifies the private information from the text representation [3]. At the same time, learning a representation that minimizes the leakage of private information by fooling the discriminator. The adversarial learning can be formally written as:

$$\min_{\left\{\theta_{D_P^t}\right\}_{t=1}^{T}} \max_{\epsilon} \mathcal{L}_{D_P} = \min_{\left\{\theta_{D_P^t}\right\}_{t=1}^{T}} \max_{\epsilon} \frac{1}{K.T} \sum_{t=1}^{T} \sum_{k=1}^{K} \mathcal{L}_{D_P^t}\left(\hat{p}_t^k, p_t\right), \text{ s.t. } \epsilon \le c_1 \quad (6)$$

where $\theta_{D_P^t}$ is demonstrates the parameters of discriminator model D_P, $\mathcal{L}_{D_P^t}$ represents the cross entropy loss function, $c1$ is a predefined privacy budget constraint, T is the number of private-attributes, $\hat{p}_t^k$ is the predicted t-th private-attribute using K-th sample and it is defined as follows:

$$\hat{p}_t^k = \text{softmax}\left(\tilde{\mathbf{z}}^k, \theta_{D_P^t}\right) \quad (7)$$

to calculate the predicted t-th sensitive attribute using the k-th sample. The outer minimization of equation (Eq. 6) finds the strongest private-attributes inference attack and the inner maximization seeks to fool the discriminator by obscuring private information.

3.5 DP_{BERT} - Learning the Text Representation

In the previous sections, we show how we can add noise to prevent the adversary from regenerate the original text from the embedding representation and minimize the chance of privacy leakage by achieving differential privacy (Eq. 2), control the amount of the added noise to preserve the semantic meaning of the textual information (Eq. 5), and to protect the private-attributes (Eq. 6). Inspired by the idea of adversarial learning, we model the objective function as a minmax game among the two discriminators D_P and D_S. Assume that $\mathcal{L}_{D_P^t}$ and $\mathcal{L}_{D_S}$ denotes cross entropy loss function for the private-attributes discriminator and cross entropy loss function for semantic task, respectively. Our goal is to maximization $\mathcal{L}_{D_P^t}$ that finds the strongest private-attributes inference attack and minimize $\mathcal{L}_{D_S}$ that measured by the incorrect label in the sentiment prediction. We can write the objective function as follows:

$$\min_{\theta_{D_S}, \epsilon} \max_{\left\{\theta_{D_P^t}\right\}_{t=1}^{T}} \mathcal{L}_{D_S} - \alpha \mathcal{L}_{D_P^t} + \lambda \Omega(\theta) \quad \text{s.t.} \quad \epsilon \le c_1 \quad (8)$$

where α controls the contribution of the private-attributes discriminator D_P in the learning process, and $\Omega(\theta)$ is the parameters regularizer.

4   Experiments

4.1   Data

We use a dataset from TrustPilot from Hovy et al. [9]. In this collected dataset, there are reviews on different products and ratings from one to five star.

Each review is associated with three private-attributes: gender (male, female), age, and location (Denmark, France, United Kingdom, and the United States). The same approach of [3] is followed in this work. We follow the sitting of [10] by categorizing age attributes into three groups, above 45 years, under 35 years, and between 35 years and 45 years. For the sentiment ground truth, we consider the review's rating score as a sentiment class.

4.2   Experimental Design

We compare DP_{BERT} with its variant $ORIGINAL_{BERT}$ which uses the original text representation without adding noise to it. We report accuracy score to evaluate the utility of the given text for a sentiment analysis [14]. Specifically, we use the rating score of each review for the sentiment prediction task. In addition, we report the examination of the text representation using F1 score for predicting the private-attributes. It is worth mentioning that the higher accuracy score for the semantic discriminator shows high utility for the sentiment task and a lower F1 score for the private-attributes discriminator demonstrates high privacy in the given text.

Since the maximum length of text for BERT is 512. We follow in this work the head-only method which keeps the first 512 tokens of the given text [15]. This head-only method considers that the important information that we need to capture from the text will be in beginning of a document.

4.3   Experimental Result

We have conducted experiments to answer three questions:

- **Q1-** *Utility:* Does the learned text representation preserve the utility of the original text by keeping the same sentiment preserved?
- **Q2-** *Privacy:* Does the learned text representation hide the private information?
- **Q3-** *Utility-Privacy trade-off:* Does the trade-off between the utility and privacy reach the optimal point without sacrificing any of them?

The experimental results are demonstrated in Table 1.

To answer the first question (Q1), we report experimental results for our proposed model using the sentiment analysis task. We predict sentiment of the textual information and measure the performance using the metric accuracy. The result of sentiment prediction for DP_{BERT} shows that the representation preserves the sentiment meaning of the textual data which means high utility. $ORIGINAL_{BERT}$ performs better than DP_{BERT} and the reason is that the first one uses the original text representation without adding noise to it which means high utility.

To answer the second question (Q2), we consider three different private information, i.e., age, location, and gender. We examine efficiency of the model in privacy protection using its performance in predicting values of different private-attributes. We measure performance of private-attributes predictor using F1 metric. Our proposed model has a remarkably lower F1 score which indicates higher privacy in terms of hiding the private-attributes. In addition, DP_{BERT} exceeds $ORIGINAL_{BERT}$ in terms of hiding the private information and does not sacrifice the utility significantly.

To answer the third question (Q3), we evaluate the utility loss against privacy improvement of the given text. DP_{BERT} has achieved a better trade-off results which are shown in high privacy and low utility loss comparing with $ORIGINAL_{BERT}$. $ORIGINAL_{BERT}$ achieves high accuracy in the sentiment task while suffering from significant privacy loss.

The results have shown that DP_{BERT} learns the textual representation of a given text that does not leak private information and preserve the semantic meaning of the text by achieving higher accuracy score for semantic discriminator D_S which indicates that representation has high utility for the semantic meaning, and lower F1 score for private-attributes discriminator D_P which demonstrates that the textual representation has higher privacy for users due to obscuring their private information.

Table 1. Accuracy for sentiment prediction and F1 for evaluating private attribute prediction task. Higher accuracy shows higher utility, while lower F1 demonstrates higher privacy.

Model	Sentiment (ACC)	Private attribute (F1)		
		Age	Location	Gender
$ORIGINAL_{BERT}$	0.3870	0.4330	0.3072	0.5623
DP_{BERT}	0.2446	0.1951	0.0671	0.4047

4.3.1 Parameter Analysis

Our proposed model, DP_{BERT}, has a parameter α which controls the contribution from the private attribute discriminator D_P. We do the experiments and check the effects of different values of α. We investigate varying α values as: 0.125, 0.25, 1, 2, 4. Figure 2(a) shows the accuracy for sentiment prediction for different values of α. We can see clearly that the increase of α will decrease the accuracy of sentiment prediction task for DP_{BERT} and $ORIGINAL_{BERT}$. The performance of private attributes discriminator for age, location and gender is shown in Fig. 2(b) using F1 score. We observe that setting $\alpha = 1$ will improve the accuracy for the sentiment prediction task and preserve the privacy by keeping the F1 score low.

5 Related Work

This section describes related work on the following areas: (1) sentence embedding; (2) Textual Data Privacy; and (3)Protecting Private-Attributes Information.

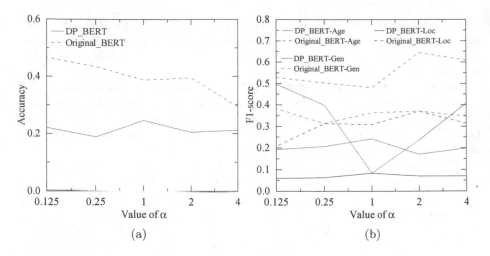

Fig. 2. Performance results for (a) sentiment prediction tasks for different values of α and (b) private attribute for different values of α. Higher accuracy shows higher utility, while lower F1 demonstrates higher privacy.

Sentence Embedding: is the process of converting a linguistic sentence to a numerical representation taking into account its meaning. The aim of sentence embedding is being able to use the logistic features for downstream tasks. There are two categories for sentence embedding techniques:non-parameterized and parameterized models [17]. Non-parameterized techniques such as tf-idf and uSIF [7] which depend on high-quality pre-trained word embedding techniques. On the other hand, parameterized techniques are more convoluted than non-parameterized techniques. SBERT [13] technique is based on BERT foundation. The work of [17], called SBERT-WK, proposed a new sentence embedding method by using geometric analysis of the space learned by deep contextualized models.

Textual Data Privacy: User-generated data has been used by researchers and service providers to better understand users' behaviors and offer them personalized services. However, publishing user-generated data may cause the problem of user privacy as this data includes information about users' private information. Several methods are reported in the literature to address how to preserve the textual information privacy. Beigi et al. [3] used adversarial learning to anonymize the textual information by proposing a privacy-preserving text representation learning framework, called DPText. The final output of this model is the text that obscures the private-attribute information by making sure the sensitive attribute does not capture in the latent representation so the adversary will not be able to infer these attributes. The previous study on creating private representation in the text field by [12] proposed a novel approach for privacy-preserving learning based on generative adversarial network GAN to train deep models with adversarial learning to improve the robustness and privacy of the neural representation. Their method has been evaluated on the tasks of part of speech tagging (POS) and sentiment analysis, protecting several demographic private-attributes such as gender, age and location.

Protecting Private-Attributes Information: Private-attribute information can be defined as the information that individuals do not want to explicitly disclose such as marital status, location, occupation, age, and gender. Numerous research studies have been mitigated the problem of privacy leakage. Recent works [1,2] identify different aspects of user privacy. In particular, [1,2] illustrate the privacy risks and compare different traditional privacy models for protecting user private-attributes.

Our work is distinct from the previous works in that we use BERT to extract the sentence embedding in our proposed model. Our model can be useful in preserving privacy in published datasets that are used for different tasks.

6 Conclusion

In this paper, we proposed a privacy-preserving text representation learning framework, DP_{BERT}, which learns a text representation that is differential private, preserves users' private information, and maintains high utility by keeping the sentiment meaning of the given text. It has four main components which are BERT, differential-privacy-based noise adder, utility checking which is semantic meaning discriminator, and privacy checking which is private-attributes discriminator. Our results showed the effectiveness of our proposed framework, DP_{BERT}, in minimizing chances of privacy leakage of the private-attributes while preserving text semantic meaning in the same time.

Acknowledgement. This work, in part, is supported by the Saudi Arabian Cultural Mission (SACM) in the United States.

References

1. Alnasser, W., Beigi, G., Liu, H.: An overview on protecting user private-attribute information on social networks. In: Cruz-Cunha, M.M., Mateus-Coelho, N.R. (eds.) Handbook of Research on Cyber Crime and Information Privacy, Chap. 6 (2020)
2. Beigi, G., Liu, H.: A survey on privacy in social media: identification, mitigation, and applications. ACM/IMS Trans. Data Sci. **1**(1) (2020). https://doi.org/10.1145/3343038
3. Beigi, G., Shu, K., Guo, R., Wang, S., Liu, H.: Privacy preserving text representation learning, pp. 275–276 (2019). https://doi.org/10.1145/3342220.3344925
4. Chaudhuri, K., Monteleoni, C., Sarwate, A.D.: Differentially private empirical risk minimization. J. Mach. Learn. Res. **12**(29), 1069–1109 (2011). http://jmlr.org/papers/v12/chaudhuri11a.html
5. Devlin, J., Chang, M.W., Lee, K., Toutanova, K.: BERT: pre-training of deep bidirectional transformers for language understanding (2019)

6. Dwork, C.: Differential privacy: a survey of results. In: Agrawal, M., Du, D., Duan, Z., Li, A. (eds.) TAMC 2008. LNCS, vol. 4978, pp. 1–19. Springer, Heidelberg (2008). https://doi.org/10.1007/978-3-540-79228-4_1

7. Ethayarajh, K.: Unsupervised random walk sentence embeddings: a strong but simple baseline. In: Proceedings of the Third Workshop on Representation Learning for NLP, pp. 91–100. Association for Computational Linguistics, Melbourne, July 2018. https://doi.org/10.18653/v1/W18-3012. https://www.aclweb.org/anthology/W18-3012

8. Fung, B.C.M., Wang, K., Chen, R., Yu, P.S.: Privacy-preserving data publishing: a survey of recent developments. ACM Comput. Surv. 42(4) (2010). https://doi.org/10.1145/1749603.1749605

9. Hovy, D., Johannsen, A., Søgaard, A.: User review sites as a resource for large-scale sociolinguistic studies. In: Proceedings of the 24th International Conference on World Wide Web, WWW 2015, pp. 452–461. International World Wide Web Conferences Steering Committee, Republic and Canton of Geneva, CHE (2015). https://doi.org/10.1145/2736277.2741141

10. Hovy, D., Søgaard, A.: Tagging performance correlates with author age. In: Proceedings of the 53rd Annual Meeting of the Association for Computational Linguistics and the 7th International Joint Conference on Natural Language Processing (Volume 2: Short Papers), pp. 483–488. Association for Computational Linguistics, Beijing, July 2015. https://doi.org/10.3115/v1/P15-2079. https://www.aclweb.org/anthology/P15-2079

11. Kingma, D.P., Welling, M.: Auto-encoding variational bayes (2014)

12. Liu, P., et al.: Local differential privacy for social network publishing. Neurocomputing 391, 273–279 (2020). https://doi.org/10.1016/j.neucom.2018.11.104. http://www.sciencedirect.com/science/article/pii/S0925231219304229

13. Reimers, N., Gurevych, I.: Sentence-BERT: sentence embeddings using Siamese BERT-networks. In: Proceedings of the 2019 Conference on Empirical Methods in Natural Language Processing and the 9th International Joint Conference on Natural Language Processing (EMNLP-IJCNLP), pp. 3982–3992. Association for Computational Linguistics, Hong Kong, China, November 2019. https://doi.org/10.18653/v1/D19-1410. https://www.aclweb.org/anthology/D19-1410

14. dos Santos, C., Gatti, M.: Deep convolutional neural networks for sentiment analysis of short texts. In: Proceedings of COLING 2014, The 25th International Conference on Computational Linguistics: Technical Papers, pp. 69–78. Dublin City University and Association for Computational Linguistics, Dublin, August 2014. https://www.aclweb.org/anthology/C14-1008

15. Sun, C., Qiu, X., Xu, Y., Huang, X.: How to fine-tune BERT for text classification? In: Sun, M., Huang, X., Ji, H., Liu, Z., Liu, Y. (eds.) CCL 2019. LNCS (LNAI), vol. 11856, pp. 194–206. Springer, Cham (2019). https://doi.org/10.1007/978-3-030-32381-3_16

16. Sweeney, L.: K-anonymity: a model for protecting privacy. Int. J. Uncertain. Fuzziness Knowl.-Based Syst. 10(5), 557–570 (2002). https://doi.org/10.1142/S0218488502001648

17. Wang, B., Kuo, C.C.J.: SBERT-WK: a sentence embedding method by dissecting BERT-based word models (2020)

GPU Accelerated PMCMC Algorithm with System Dynamics Modelling

Lujie Duan$^{(\boxtimes)}$ and Nathaniel Osgood$^{(\boxtimes)}$

Department of Computer Science, University of Saskatchewan,
Saskatoon, SK, Canada
{lujie.duan,nathaniel.osgood}@usask.ca

Abstract. Recent work demonstrates that coupling Bayesian computational statistics methods with dynamic models can facilitate analysis and understanding of complex systems associated with diverse time series, those involving social and behavioral dynamics. Particle Markov Chain Monte Carlo (PMCMC) is a particularly powerful class of Bayesian methods combining aspects of batch Markov Chain Monte Carlo (MCMC) and the sequential Monte Carlo method of Particle Filtering (PF). PMCMC can flexibly combine theory-capturing dynamic models with diverse empirical data streams. PMCMC has demonstrated great potential for broad applicability across social and behavioral domains. While PMCMC offers high analytic power, such power imposes a high computational burden. In this work, we investigated the effectiveness of using Graphical Processing Units (GPUs) in reducing run times. Specifically, we designed and implemented a GPU-enabled parallel PMCMC version with compartmental simulation models. Evaluating this work's impact with a realistic PMCMC health application showed that GPU-based acceleration achieves up to 160× speedup compared to a corresponding sequential CPU-based version. Use of the GPU accelerated PMCMC algorithm with dynamic models offers researchers a powerful toolset to greatly accelerate learning and secure additional insight from the high-velocity data increasingly prevalent within social and behavioral spheres.

Keywords: PMCMC · System dynamics · GPU · Markov chain
Monte Carlo · MCMC

1 Introduction

The growing availability of high-variety, high-velocity, higher-veracity data sources related to human behavior raises prospects for grounding increasingly articulated dynamic models emerging from research into human behavior. However, doing so requires effectively addressing two important challenges. Firstly, even the richest single data source typically offers evidence regarding only a confined subset of intrinsically multidimensional human behavior; grounding richer behavioral models typically requires drawing on more than one such dataset. Secondly, there is a need for a principled meshing of such diverse sources of temporal

© Springer Nature Switzerland AG 2021
R. Thomson et al. (Eds.): SBP-BRiMS 2021, LNCS 12720, pp. 101–110, 2021.
https://doi.org/10.1007/978-3-030-80387-2_10

behavioral data with theory captured in a model, one that takes into account the fact that often such data relates to emergent model and system behavior.

PMCMC is a set of particularly powerful Bayesian inference frameworks suitable for a broad range of inference problems, including those that combine multiple time series with dynamic models. The Marginal Metropolis Hastings (MMH) PMCMC method on which this paper specifically focuses allows for estimating the joint posterior distribution of both static model parameters and latent state of the posited underlying system over time. This method provides a means of arriving at a probabilistic consensus picture of the underlying system and parameter values that reflects both theory captured in a dynamic model and unfolding data. While there are many successful applications of the PMCMC method, the computational resources and prolonged runtimes constitute top challenges.

Although implementations of the PMCMC method in different programming languages are available, they suffer from several notable disadvantages. Implementations in statistical languages such as R support ready integration and interaction, but can suffer from slow execution that limits them to smaller models and simpler inference tasks. Others taking advantage of GPU or special hardware are challenging to integrate with large compartmental simulation models. Previous work from the authors includes a PMCMC library in language C and R to work specifically with compartmental System Dynamics (SD)/Ordinary differential equation (ODE) models [9]. While the high-efficient C implementation secures notable performance gain, running times remain high, particularly with large particle complements, larger MCMC chains and due to the numerical integration overhead in larger SD models. Given the larger particle complements frequently required for effective sampling in large compartmental models, it is not unusual for such run times to extend over days or weeks. This work demonstrates a high-performance graphic processing units (GPUs)-based implementation accelerating PMCMC use with SD simulations while retaining modularity with respect to model choice. We demonstrate a significantly reduction in the execution time, resulting in a quick convergence of the algorithm and a markedly reduced interval between empirical data availability and model results.

2   Background and Related Work

2.1   Particle Markov Chain Monte Carlo Methods and Parallelization

Particle Markov Chain Monte Carlo [1] is a Bayesian method of rising popularity that combines the power of PF and MCMC and aims to provide solutions for joint inference of state and parameter values in even high dimensional complex systems. This sampling-based estimation algorithm uses PF to simplify the likelihood calculation of the MCMC process, enabling it to explore high-dimensional parameters space efficiently with time series data [3], making the MCMC algorithm general and universal. While of relatively recent origina, PMCMC has been applied across many areas including bioinformatics, hydrologic and environmental modelling and inference, robotics target tracking and mobile wireless channel tracking [5,7,13,16]. One rising sphere of application of PMCMC lies in estimation of states and parameter values of compartmental models in health, such as SIR transmission models. This was shown to work well with complex systems characterized as including many latent variables and multi-dimensional observation data [9]. Practical PMCMC use requires addressing several challenges, including effective achievement of convergence, tuning parameters and hyperparameters, choice of priors, and challenges in specification of models. But a foremost pragmatic one lies in managing the heavy associated computational burden. Fortunately, because of the modular way in which PF is combined with the MCMC sampling within PMCMC, many methods developed to speed up those methods individually can also be readily adapted to PMCMC. Past efforts have contributed many innovations to shorten the execution time for PF or MCMC separately or together; some have tried to parallelize such algorithms with multi-core machines, distributed clusters, GPU and FPGAs [6,8,10,11] while others sought to simplify and modify the algorithm so that the computation can be reduced or become easy to parallelize, while maintaining the accuracy of the algorithm [2,12,14]. This paper sought to accelerate the use of PMCMC with complex SD models by migrating the algorithm to special hardware without changing the mathematical structure of the PMCMC algorithm.

2.2 GPU Programming and CUDA

Graphic Processing Units (GPU) can accelerate computationally intensive algorithms such as machine learning and computational statistics methods effectively. For example, Deep Learning has benefited from large parallel hardware such as GPUs for fast learning, processing, interpreting and parameter tuning. Compute Unified Device Architecture (CUDA) is a general-purpose heterogeneous programming language for both graphic and non-graphic calculations on easily accessible NVIDIA GPUs. Programming with GPU differs from multi-threading CPU programming. The dramatically distinct core counts and features between the host CPU and GPU devices require use of a different programming model.

Much optimization for host CPUs focuses on latency when completing tasks, while GPUs optimize for throughput, reflective of their purposeful design for rendering images and videos, where the computing time for individual pixels is crucial for finishing the full image containing thousands of pixels. GPUs have many simple processing cores that support single-instruction-multiple-data (SIMD) semantics. To accommodate I/O needs, GPUs are usually equipped with high-bandwidth, high-clock speed and large memory right on the chip to address increasing demand for high graphic resolution and frame rates. Even though GPUs were not built for computer vision and image processing tasks initially, GPU characteristics make it perfect for those non-sequential high-throughput applications. GPU structure allows programmers to spawn up thousands of threads simultaneously. The memory hierarchy and management are also distinct. Instead of the notation of stack, heap, and cache, GPUs have global, constant, shared memory, etc., each for designated purposes; poor usage can severely impair performance. CUDA virtualizes the physical hardware, such that a thread is a virtualized scalar processor, with registers, program counter and state, and a block is a virtualized multiprocessor with shared memory.

3   Implementation

Starting from the central computational task of the algorithm, the GPU implementation's first step involved moving the differential equation calculation of the PF process to the GPU. With a medium-sized model's ODE, when integrating over one time step with 16384 particles, the CPU takes 18.6 s while only 0.194 s for GPU with 1920 CUDA cores, a 95 fold improvement for this core function. The next step was moving the likelihood calculation, which is executed right after each single time step. Both steps are straightforward; because particles do not interact during numerical integration, no inter-thread communication or synchronization is required. A key step requiring cross-thread synchronization within PF (and, by extension, PMCMC) concerns the observation step, during which weights must be updated for each particle, normalized, and resampling performed. After weights have been normalized and indices of the particles to carry over to the next time step have been decided following resampling, a swap of particles will copy all the state variables of the parent particle to the new child particle locations. To reduce memory usage on the device and the host, this swap occurs in-place by reusing the state variable array; a synchronization call is needed to make sure all *READ*s have been finished before any *WRITE*s to the array.

The heterogeneous programming model means that the memory management operations can be expensive; these include allocating memory, copying variables between the device and the host, and reading by GPU chips from GPU memory or cache. Whenever possible, we should reuse memory allocation and minimize memory operations; we investigated the trade-offs between memory allocations and memory copying when achieving both concurrently is impossible. Since each PF step would only read the empirical values observed at the current time step, naturally we would seek to reuse the same allocated memory across all PF steps, and then copy the values to the device. Alternatively, we can allocate the entire observation matrix and copy the data over at the start. We would then trade performance with space, as combining memory copy actions can better utilize the host-device communication bandwidth. We observed 2.3% to 4.7% performance improvement by allocating all memory for the observation matrix compared to repeated copying from the reused allocated memory.

Constant memory lives in the device DRAM but is cached in several layers. In other words, once fetched, the read-only constant memory is stored very close to the processing units, and the cost of reading from constant memory is minimal, similar to reading from per-block shared memory. The empirical observation array would be a perfect candidate to move to the constant memory: firstly, no modifications are required for the observations; secondly, all particles are trying to access the same observation at the same time point of the entire time series; finally, because all the particles need to access the observations, it is better to use constant memory than shared memory or texture memory. We observed a 0.16%

to 3% performance improvements. While this relatively small because of the smaller length of the observation array we used, it bears emphasis that not only does this mean a smaller allocation overall, but the allocation and initialization only happens once for the entire PMCMC chain, so the per-MCMC-iteration overhead will also diminish with a longer chain.

4   Using Particle MCMC in Influenza Model Inference

Bayesian computation methods enable the ready combination of empirical data with SD models to characterize communicable disease outbreaks; the availability of information-rich and easy-to-access social media and other online data further enhance such models' predictive accuracy. [15] found adding Google Search volume on related topics could substantially elevate predictive accuracy of an influenza model used with PF beyond what is possible using clinical data alone. It also demonstrated how multiple data sources could be used together in a SD model to characterize influenza spread, and supports the hypothesis that fear of getting infected is also contagious. Work described in this section adapted this model to PMCMC and compared how the CUDA-accelerated PMCMC code can be accelerated by such simulations.

4.1   Model Description

As shown in Fig. 1, the influenza model adapted from [4] has eight stocks. Besides the commonly known susceptible, exposed, infectious and recovered stocks, four more stocks were introduced to simulate the portion of the population who are anxious about the pandemic and eventually will become self-isolated from the general population. Our PF model described in [15] was built in AnyLogic with the PF method and has been run with the Canadian populations of Quebec and (separately) Manitoba, respectively. We modified the model to target the population of Saskatchewan and added one more data source: The tweet counts with related words. The weekly Google Search volume was collected for this province and the clinical data was available weekly[1] for the flu season from October to March each year. When calculating the total likelihood, we used the product of the likelihoods calculated from the three time series.

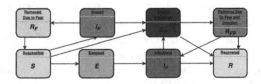

Fig. 1. Stock-and-flow diagram for the influenza model (adapted from [15])

---

[1] Sask. Weekly Influenza Surveillance Reports: https://tinyurl.com/SKFluReports.

4.2   Model Inference Results

Use of PMCMC with the adapted model allow inference of parameters and states over time, based on the empirical data. Figure 2 depicts the distribution (across MCMC iterations) of sampled states over time as a 2D histogram. The x-axis shows model time (in weeks), as the data extends over 20 weeks. The y-axis is the values of model state variable. The density represents the distributions of the particles, where particles are used to approximate the statistic possibilities. The state values are directly comparable to the empirical data, and the empirical data are shown plotted on top of the density. Essentially, the sampled values' distributions represent the density of the model's belief of the current state, while the empirical data at that step will update the belief using Bayes' theorem. We can see that the empirical data is mostly falling in the highest posterior density region, which suggests that the model performs well in inferring accuracy.

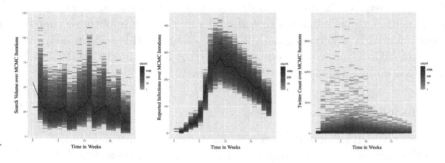

Fig. 2. Empirical data compared to corresponding model posterior distribution.

5   Experiments and Results

This work performed experiments to assess the performance improvements, to better understand the computational patterns, and investigate hardware utilization. By comparing the running time between the original serial version of the algorithm and the new GPU-accelerated parallel implementation on different testing environments, we sought to identify variables impacting the running time most significantly and locate bottlenecks. We also sought to uncover the scalability of the implementation by varying parameters such as the counts of MCMC iterations and particles, trying to identify the pattern and anticipate the potential improvements of GPU version over CPU version if more powerful GPUs become available. The hardware specifications are important factors moderating performance improvements and need to be considered when looking at the results. As shown in Table 1, test machines and environments were selected to represent a wide range of machines, including CPU-, memory-, or GPU- bounded single machines, or high performance computing environments and remote servers. Each set of experiments has been repeated 20 times, to decrease the effects of resource competition of other applications. All used an ODE integration time step of 0.01.

Table 1. Testing machines GPU specifications

ID	Name	CPU	GPU	GPU memory
1	D1	Intel i7-4770k	GTX 1050 Ti 768 cores	4 GB 112 GB/s
2	D2	Intel i3-6100	GTX 1070 1920 cores	8 GB 224 GB/s
3	D3	AMD A6-5400K	GTX 1050 Ti 768 cores	4 GB 112 GB/s
4	Cedar	Intel Xeon E5-2650 v4	Tesla P100 Pascal 3584 cores	12 GB 549 GB/s
5	Graham	Intel Xeon E5-2683 v4	Tesla P100 Pascal 3584 cores	12 GB 549 GB/s
6	Beluga	Intel Gold 6148	Tesla V100 5120 cores	16 GB 900 GB/s
7	Kepler2	Intel Xeon E5-2620 v2	Tesla K20c 2496 cores	4 GB 208 GB/s
8	Azure NC6	Intel Xeon E5-2690 v3	Tesla K80 4992 cores	24 GB 480 GB/s

With baseline settings for the performance experiment, with 4096 particles and 800 MCMC iterations, the GPU-accelerated version was faster across testing machines, and the fastest was when running on *Beluga*, benefiting from the largest CUDA core count, abundant memory and fastest memory bandwidth. By contrast, the higher CPU clock speed and larger system memory contributed to *D2* achieving the fastest CPU version running speed. Scaling up the count of particles by a factor of eight to 32768 and reducing MCMC iterations down by the same factor to 100, we evaluated the relative impact of those critical settings. The running times were compared for each machine between the two configurations and presented in the left plot in Fig. 3. They exhibited similar execution time running sequentially, reflecting the similar amount of tasks are involved. With parallel code, the difference was unnoticeable for a relatively capable host and device. The count of particles would decide the task size at each MCMC iteration, and those tasks would be launched together and scheduled to run by the device. The not-so-different running time suggested that our implementation handled it well and did not require extra overhead to process a larger count of particles. On resource-restricted machines, such as *D2*, *D3*, and *K2*, the difference between the performance of these GPU configurations is far more significant. It is possible that the host and device bus speed prevented transferring a larger count of particles efficiently; the limited count of cores means it would take longer to finish threads synchronization when needed.

The middle plot in Fig. 3 displays the CPU over GPU run time ratio distribution across different testing environments. For current hardware, it is easier to parallelize a low particle count, yielding a constant proportional improvement for low particle counts. Even though a larger particle count yields smaller improvements on most lower-end GPU devices due to the overhead associated with mapping different groups of particles directly to the GPU cores, high-end GPUs can be fully utilized by all those particles, achieving up to 160× speedup.

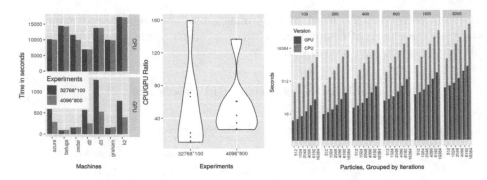

Fig. 3. Left: running time differences between configurations on different machines' CPU and GPU; middle: violin plot for speed up ratio across all machines; right: running time in seconds when varying different counts of iterations and particles for influenza model, on Log2 scale on testing machine Cedar

Next, we investigated scalability when keeping invariant the testing environment. The right plot in Fig. 3 shows the running time in seconds for both parallel and non-parallel versions when changing count of iterations and (separately) particles. Arranging the results by ascending particle count and grouping them by iteration count on a $log2$ scale, it is clear that the CPU version of the code exhibits a log-linear escalation in running time with iteration counts; for the GPU implementation, the running time increases faster when the count of particles increases within each iteration group. This reflects the fact that as the count of particles increases, the task size at each MCMC iteration grows and extra time is required for host-device communication and thread synchronization, which scales non-linearly with the task size.

Across the particle and iteration counts parameter settings, we can observe marked speedup; the actual improvements secured depends on the configuration. Figure 4 shows the CPU over GPU running time ratio with different particles and iteration counts on individual testing machines, *Cedar* and *D1*, as samples of HPC and personal workstations are grouped by the count of particles. Within a given particle count, the performance gain (improvement ratio) remains close to constant across different iteration settings. By contrast, with increasing particles count, the improvement ratio initially rapidly increases, and then reaches a plateau at which hardware is fully utilized. For *D1* compared to the first testing machine, the peak improvement ratio was reached much earlier, with around 2048 to 4096 particles, due to hardware limitations, from memory size, cores and SMs, to core and memory clock speed. On this more resource-constrained machine, the performance improvement secured by GPUs starts to decrease due to a struggle in handling the excess of concurrent tasks. With the maximum particle count (16384), this overhead leads to an improvement ratio the same as that secured when the device is underutilized with just 512 particles.

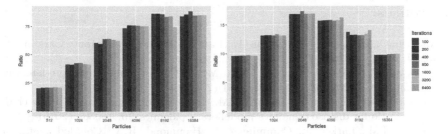

Fig. 4. Running time CPU over GPU ratio, with different iterations, on testing machine Cedar (left) and D1 (right)

6 Discussion

The performance experiments on various testing environments makes it relatively easy to predict performance improvement as hardware advances. One core design philosophy of CUDA that provision of a higher-level abstract programming interface independent of the actual scale and layout of parallel cores in the underlying hardware, can allow an efficient and scalable implementation of an application to take advantage of more cores and memory right away and without the need of any major code changes. Our testing results, carried out among several generations of CUDA-enabled GPUs, show that the implementation reflects such aspirations: We can scale up the experiments and particle count for more timely receipt of accurate results.

A key need of social and behavioral modeling is to mesh diverse high-velocity datasets with theory. With its capacity to derive a joint posterior distribution over the latent state of a system over time together with the value of associated static parameters, PMCMC represents an important and rapidly rising method capable of integrating diverse longitudinal behaviorally relevant data with theory captured in the form of a dynamic model. In this work, we provided a novel implementation of the GPU-accelerated PMCMC algorithm with dynamic modelling, demonstrated it with an influenza transmission simulation enriched with social media and online activity data, and performed various experiments targeting the performance pattern of our work on diverse testing environments. This work is already starting to facilitate effective adoption of PMCMC to build state-of-the-art simulation models taking advantage of today's tremendous amount of high-velocity social and behavioral data available online, through mobile devices, and elsewhere, while reducing the time needed to develop, tune and deliver such models with a well-considered parallel implementation on accessible and widespread commodity graphic processing units.

Acknowledgements. The authors gratefully acknowledge Dr. Juxin Liu's methodological guidance on PMCMC. NDO further expresses his gratitude to SYK for inspiring and making sustainable delivery of this work. This research was enabled in part by NSERC support and Compute Canada computing resources (www.computecanada.ca).

References

1. Andrieu, C., Doucet, A., Holenstein, R.: Particle Markov chain Monte Carlo methods. J. R. Stat. Soc. B **72**(3), 269–342 (2010)
2. Brockwell, A.E.: Parallel Markov Chain Monte Carlo simulation by pre-fetching. J. Comput. Graph. Stat. **15**(1), 246–261 (2006)
3. Endo, A., van Leeuwen, E., Baguelin, M.: Introduction to particle Markov-chain Monte Carlo for disease dynamics modellers. Epidemics **29**, 100363 (2019)
4. Epstein, J.M., Parker, J., Cummings, D., Hammond, R.A.: Coupled contagion dynamics of fear and disease: mathematical and computational explorations. PLoS ONE **3**(12) (2008)
5. Golightly, A., Wilkinson, D.J.: Bayesian parameter inference for stochastic biochemical network models using particle Markov chain Monte Carlo. Interface Focus **1**(6), 807–820 (2011)
6. Henriksen, S., Wills, A., Schön, T.B., Ninness, B.: Parallel implementation of particle MCMC methods on a GPU. In: 16th IFAC Symposium on System Identification, vol. 45, pp. 1143–1148 (2012)
7. Kattwinkel, M., Reichert, P.: Bayesian parameter inference for individual-based models using a Particle Markov Chain Monte Carlo method. Environ. Model. Softw. **87**, 110–119 (2017)
8. Lee, A., Yau, C., Giles, M.B., Doucet, A., Holmes, C.C.: On the utility of graphics cards to perform massively parallel simulation of advanced Monte Carlo methods (2010)
9. Li, X., et al.: Illuminating the Hidden Elements and Future Evolution of Opioid Abuse Using Dynamic Modeling. Big Data and Particle Markov Chain Monte Carlo, SBP-BRiMS (2018)
10. Lovell, D., Malmaud, J., Adams, R.P., Mansinghka, V.K.: ClusterCluster: Parallel Markov Chain Monte Carlo for Dirichlet Process Mixtures, April 2013
11. Mingas, G., Bottolo, L., Bouganis, C.S.: Particle MCMC algorithms and architectures for accelerating inference in state-space models. Int. J. Approx. Reasoning **83**, 413–433 (2017)
12. Mingas, G., Bouganis, C.S.: Population-based MCMC on multi-core CPUs, GPUs and FPGAs. IEEE Trans. Comput. **65**(4), 1283–1296 (2016)
13. Nevat, I., Peters, G.W., Yuan, J.: Channel tracking in relay systems via particle MCMC. In: IEEE Vehicular Technology Conference (2011)
14. Quiroz, M., Kohn, R., Villani, M., Tran, M.N.: Speeding up MCMC by efficient data subsampling. J. Am. Stat. Assoc. **114**(526), 831–843 (2019)
15. Safarishahrbijari, A., Osgood, N.D.: Social media surveillance for outbreak projection via transmission models: longitudinal observational study. JMIR Public Health Surveill. **5**(2), e11615 (2019)
16. Wang, S., Huang, G.H., Baetz, B.W., Ancell, B.C.: Towards robust quantification and reduction of uncertainty in hydrologic predictions: integration of particle Markov chain Monte Carlo and factorial polynomial chaos expansion. J. Hydrol. **548**, 484–497 (2017)

An Analysis of Global News Coverage of Refugees Using a Big Data Approach

Veysel Yesilbas⬚, Jose J. Padilla⬚, and Erika Frydenlund⁽⊠⁾⬚

Virginia Modeling, Analysis, and Simulation Center, Suffolk, VA, USA
efryden1@odu.edu

Abstract. The number of refugees and other displaced persons is at an all-time high globally. News coverage of displacement often evokes negative and even dehumanizing portrayals of the situations, politics, and economics of refugees and their host communities. In this study, we use GDELT to collect a large representation of global news across languages and countries to understand the volume, tone, and topics of news media coverage of refugees. Looking at specific events that trigger peaks in news coverage, we find that while the tone of news is always negative, this negativity reflects both empathy and sorrow for refugees as well as anti-migrant sentiment. World Refugee Day is a significant annual event that drives up news coverage at the international level, even though this also causes an increase in negative tone of the articles, reflecting the empathetic aspects of negative tone. At the international level, talking rather than action codes dominate the news. We suggest this as an area of future work, but it points to the large role that international diplomacy and international appeals for aid play in the humanitarian response of refugee situations.

Keywords: Big data · Refugees and migrants · Media analysis

1 Refugees and Other Displaced Persons in the Media

An unprecedented number of people are currently displaced in this world, nearly 80 million people as of June 2020 [1]. While most of these displaced persons—over 86%—live in developing countries, news stories of their journeys and situations reach every corner of the world. Displacement is not a monolith. United Nations definitions and data encompass different types of displacement such as internal to one's own country (IDPs), across international borders (refugees and asylum seekers), those in "refugee-like" situations, and those who have found themselves with no nationality for various reasons (stateless). These categories can overlap (stateless and asylum seeker) and change over time (IDP then refugee). In this study, we use the term "refugee" and "migrants" to speak of displacement in the media but recognize that labeling displacement is complex and changing.

Understanding about the news coverage related to refugees and migrants is emerging as more data and tools become available. Some studies suggest that it is too soon to definitively suggest securitized narratives about refugees in certain country's coverage

© Springer Nature Switzerland AG 2021
R. Thomson et al. (Eds.): SBP-BRiMS 2021, LNCS 12720, pp. 111–120, 2021.
https://doi.org/10.1007/978-3-030-80387-2_11

of refugee-related issues [2], while others are more definitive about the "dehumanizing" nature of media coverage that positions refugees as threats to their host communities [3]. The nature of news coverage may vary by region by the type of framing used (e.g., securitization, political, economic), and by the fact that most stories about refugees do not include the voices of refugees themselves, leading to a global conversation *about* but not *with* those experiencing displacement [4]. Some argue that visual depictions of refugees in the news creates a dehumanizing representation of those forcibly displaced [5]. In this study, we add to the conversation about media coverage of refugees and migrants by taking a big data approach, utilizing over six years' worth of global news coverage across over 100 languages and countries to understand what kinds of insights this global-level view can provide.

2   Methodology

We use GDELT to understand how the tone of refugee and migrant-related news coverage changes between countries. Beyond other databases, GDELT has larger coverage, more detail, and better structured data that allows deeper analysis. We mined 873,755 events representing coded news articles from the GDELT Events Table through Google BigQuery in the Google Cloud Platform using SQL syntax to collect data on the tone and numbers of articles about refugees.

2.1   Dataset

The GDELT Project is an online data repository of news collected across many news outlets and languages that updates every 15 min [6]. To structure this data, GDELT relies on CAMEO and other systematic means of coding event data. CAMEO is an event coding structure that enables GDELT to automatically label news articles according to a structure originally designed to understand different aspects of international disputes [7]. While this study is not about international disputes, the framework provides a useful overview of the types of events—from public statements to fighting and conflict about a topic of interest.

There are other news aggregators, such as Factiva [8], that offer a searchable collection of current news articles covering many countries and languages of origin. In addition to its highly structured coding scheme, GDELT has additional built-in variables for tracking article tone that calculate the difference between the percentage of positive and negative words in an article compared against a standard lexicon to give an overall score [6]. This provides an additional layer of insight about the data beyond simply the quantity of articles filtered by topic and location.

2.2   Data Query

GDELT Event records are stored in an expanded version of the dyadic CAMEO Event and Actor Codebook format, capturing actions performed by Actor1 upon Actor2 [9]. These actors can be specified according to CAMEO attributes. In our query, Actor1 is the country and Actor2 is a refugee or displaced person ("REF"). This means that, in this study, refugees are the ones being spoken about or acted upon, and countries are the actors doing the action.

The GDELT codebook, which relies heavily on the CAMEO structure, provides a long list of codes associated with each article from which to query [9]. Those selected for this study are described in Table 1. For our query, we specified that Actor2Code is "Ref," for refugees. These actors can be further disaggregated by Actor2Name and, according to the GDELT codebook, can include: "Asylum seeker, boat people, displaced families, displaced family, displaced people, displaced person, displaced residents, expatriate, indigenous refugees, refugee, returnee, refugee returnee, stateless person" [9]. As can be seen from that list, this means that the article coverage is about displacement broadly, but not so broad as to cover related topics such as immigration.

Table 1. Queried codes from GDELT codebook

Code	Input	Output
Actor1Country		Country doing the event related to refugees
Actor2Code	Ref	
Actor2Name		Specific name of actor within the *Actor2Code*
EventRootCode		All CAMEO event codes attached to the article
AvgTone		Values ranging from ($-100,100$) indicating negative to positive tone over the entire article

2.3 Limitations

The main limitations of this study come from the way that GDELT classifies news articles. It was not possible, though we tried different methods, to determine the amount of noise or proportion of false news classification in GDELT. We noted cases where news was possibly related to refugees through spot-checking, but not news in the way we expected. For example, one article coded as Actor1Country = "Germany" and Actor2Name = "Refugee" was for a university event on the Holocaust held in the United States of America [10]. In this study, the number or ratio of such "noise" in the dataset is unknown. The spot-checking we conducted suggests that this number may not be high enough to impact the high-level view of the data trends.

Another limitation is that "refugee" is not a universal term for displaced populations about whom we are interested, and terms such as "migrant" or "foreigner" common in South Africa, or "migrantes Venezolanos" in Colombia may not be covered very well in our query. We are still investigating this issue. Finally, though GDELT has impressive geographic and linguistic reach, it is missing parts of the global news landscape that is accessible online, particularly in developing countries [10], though news sources and codebooks are being added all the time.

3   Global News About Refugees

At the global level, we can draw some high-level insights about the collection of articles from January 2015 to March 2021. The number of countries reporting news related to refugees has been on the decline in the last six years (Fig. 1, top), likely losing some momentum as the Syrian refugee and European migration crises have somewhat stabilized. This corresponds to the data showing that the total number of articles peaked in the latter half of 2015 and has slowly tapered off (Fig. 1, middle). In general, these articles have had a negative tone, though most of this negative tone is centered around specific peaks in news coverage. Notice in Fig. 1 that graphs for the total number of articles and total tone mirror one another. The average tone, however, has stayed relatively constant until it became slightly more negative after April 2020, possibly in response to COVID-19. What is unclear from this is whether negative tone indicates anti-refugee coverage or empathy for the dire conditions that refugees found themselves in. For that, we must first understand what drives the peaks in article volume related to refugees and then try to reconcile "tone" as a variable with the types of CAMEO event codes attached to each article.

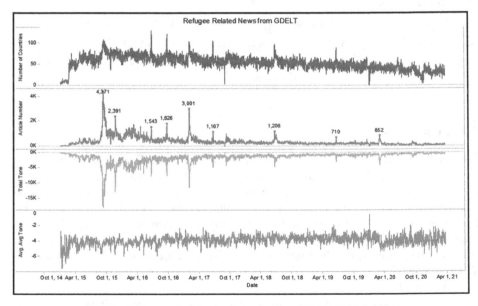

Fig. 1.   Refugee related news from January 2015 to March 2021

3.1   Refugee-Related Events Associated with Peak Coverage

Looking first at events associated with peaks, we used Total Number of Articles, Average Article per Country and Sum of Average of Tones. This revealed 14 major events (Fig. 2 and Table 2). Four of these were connected to the United Nation's annual World Refugee Day [11] (not included in Table 2), which brings to light the conditions and stories of refugees around the world to advocate for refugees' rights and inclusion. From this, we can gather that even positive events can have a negative tone. This negative tone may then represent moments of collective empathy in the news media in addition to anti-migrant coverage.

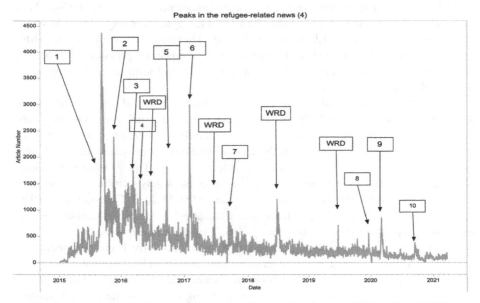

Fig. 2. Peaks Identified in the refugee related news, from January 2015 to March 2021

Table 2. Major events by date and tone

	Date	Major associated event	AvgTone
1	September 2, 2015	Death of Alan Kurdi	−4.3
2	November 13, 2015	Paris Attacks	−3.7
3	March 7, 2016	EU-Turkey Refugee Summit	−3.8
4	April 16, 2016	Pope Francis visit to a refugee camp in Greece	−3.5
5	September 20, 2016	UN Refugee Summit	−4.1
6	January 29, 2017	Mass shooting at Canadian Mosque	−4.3
7	September 6, 2017	EU court upholds member states' refugee quotas	−4.3
8	December 17, 2019	First Global Refugee Forum	−3.3
9	March 1, 2020	Turkey/Greece Border Clashes	−4.5
10	September 10, 2020	Fire at Moria refugee camp in Lesvos, Greece	−4.6

Five of the events included organized international diplomatic meetings, visits, and court decisions about refugees. The first was a diplomatic meeting between leaders of the European Union and Turkey to try to stem the flow of migrants and solidify relationships for managing migration in the region [12]. The second was Pope Francis's visit to Lesvos, Greece and bringing 12 refugees back with him for resettlement in Italy [13]. The third event was the United Nations Summit for Refugees and Migrants 2016. This was a massive moment in for international refugee protections where, for the first time [14], the UN assembled international Heads of State and unanimously adopted The New York Declaration that began the path towards the Global Compacts for migration [15]. In the fourth event of this type, the EU courts ruled against Hungary and Slovakia's complaints against the EU migration policy and member states' refugee reception quotas [16]. The final event was the first global refugee forum, a meeting of ministers following up on the Global Compacts for migration [17].

The remaining five events included accidents, clashes, fire, and attacks. The highest peak, and probably most globally recognized event related to refugees in recent years, was the death of a Syrian boy, Alan Kurdi, as his family attempted to cross the Aegean Sea from Turkey into Greece—the path that over one million people made in that year alone. The next event was the Paris attacks that left 130 dead. This was linked to at least one refugee who may have passed through Greece, immediately sparking discussions of stricter border control and surveillance of migrants [18]. In 2017, Canada was shocked by a mass shooting in a mosque. Some attributed the attack to the Prime Minister's recent welcome of Syrian refugees to the country [19]. Though refugees were not involved in the incident in any way, Canada's migration policies may have been the impetus for the attack, so discussions about refugees featured prominently in global coverage of the event. The fourth event, clashes between Turkey and Greece involving migrants, involved migrant casualties of those who were being used to incite political tension between the two countries [20]. The final event involved a fire at Moria refugee camp in Lesvos, Greece [21].

Looking more closely at the most dramatic article volume peak in Fig. 3, the death of Alan Kurdi—where news media shows images of his small body washed up on the shore—caused a 200% increase in the number of refugee-related news (from 1,461 to 4,371 articles). The total tone across all articles went from −6,292 to −17,490, approximately 178% decrease in the days following the event, peaking five days after his death on September 7. It took about 23 days for the news to reach pre-event levels (tone = −6,320; articles = 1,641 by September 25). The story also had wide global coverage, where the number of countries covering the news increased from 76 to 104. Figure 3 and Table 3 depict how the coverage of this event evolved in article numbers and tone. This event caused both the largest surge in articles and the biggest drop in tone, but the coverage was a global outcry about refugees' conditions. This is further evidence that the negative tone can be empathy for refugees and not simply anti-migrant sentiment.

3.2 Observing the Content CAMEO Codes of Refugee Related News

Thus far, we have established that tone and the number of articles can help to identify key events in a global media conversation about refugees. We have also demonstrated that "negative tone" can encompass both empathy and disdain for refugees or other communities associated with refugees and migrants. GDELT's use of CAMEO codes

Table 3. Event summary for the death of Alan Kurdi in global news coverage

Event Phase →	Event	Peak	End
Date in 2015 →	*Sept 2*	*Sept 7*	*Sept 25*
Total Number of Articles	1,461	4,371	1,641
Total Number of Countries	76	104	82
Average Number of Articles per Country	19	42	20
Sum of Average of Tone of Countries	-6,292	-17,490	-6,320
Mean of the Average Tone	-4.3	-4	-3.85

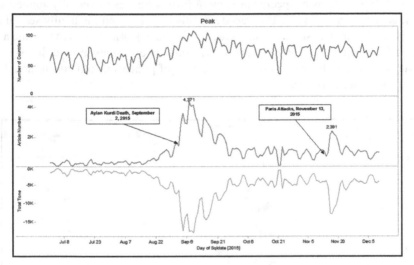

Fig. 3. Countries covering news, number of articles globally, and total tone related to media coverage of the death of Alan Kurdi

can offer an additional layer to examine the drivers behind those changes; that is, what are the actions or events being covered in these articles?

Breaking down the most impactful event in our dataset, the death of Alan Kurdi, we can look at the distribution of CAMEO codes across the event timeline. There had been several small peaks in news articles leading up to his death. In the global context, it was becoming clear at that time that migration into Europe from Turkey was on the rise. Even at that time, "Consult," was the most represented CAMEO code in the dataset. This code refers to different types of diplomatic engagement (e.g., meetings, telephone discussions, visits, and mediation) [22]. "Consult" dominates the refugee-related articles before, during, and after Alan Kurdi's death (Fig. 4), but the cross-border nature of migration, particularly forced migration and humanitarian response, necessitates extensive diplomatic engagement. This includes everything from diplomatic meetings between countries to diplomatic visits to refugee camps or transit areas.

After Alan Kurdi's death, the article coverage begins to peak. Beyond "Consult," the CAMEO codes that are most represented are "Appeal," "Provide Aid," and "Engage in Diplomatic Cooperation." In the CAMEO codebook, "Appeal" covers appeals for cooperation, aid, and peacekeeping, as well as appeals for political change [22]. In this context, this confirms our understanding of the situation where Alan Kurdi became a symbol for the plight of refugees and mobilized many who would not otherwise have known about humanitarian situations. "Provide Aid" includes humanitarian, economic, and military aid, but additionally includes "granting asylum" that ranges from political to diplomatic asylum. Both aid and asylum are highly relevant to this migration context as appeals for aid and asylum came in not just from traditional international actors like countries or UN agencies, but also from celebrities and individuals in countries who volunteered to take refugees into their homes. "Engage in Diplomatic Cooperation" refers broadly to actions as specific as signing formal treaties to verbally supporting a "policy, action, or actor" [22]. The tragedy of this little boy and his family served as a global symbol of human suffering and refugees' arduous journeys, which became a call to action for politicians, celebrities and other elites, and citizens around the world. It is no surprise that at the very least, verbal support of policies would constitute a large portion of the article coverage.

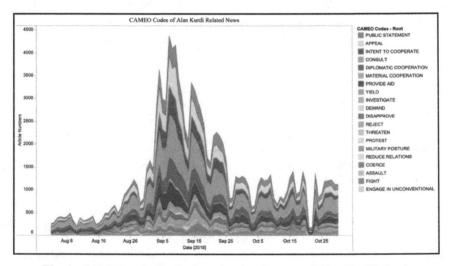

Fig. 4. CAMEO code distribution related to the news of Alan Kurdi's death

Combined, these major CAMEO codes align with our understanding of the context and situation through other ongoing studies using qualitative analysis of news articles and ethnographic fieldwork in Greece. The Greece/Turkey border is a hotly contested zone in the European migration situation. With a long history of contention between the two countries, and Greece's position at the frontiers of Europe, a significant amount of political activity—from diplomatic meetings and agreements to appeals for better conditions for refugees and migrants—consistently feature in local, national, and global news. In fact, these general trends hold for the entire database of refugee-related articles, except that at the international level, "Make Public Statement" rises to prominence. This code is extremely broad, covering topics ranging from "make optimistic comment" to "consider policy option." Investigating the role of the subcategories of each of the CAMEO codes is a direction for future work.

4 Discussion

The GDELT dataset allows us to see a global view of media for particular topics. In this study, we focused on refugees and displaced persons, looking at the evolution of article coverage from January 2015 until March 2021. In that time, the dataset revealed 14 major events (see Table 2) that were driving the global conversation about refugees. Several interesting findings come out of this analysis. First, "negative tone" in the GDELT dataset, when referring to refugees, can cover both moments of empathy and moments of anti-migrant sentiment. Since GDELT is automatically comparing against pre-established lexicons, it cannot make this distinction; though both manifest as "negative" language, the results yielded a negative tone of -3.97 for the average tone of all refugee-related events included in the analysis. This may not be the case in all topic areas, but refugees and migrants often evoke both anti-migrant, nationalistic rhetoric as well as feelings of despair for their tragic conditions. This was demonstrated as even positive events, like World Refugee Day, increased article volume globally, but decreased tone, indicating more "negativity." Second, the CAMEO codes reveal a rich area of research for future exploration, adding a significant amount of nuance to the types of global conversations about refugees. We know that "Consult" dominates the article coding, comprising 17.5% of our refugee query, but this is largely a "talk" and not an "action" code. This aligns with criticisms of many about the international humanitarian response to displacement but doesn't seem to account for the vast amount of humanitarian action on the ground. Perhaps this is because these activities are not "sensational" enough to be well-covered on a regular basis in the global news. This presents an additional direction of research where we plan to dig deeper into the sub-fields of these CAMEO codes to derive further insights. At further levels of disaggregation of the data, we are also interested to see how these analyses break down by country, particularly to compare refugee news coverage in countries that host large numbers of refugees and migrants versus those that do not. Another important finding is the signature found within the time series that helps to discriminate the different types of events. For example, the dissemination rate of the Alan Kurdi event (see Table 3 and Fig. 3), diffusing 18 days after the peak day, differs from the summit-type events, which occur and recede on the same day.

Acknowledgments. This research is funded by grant number N000141912624 by the Office of Naval Research through the Minerva Research Initiative; none of the views reported in the study are those of the funding organization.

References

1. UNHCR: 1 per cent of humanity displaced: UNHCR Global Trends report. UNHCR, Geneva (2020)
2. Caviedes, A.: An emerging 'European' news portrayal of immigration? J. Ethn. Migr. Stud. **41**, 897–917 (2015)

3. Esses, V.M., Medianu, S., Lawson, A.S.: Uncertainty, threat, and the role of the media in promoting the dehumanization of immigrants and refugees. J. Soc. Issues **69**, 518–536 (2013)
4. Cooper, G., Blumell, L., Bunce, M.: Beyond the 'refugee crisis': how the UK news media represent asylum seekers across national boundaries. Int. Commun. Gaz. **83**, 195–216 (2021)
5. Chouliaraki, L., Stolic, T.: Rethinking media responsibility in the refugee 'crisis': a visual typology of European news. Media Cult. Soc. **39**, 1162–1177 (2017)
6. Leetaru, K., Schrodt, P.A.: GDELT: Global data on events, location, and tone, 1979–2012. ISA annual convention, vol. 2, pp. 1–49. Citeseer (2013)
7. Abu-Jabr, R., Gerner, D.J., Schrodt, P.A., Yilmaz, O.: Conflict and Mediation Event Observations (CAMEO): A New Event Data Framework for the Analysis of Freign Policy Interactions. Annual Meeting of the International Studies Association, New Orleans (2002)
8. https://professional.dowjones.com/factiva/
9. The GDELT Project: The GDELT Event Database Data Format Codebook v2.0. (2015)
10. Pogorelov, K., Schroeder, D.T., Filkukova, P., Langguth, J.: A system for high performance mining on GDELT data. In: 2020 IEEE International Parallel and Distributed Processing Symposium Workshops (IPDPSW), pp. 1101–1111. IEEE (2020)
11. https://www.un.org/en/observances/refugee-day
12. The European Council: Meeting of the EU heads of state or government with Turkey, 7 March 2016. In: Council of the European Union (ed.), European Council website (2016)
13. Yardly, J.: Pope Francis Takes 12 Refugees Back to Vatican after Trip to Greece. The New York Times, New York (2016)
14. https://refugeesmigrants.un.org/summit
15. UN General Assembly: Resolution adopted by the General Assembly on 19 September 2016: 71/1. New York Declaration for Refugees and Migrants. United Nations, Geneva (2016)
16. Rankin, J.: EU court dismisses complaints by Hungary and Slovakia over Refugee Quotas. The Guardian, UK (2017)
17. UNHCR: First Global Refugee Forum, 17 and 18 December 2019, Palais des Nations, Geneva (2019). https://www.unhcr.org
18. Hewitt, G.: Paris attacks: Impact on border and refugee policy. BBC News, online (2015)
19. Austen, I., Smith, C.S.: Quebec Mosque Shooting Kills at Least 6, and 2 Suspects Are Arrested. The New York Times, online (2017)
20. Stevis-Gridneff, M., Kingsley, P.: Turkey Steps Back from Confrontation at Greek Border. The New York Times, online (2020)
21. Moria migrants: Fire destroys Greek camp leaving 13,000 without shelter. BBC News, online (2020)
22. Schrodt, P.A.: CAMEO: Conflict and Mediation Event Observations Event and Actor Codebook. Pennsylvania State University, Codebook (2021)

An Identity-Based Framework for Generalizable Hate Speech Detection

Joshua Uyheng$^{(\boxtimes)}$ (iD) and Kathleen M. Carley (iD)

CASOS Center, Institute for Software Research, Carnegie Mellon University,
Pittsburgh, PA 15213, USA
{juyheng,kathleen.carley}@cs.cmu.edu

Abstract. This paper explores the viability of leveraging an identity-based framework for generalizable hate speech detection. Across a corpus of seven benchmark datasets, we find that hate speech consistently features higher levels of abusive and identity terms, robust to social media platforms of origin and multiple languages. Using only lexical counts of abusives, identities, and other psycholinguistic features, heuristic and machine learning models achieve high precision and weighted F1 scores in hate speech prediction, with performance on a three-language dataset comparable to recent state-of-the-art multilingual models. Cross-dataset predictions further reveal that our proposed identity-based models map hate and non-hate categories with each other in a conceptually coherent fashion across diverse classification schemes. Our findings suggest that conceptualizing hate speech through an identity lens offers a generalizable, interpretable, and socio-theoretically robust framework for computational modelling of online conflict and toxicity.

Keywords: Hate speech · Social media · Identities · Machine learning

1 Introduction

Developing computational methods for the detection of hate speech is an important task for the emerging science of social cyber-security [2,14]. While a vast literature in machine learning and allied fields has sought to develop cutting-edge tools to address this problem [4,6,16], some consequences of this proliferation of work include a divergence in hate speech definitions, and with the increasing traction of complex deep learning models, the interpretability of model predictions [15].

This work was supported in part by the Knight Foundation and the Office of Naval Research grants N000141812106 and N000141812108. Additional support was provided by the Center for Computational Analysis of Social and Organizational Systems (CASOS) and the Center for Informed Democracy and Social Cybersecurity (IDeaS). The views and conclusions contained in this document are those of the authors and should not be interpreted as representing the official policies, either expressed or implied, of the Knight Foundation, Office of Naval Research or the U.S. government.

© Springer Nature Switzerland AG 2021
R. Thomson et al. (Eds.): SBP-BRiMS 2021, LNCS 12720, pp. 121–130, 2021.
https://doi.org/10.1007/978-3-030-80387-2_12

Table 1. Summary of datasets used. Collapsed classes are specified in parentheses. Frequencies reflect retrievable data, as some data points may be unavailable.

Dataset	Class	Frequency
Chung [3]	Hate	857 (11.19%)
	Counter	6804 (88.81%)
Davidson [4]	Hate	1430 (5.77%)
	Offensive	19190 (77.43%)
	Regular	4163 (16.80%)
De Gibert [5]	Hate	1196 (10.93%)
	Regular (Relation/Non-Hate)	8748 (89.07%)
Founta [6]	Hate	541 (7.28%)
	Abusive	1956 (26.30%)
	Spam	848 (11.40%)
	Regular	4089 (5499%)
Mathew [9]	Hate	6854 (34.02%)
	Offensive	5480 (27.20%)
	Regular	7814 (38.78%)
Qian [10]	Hate	25344 (40.80%)
	Regular	36779 (59.20%)
Waseem [16]	Hate (Racism/Sexism)	2161 (27.41%)
	Regular	5723 (72.59%)

In this work, we use social scientific theory about *identities* to build grounded computational models of hate speech. Utilizing a multilingual lexicon of known terms which reference identities [7], we show that it is possible to detect hate speech in an accurate, interpretable, and generalizable way across a variety of datasets based on diverse definitional taxonomies, languages, and social media platforms [3–6,9,10,16]. Through this work, we contribute an enhanced social scientific understanding of hate speech that may inform future modelling efforts, as well as a general and scalable method in its own right that may be deployed for hate speech detection in emergent and applied settings [2,12,13].

2   Data and Methods

2.1   Dataset Curation

To facilitate systematic analysis, this study relied on a curated collection of hate speech datasets for systematic analysis. Beginning with the online repository[1] of Vidgen and Derczynski [15], we filtered out datasets with fewer than 1000 examples, which included only hate (and no negative examples), or which conflated

---

[1] https://hatespeechdata.net.

hate with bullying (outside the current scope). We settled on seven datasets summarized in Table 1.

Because several datasets involved social media data (e.g., Twitter), Table 1 reports the frequency of each class label as retrievable from each dataset as of February 2021. Due to possible suspensions of hateful utterances online, these reported frequencies may differ from statistics reported in their original papers. Additionally, for classes which had few examples (e.g., less than 100), we collapsed several classes into a single class, made explicit in Table 1. For each dataset, we uniformly identify the "hate" category. We note that non-hate categories across datasets may vary widely, including regular speech [4–6,9,10,16], abusive or offensive but not hateful speech [4,6,9], spam [6], and counter-hate [3]. While most datasets originate from Twitter, others also include utterances generated offline [3], and from other platforms like Reddit [10], Gab [9,10], and Stormfront [5]. One dataset moreover includes multilingual data [3].

2.2   An Identity Lexicon with Psycholinguistic Features

To study the use of identities in the collected hate speech datasets, we drew upon social scientific theorizing around identities, or concepts of socially embedded selves and groups. We leverage recent work which expands and validates a lexicon of *identity terms* for computational modelling [7]. We specifically use the Netmapper software[2] which counts these identity terms across several dozen languages, and has been previously used in applied settings of psycholinguistic analysis for social cyber-security [12,13].

We additionally use Netmapper to measure various other psycholinguistic features that have been associated with a wide variety of cognitive and emotional states [11]. These include pronoun usage, various positive and negative emotion words, and patterns of punctuation. Of particular interest, however, we also probe the joint presence of identity terms alongside known *abusive terms*. Taken together, we propose that identities and abusives constitute general, reliable, and theoretically grounded empirical touchstones for hate speech detection.

2.3   Problem Formulation and Experimental Setup

Utilizing the foregoing psycholinguistic measurements across the seven datasets in this study, we examine several research questions of interest. We divide the results that follow into three stages, beginning with a statistical analysis of identities and abusives across the curated datasets (RQ1), a predictive modelling analysis that evaluates the use of psycholinguistic features for hate speech detection within individual datasets (RQ2), and a generalizability analysis that maps out the quality and consistency of cross-dataset predictions (RQ3).

---

[2] https://netanomics.com/netmapper/.

RQ1: Does hate speech contain more abusive and identity terms? To answer our first question, we perform two types of regression analysis. We begin with separate analyses of abusives and identities for each dataset. Here, we perform linear regression to predict the number of abusives and identities in a given text based solely on its label. Based on extant definitions of hate speech, we expect that, across datasets, hate speech will indeed have higher levels of *both* abusive and identity terms.

Next, we perform a binary logistic regression analysis over a consolidated dataset, where we predict whether a given text is classified as hate or not hate. Here, the predictors are now the number of abusive and identity terms in a given text. In additional, we control for each dataset's platform of origin and its multilingual focus by including them as covariates in the regression model. In this case, we also hypothesize positive effects for both abusives and identities, robust to the explored controls.

RQ2: How can abusive and identity terms be used to detect hate speech? In the second stage of this work, we shift from statistical analysis to predictive modelling. Here, our objective is now to evaluate whether hate speech can be accurately detected using abusives and identities. We begin by evaluating the precision of a heuristic model that simplistically predicts that a text is hate speech if it contains at least one abusive term, and at least one identity term. Given only two features, this heuristic may not achieve particularly high precision. However, we do expect it to significantly outperform a random baseline, as it aligns with our identity-based theoretical understanding of hate speech.

We then assess the use of psycholinguistic counts more broadly - which include abusives and identities - for hate speech detection using a range of machine learning models, including logistic regression, random forests, and support vector machines. We augment feature inputs with their squared values and pairwise products to capture interactions between psycholinguistic measures, since these second-order terms capture signals from when pairs of variables have simultaneously high values.

We compare these results against a deep learning benchmark that uses the full text. We specifically use a classical convolutional neural network (CNN) for sentence classification [8]. We use an Adam optimizer and perform a grid search over word embedding dimension, filter size, and dropout rate. We use the average weighted F1 measure obtained from five-fold cross-validation for evaluation. Here, we expect that the deep learning model will consistently achieve the best performance, but we also expect that the psycholinguistically informed machine learning models will not be severely worse.

RQ3: How generalizable is identity-based hate speech detection? Finally, we consider the generalizability of the proposed identity-based framework for hate speech detection. Here, we adopt both confirmatory and exploratory tools to flesh out a rich examination of cross-dataset dynamics.

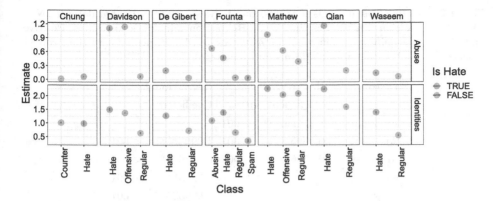

Fig. 1. Results of linear regression on single datasets, predicting abusives and identities based on class labels. In general, hate speech has higher (red) values of *both* abuse and identities than non-hate (blue). (Color figure online)

We first take the best-performing machine learning model from the previous stage of analysis, and use it to generate predictions for all other datasets in our corpus. For instance, a model trained using the Davidson dataset [4] would generate predictions for the Founta dataset [6]. So all instances in the Founta dataset of Hate, Abusive, Spam, and Regular, will be assigned a Davidson-based label of Hate, Offensive and Regular. Given the diversity of definitional taxonomies, we assess the extent to which each model accurately maps its own associated hate class to the hate class of other datasets.

From an exploratory standpoint, we also perform principal component analysis on the cross-dataset predictions. This allows us to qualitatively assess a shared, low-dimensional visualization of text classes across datasets. In both analyses of generalizability, we expect that if our identity-based psycholinguistic approach is successful, hate categories will be mapped to each other across datasets, and non-hate categories likewise.

3 Results

3.1 Hate Speech Consistently Features Identity Abuse

Figure 1 visualizes the levels of abusive and identity-based language in hate speech (red) and other classes (blue) in each dataset. In the case of the De Gibert [5], Mathew [9], Qian [10], and Waseem [16] datasets, comparisons are straightforward, as hate speech shows higher levels of both measures. Interestingly, however, we note that in both Davidson [4] and Founta [6] datasets, the Offensive and Abusive classes respectively have more abusive terms. Yet in both cases, hate speech features more identity terms than these classes, while *also* containing more abusives than the designated Regular classes. Conversely, in the Chung [3] dataset, hate speech shows fewer identity terms than the Counter

Table 2. Results of binary logistic regression analysis over consolidated dataset ($N =$ 140979). Regression 1 is the baseline, while Regression 2 includes covariates. Positive and statistical significant coefficients for Abusives and Identities point to their robust associations with hate across platforms and languages.

Factors	Regression 1		Regression 2	
	Coefficient	p	Coefficient	p
Intercept	-1.7436***	$<.001$	–	–
Abusives	0.8895***	$<.001$	1.0909***	$<.001$
Identities	0.1196***	$<.001$	0.0156***	$<.001$
Gab	–	–	1.7283***	$<.001$
Reddit	–	–	-2.7653***	$<.001$
Stormfront	–	–	-2.2021***	$<.001$
Twitter	–	–	-3.1943***	$<.001$
Multilingual	–	–	-2.1225***	$<.001$

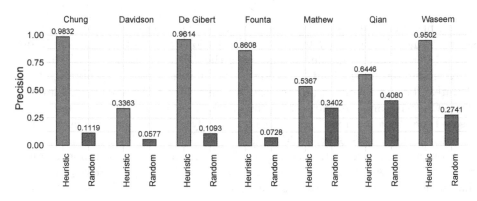

Fig. 2. Evaluation of heuristic predictive model precision against random baselines.

class; but as expected, hate speech has more abusive terms. Thus, our hypothesis for this analysis holds: hate speech systematically features *both* abusive and identity terms, at significantly higher levels than other kinds of text.

We strengthen this per-dataset observation by analyzing the consolidated corpus ($N = 140979$). Table 2 shows that regardless of dataset, indeed, higher levels of abusive (Model 1: $\beta = 0.8895, p < .001$; Model 2: $\beta = 0.1196, p < .001$) and identity (Model 1: $\beta = 1.0909, p < .001$, Model 2: $\beta = 0.0156, p < .001$) terms predict higher likelihood of a text being hate speech, with and without controlling for platforms of origin and multilingualism suggest the robustness of these effects.

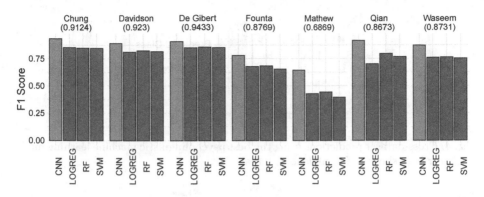

Fig. 3. Cross-validated weighted F1 scores of machine learning models versus CNNs. Datasets are labelled with the best relative performance of machine learning models.

3.2   Evaluating Identity Abuse for Hate Speech Detection

Predictive modelling results bolster our statistical findings by showing the practical utility of abusive and identity terms in detecting hate speech. Figure 2 shows that a heuristic that predicts hate speech solely using these two features achieves 86.08–98.32% precision for four of the seven datasets, with the remaining three datasets showing 33.63–64.46% precision. Note we only evaluate precision because the two features alone do not account for all non-hate classes.

These results indicate that for many datasets of hate speech, our proposed theoretical identity-based framework may have predictive utility. However, more features need to be accounted for in other hate taxonomies, in view of the latter three datasets. Yet despite their relatively low performance, we note with interest that across all datasets, the heuristic vastly outperformed all random baselines. Two-sample proportion tests with continuity correction affirm these differences to be highly statistically significant ($p < .001$).

Pushing our psycholinguistic framework further, we now evaluate various machine learning models against a deep learning benchmark [8]. On average, the F1 scores of psycholinguistic machine learning models are 12.11% higher than the precision scores of the two-feature heuristics using only identities and abusives. Yet as expected, a CNN with high-dimensional representations of the entire text achieves better performance compared to machine learning models only given psycholinguistic counts as inputs.

However, in all cases but one [9], the best machine learning model performance is 86.73–94.33% that of the CNN model. That means using our approach there is practically between a 5.67–13.27% performance reduction from a deep learning model, even with significantly fewer parameters, and with much greater interpretability. We also note that in the multilingual dataset [3], the F1 scores of the machine learning models (0.8445–0.8523) were comparable to a state-of-the-art analysis of deep learning models in a multilingual setup (0.6651–0.8365) [1]. This may be crucial to systematically explore further.

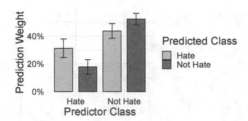

Fig. 4. Cross-dataset predictions using the best psycholinguistic machine learning models. Error bars show 95% confidence intervals.

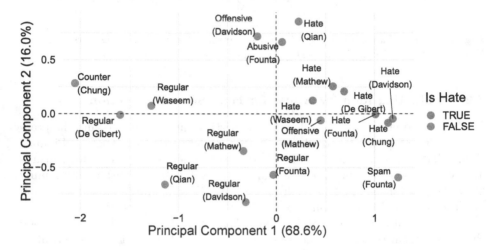

Fig. 5. Principal component analysis of cross-dataset predictions, with hate in red, and non-hate in blue. Principal components are presented with variance explained. (Color figure online)

For the outlier case [9], though machine learning models at best achieve 68.69% of the CNN performance, the CNN itself only obtains a weighted F1 score of 64.27%. As the dataset was used to train explainable models with human-generated context, these examples may be difficult to classify without this additional information, as here we only leverage the raw text.

3.3 Mapping Cross-dataset Generalizability

Finally, we turn to generalizability analysis. Figure 4 shows the accuracy of cross-dataset predictions for all hate and non-hate classes. Across datasets, we find that our proposed identity-based framework reliably maps hate to hate, and non-hate to non-hate. However, non-negligible discrepancies may still arise, likely due to intermediate categories like Offensive, Abusive, and Spam classes.

To explore these nuances further, Fig. 5 shows the results of a principal component analysis on cross-dataset prediction weights. Remarkably, we find that the Hate classes for all seven datasets tightly cluster on the top-right corner of a low-dimensional visualization of the first two principal components (84.6% of the variance). As might be expected, the other classes plotted closest to hate speech are the harmful yet non-hate classes of concern. Notably, Spam is distinctly located in a separate location from Offensive and Abusive, which are conceptually more similar to each other. Furthermore, most Regular classes are clustered near the left and bottom-left corner of Fig. 5, with the Counter class occupying the left-most coordinate. Conceptually, the Counter class, which actively combats hate [3], may also be understood as the most distinct form of speech relative to both hate and other irregular yet non-hateful texts.

Collectively, then, we find a highly theoretically coherent mapping of the classes considered in our corpus, with cross-dataset model predictions empirically resonating with latent conceptual relationships. This suggests that our proposed identity-based framework may valuably capture shared, underlying features of hate speech and related constructs that cut across datasets.

4   Conclusions and Future Work

This paper showed that an identity-based approach reflects key features of hate speech across several datasets [7]. Hate speech not only features more identity and abusive terms (RQ1), but these terms may also detect hate speech with enhanced interpretability relative to state-of-the-art models (RQ2) [1]. Identity-based representations further produce generalizable models that map different forms of hate and non-hate in a theoretically coherent fashion (RQ3).

This work shows the conceptual benefit of an identity perspective for understanding hate as abuse targeted against social groups, with practical benefits for generalizable, interpretable, and scalable computational modelling [2,14]. Future applied work may leverage the multi-dataset effort presented here for even richer maps of hate dynamics - alongside linked phenomena like spam and counter-hate [3,6] - to capture the multi-faceted nature of online toxicity, and potentially inform more nuanced policy and platform responses [14]. Our theory-based method could also be potentially used to classify the targets of hate and measure the coordination of hate in information operations [2,12,13].

To this end, extensions to our work may readily be pursued through an expanded data corpus and set of models. Psycholinguistic and identity-based features could also be utilized alongside - rather than independent from - word or sentence embeddings already prevalent in cutting-edge hate speech detection applications [15]. It is also extremely promising to more systematically compare - or combine - our approach against more state-of-the-art models in multilingual hate speech prediction [1]. Finally, from a qualitative standpoint, it is crucial to consider kinds of hate that may not explicitly abuse identities, yet take on silent but salient and harmful forms - these are issues outside the framework we propose, and form vital directions for further, multidisciplinary research [2,14].

References

1. Aluru, S.S., Mathew, B., Saha, P., Mukherjee, A.: Deep learning models for multilingual hate speech detection. arXiv preprint arXiv:2004.06465 (2020)
2. Carley, K.M.: Social cybersecurity: an emerging science. Comput. Math. Org. Theory **26**(4), 365–381 (2020)
3. Chung, Y.L., Kuzmenko, E., Tekiroglu, S.S., Guerini, M.: CONAN-COunter NArratives through Nichesourcing: a multilingual dataset of responses to fight online hate speech. In: Proceedings of the 57th Annual Meeting of the Association for Computational Linguistics, pp. 2819–2829 (2019)
4. Davidson, T., Warmsley, D., Macy, M., Weber, I.: Automated hate speech detection and the problem of offensive language. In: Proceedings of the International AAAI Conference on Web and Social Media, vol. 11 (2017)
5. De Gibert, O., Perez, N., García-Pablos, A., Cuadros, M.: Hate speech dataset from a white supremacy forum. In: Proceedings of the 2nd Workshop on Abusive Language Online, pp. 11–20 (2018)
6. Founta, A., et al.: Large scale crowdsourcing and characterization of Twitter abusive behavior. In: Proceedings of the International AAAI Conference on Web and Social Media, vol. 12 (2018)
7. Joseph, K., Wei, W., Benigni, M., Carley, K.M.: A social-event based approach to sentiment analysis of identities and behaviors in text. J. Math. Sociol. **40**(3), 137–166 (2016)
8. Kim, Y.: Convolutional neural networks for sentence classification. In: Proceedings of the 2014 Conference on Empirical Methods in Natural Language Processing (EMNLP), pp. 1746–1751, October 2014
9. Mathew, B., Saha, P., Yimam, S.M., Biemann, C., Goyal, P., Mukherjee, A.: HateXplain: a benchmark dataset for explainable hate speech detection. arXiv preprint arXiv:2012.10289 (2020)
10. Qian, J., Bethke, A., Liu, Y., Belding, E., Wang, W.Y.: A benchmark dataset for learning to intervene in online hate speech. In: Proceedings of the 2019 Conference on Empirical Methods in Natural Language Processing and the 9th International Joint Conference on Natural Language Processing, pp. 4757–4766 (2019)
11. Tausczik, Y.R., Pennebaker, J.W.: The psychological meaning of words: LIWC and computerized text analysis methods. J. Lang. Soc. Psychol. **29**(1), 24–54 (2010)
12. Uyheng, J., Carley, K.M.: Bots and online hate during the COVID-19 pandemic: case studies in the United States and the Philippines. J. Comput. Soc. Sci. **3**(2), 445–468 (2020)
13. Uyheng, J., Carley, K.M.: Characterizing network dynamics of online hate communities around the COVID-19 pandemic. Appl. Network Sci. **6**(1), 1–21 (2021). https://doi.org/10.1007/s41109-021-00362-x
14. Uyheng, J., Magelinski, T., Villa-Cox, R., Sowa, C., Carley, K.M.: Interoperable pipelines for social cyber-security: assessing Twitter information operations during NATO Trident Juncture 2018. Comput. Math. Organ. Theory **26**, 1–19 (2019)
15. Vidgen, B., Derczynski, L.: Directions in abusive language training data, a systematic review: garbage in, garbage out. PLoS ONE **15**(12), e0243300 (2020)
16. Waseem, Z., Hovy, D.: Hateful symbols or hateful people? Predictive features for hate speech detection on Twitter. In: Proceedings of the NAACL Student Research Workshop, pp. 88–93 (2016)

Using Diffusion of Innovations Theory to Study Connective Action Campaigns

Billy Spann[1], Maryam Maleki[1], Esther Mead[1], Erik Buchholz[1], Nitin Agarwal[1(✉)], and Therese Williams[2]

[1] COSMOS Research Center, UA – Little Rock, Little Rock, AR, USA
{bxspann,mmaleki,elmead,ejbuchholz,nxagarwal}@ualr.edu
[2] Information Systems and Operations Management, University of Central Oklahoma, Edmond, OK, USA
twilliams120@uco.edu

Abstract. The S-shaped signatures from connective action events offers insights into past social science theories that support characterizing these events through a quantitative approach. The observational approach taken in this analysis builds on the diffusion of innovations theory, critical mass theory, and previous s-shaped production function research to provide ideas for modeling future campaigns. A key benefit to this approach is that these technologies have been studied extensively and the adoption curves from online social movements are platform independent and occur across a range of industries, technologies, and platforms. For this analysis we analyze 9 misinformation and public discourse hashtags which cover a range of topics related to COVID-19 such as lockdowns, face masks, and vaccines. Plotting the cumulative frequency of Tweets from January 1 to December 31, 2020 we observe s-curve signatures representing the adoption of connective action for these campaigns. We then categorize each of these campaigns by examining their affordance and interdependence relationships by assigning retweets, mentions, and original tweets to the type of relationship they exhibit. This will help researchers understand the relationships between users, characterize what actions take place, and how information flows through users as adoption occurs. The contribution of this analysis provides a foundation for mathematical characterization of connective action signatures, and further, offers ideas on how to design support or countermeasures for the type of campaign taking place. The first approach will drive future work toward developing a predictive model, while the affordance approach will help us to understand the organizational components.

Keywords: Collective action · Connective action · Social network analysis · COVID-19 · Diffusion of innovation · Critical mass theory · S-Curve

1 Introduction

In this paper, we examine connective action through the lens of information diffusion by looking at social media campaigns and their event signatures related to the COVID-19 pandemic. A study of Twitter data from nine COVID-19 related hashtags reveals s-shaped

© Springer Nature Switzerland AG 2021
R. Thomson et al. (Eds.): SBP-BRiMS 2021, LNCS 12720, pp. 131–140, 2021.
https://doi.org/10.1007/978-3-030-80387-2_13

diffusion curves that show the rate at which information spreads throughout the life cycle of nine misinformation campaigns. The s-shaped phenomenon is similar to diffusion of innovations (DOI) theory [15] that describes how new ideas and technology spread over time. We also refer to this s-shape as a production function as found in critical mass theory [12]. Our work extends previous contributions by comparing observations from our social network analysis, epidemiological studies, and connective action framework and comparing these to the DOI and s-curve socio-technical theories. We bridge the mathematical properties of the DOI and s-curve functions with the observations we are seeing from a social networking perspective to help guide our research toward developing a predictive model for detecting connective action campaigns. Using past theories that have been thoroughly researched and transferring knowledge to the connective action domain also allows us to apply this technique across different platforms and remain platform independent. Two main contributions of this study are (1) initially using the s-curve characteristics to model an emerging connective action campaign, and (2) distinguish the types of campaigns by examining the affordances and interdependencies associated with the connective action. Understanding this affordance-interdependence relationship will help researchers determine the type of countermeasures to properly respond to the type of campaign.

The information campaigns where s-shaped phenomenon is observed are consistent with the diffusions of innovation concept. Similar to the stages that influence adoption of an innovation or idea, we also describe factors that influence adoption of connective action (mis)information campaigns. Connective action campaigns use affordances through online social networks to influence rapid adoption, while reducing costs to join and providing anonymity. Since the costs to join a campaign are low, the decision to adopt the idea, or join the movement requires little exposure. As part of our analysis, we analyzed 88,776 Tweets for popular COVID-19 discourse and anti-lockdown protests that occurred throughout 2020. We examine input features used in these connective action campaigns and show how these features follow the s-curve function, leading to mathematical characterization. We can understand the type of event that is occurring by observing the production function of the message frequency plot. Some events take longer to get started and adopt, while some events have a flash mob startup feature where they start quickly and then quickly disperse. Finally, some events combine these two approaches where the action slowly starts, quickly accelerates, then slowly decreases. This is the s-curve pattern we observe. These can be observed in their production function curves described in collective action of critical mass theory.

Two fundamental research questions are studied: RQ1: Can we leverage existing social science theories on diffusion of information to provide background for a mathematical representation of these information campaigns? RQ2: After quantifying the campaign, can we apply social network analysis to the proposed affordance-interdependence relationships to offer a view of how the campaign unfolds?

The remainder of this paper is presented as follows. Section 2 provides a comprehensive review of the social science theories we use to compare to our connective action models. In Sect. 3, we provide our approach to data collection and processing of the data. Next, we discuss the roles of the affordance-interdependency relationships in Sect. 4 and how they describe certain events. We conclude with Sect. 5 discussing our findings and future work to this line of research.

2   Literature Review

S-shaped curves are applied to numerous applications including predicting population changes, projecting the performance of technologies, predicting penetration into markets, modeling ecological changes, and several others. The s-shaped curve goes by many names, but it was first introduced by Belgian mathematician Pierre-Francious Verhults in 1838 as the logistic function. The bell-shaped curve is used as an underlying template to depict the rate of growth for a particular system for a specific period of time. Hu and Wang [7] analyzed data from the Chinese social network, Wealink, to show how the number of users and friends on the network can be modeled with a logistic s-shaped curve. The social network evolves in a variety of ways, from an s-shaped pattern of adoption of users and friends to different patterns in path length, diameter, density, and assortativity. Bejan and Lorente [1] also point out that every successful innovation has this similar trajectory, which is characterized by an initial stage of low acceptance, then another stage when that acceptance begins to spread throughout a population, then at some point that acceptance reaches a critical mass, wherein there is a sharp rise in the acceptance among users, and then, finally, that rate of acceptance reaches a saturation point or ceiling and begins to trail off. Fink et al. [4] modeled user adoptions as a Bernoulli process where the likelihood of adoption after each exposure to a neighbor's use of the hashtag is a function of the number of distinct adopting neighbors. Shimogawa et al. [15] also discuss the pervasive use of the s-curve, specifically shaping their discussion around the concept of innovation diffusion. They contend that, from this perspective, the true long-term extrapolations from the s-curve are difficult to predict when it comes to the diffusion of innovations because of social differences about the final adopter population.

Everett Rogers's [13] Diffusion of Innovations (DOI) theory is defined as the process by which an innovation is communicated through certain channels over time among the members of a social system. Rogers claims that innovations must be widely adopted to reach critical mass. The categories of adopters are innovators, early adopters, early majority, late majority, and laggards. Most early DOI research focuses on the individual as the unit of adoption in technology diffusions. Our research leverages the focus on the individual but considers the individual as an organizer in connective action movements. Ma et al. [10] used the DOI theory to combine social network analysis with multiple regression analysis to demonstrate that "opinion leadership" is the most important factor that can forecast an individual's news distribution, followed by "news attribute" and "tie strength". Our connective action model quantifies these leadership and tie strength qualities by calculating centrality measurements and identifying brokers within the network that have influenced the campaign.

Marwell's et al. [11] critical mass theory evaluated which aspects are important for a critical mass to have successful contract negotiations (e.g., collective action). They evaluated the importance of cost, density, and centrality in determining how much individuals are willing to contribute to a cause. This contribution was evaluated in a variety of groups, which varied social ties, interest, and resources in three groups: homogenous groups, heterogeneous social tie groups, and heterogenous ties/interest/resource groups. Overall, while social network density improves collective action (as expected), network centralization remains consistently important (surprising). Geddes [5] proposed several factors that help achieve at least 15% network penetration so that a critical mass would be achieved, and network saturation would occur. The author also suggested that after around 15% of a community has been penetrated, the rate of acceleration of adoption dramatically increases until it plateaus at a saturation point. To enhance 15% adoption, the author suggests leveraging existing networks and relationships, restricting topics of conversation, demonstrating the value that users achieve, and shrinking the size of the pool. We propose that in connective action campaigns, localized communities as measured by modularity can achieve adoption, and then spread to other communities via brokerage nodes.

Connective action is a term first defined by Bennett and Segerburg [3] as separate from collective action in that they are more individualized and more technologically organized actions. They result in actions without the need of collective identity framing and without the need of organizational resources. This allows world-wide protests to be successfully organized and can reach smaller individualized populations that may be hard to convince to join if they must personally identify with an organization's goals. The connective action principles can be summarized as collective identity, network organization, and mobilization of resources. The role of individuals in connective actions is further expanded by interpersonal dynamics that may operate behind the scenes in a digitally networked action (DNA). These interpersonal dynamics can provide a structure to seemingly unstructured movements and can also, inadvertently, promote individuals as the face of a movement. This can be both desirable and undesirable. Hemsley [6] investigated viral information event signatures and showed that initial sharp peaks indicate a viral spread of a message that show an s-curve over time related to the diffusion of information.

3   Data Collection and Research Methodology

We begin by studying information campaigns from the lens of our connective action framework. We consider the fundamental principles of connective action, viz., collective identity, network organization, and mobilization as we observe these misinformation campaigns. The social processes associated with these connective action networks are grounded in previous social science research such as diffusion of innovations and critical mass theory.

Twitter premium APIs were used to collect tweets related to COVID-19 topics from January 1 to December 31, 2020. Our goal was to collect data for different misinformation and public discourse hashtags which cover a broad range of topics related to COVID-19 such as lockdowns, face masks, and vaccines. For each of the three categories, two separate hashtags were analyzed: one representing pro-narratives, and the other representing anti-narratives around misinformation. For example, hashtags representing anti-narrative misinformation that were analyzed include: #Lockdownskill, #Nofacemask, and #Novaccineforme. The hashtags representing the corresponding pro-narrative information include: #Lockdownswork, #Wearafacemask, and #Vaccinesaveslives. Additionally, three other hashtags representing items of misinformation were analyzed: #Covidscam, #BillGatesVirus, and #Coronascam. The number of resultant tweets for each hashtag is shown in Table 1. Our initial observations support a mathematical characterization for connective action campaigns, and then show how affordances can be used to identify the inter-user message strategy. The first approach will drive future work toward developing a predictive model, while the affordance approach will help us to understand if this is organized by multiple users or just a central group of users, at which point we can devise countermeasures or support for the campaign.

4   Analysis and Results

We analyzed 9 different datasets from Twitter Hashtag networks. We plot the cumulative tweets over time, showing the adoption as it occurs. The early amplification, adoption, and sustainment cycles of these campaigns take on an s-curve shape that is discussed in Sect. 2. As part of the development of our connective action framework, we need to create a mathematical representation of the event signatures.

Examining these 9 connective action campaigns from Twitter reveals characteristics of this s-shape curve phenomenon as seen in Fig. 1[1] below. Clearly, there is an initial growth stage, accelerating stage, and then a decelerating stage. However, we notice some campaigns continue to grow after an initial deceleration.

The previous theories of diffusion of innovations and critical mass theory can also be used to characterize these curves into mathematical representations of diffusion for connective action.

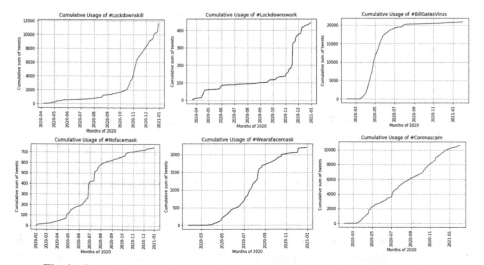

Fig. 1. Cumulative tweet frequency for 9 COVID-19 connective action campaigns

---

[1] Due to space limitations and to improve image quality, we are presenting a representative sample of the results.

S-curves can take on a logistic sigmoid function, which is the most common activation function used in machine learning algorithms using logistic regression. Using the sigmoid function gives a probability of adoption after users are exposed to information. We realize that modeling behavior change may require more than a sigmoid function, but for purposes of developing an initial framework we start here.

$$\sigma(x) = \frac{1}{1 + e^{-x}}, \quad \text{where } \sigma(x) \text{ is probability of adoption}$$

Equation 1 – Sigmoid function describing s-shaped adoptions.

When we examine "successful" information campaigns with a high rate of adoption, we also see the s-shaped production function. This is similar to the DOI model, where we observe early initiators, amplifiers, and adopters who sustain the campaign until it reaches a steady state of growth or stops growing altogether. To characterize these campaigns, we look at the rate of change of adoption within the group. The slower the adoption rate, the flatter the "S" curve will be. Whereas, if the adoption rate is fast, the "S" curve will be narrower. This is evident when we compare the anti-narrative hashtags campaigns (left graph of Fig. 2). We clearly see that some campaigns are slower to adopt than others (consider the steepness of the curve). We also see that some campaigns have reached critical mass, flattened, and then regained momentum. When we compared the pro-narrative campaigns against the anti-narrative campaigns, we saw that the misinformation campaigns were more successful over the same time period in achieving adoption (right graph of Fig. 2).

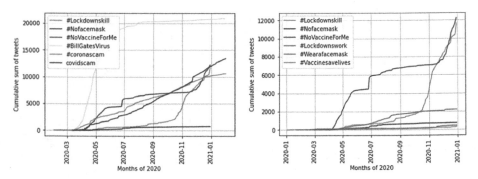

Fig. 2. Contrasting pro vs anti-narrative connective action campaigns.

The next step in our analysis is to examine the affordances of social media and how its use enables new forms of connective action. We apply labels to each tweet and graph the cumulative frequency of the campaign tweets over time to identify the affordances and interdependencies for each network (Fig. 3). Social media affordances, emerging roles, and interdependence in connective action have been studied from a qualitative perspective as presented by Vaast, et al. [16], but there is limited research on computational methods that quantify these social science concepts. Affordances correspond to "action possibilities and opportunities that emerge from actors engaging with a focal technology" [3]. Leonardi distinguished between these types of actions by labeling them as individualized, shared, and collective affordances [9]. Individual affordances are those that are realized by individual actors to fit their own specific needs that others may not share. Shared affordances correspond to affordances that are actualized by actors who share similar needs and uses of the technology. Collective affordances are actualized as various actors use a technology in a way that reflects their own needs and that, in the aggregate, generates a collective outcome. There are also different types of interdependence between users that correspond to the intensity and types of interactions and behaviors among actors that depend on each other for their own actions as they accomplish tasks together [13]. These are pooled, sequential, and reciprocal interdependence. Figure 4 summarizes the relationship between these affordances and interdependence. In pooled interdependence each user contributes to the collective action without requiring participation from other users. This is represented through the posting of *original tweets*, independent of others posting contributions to the event. Sequential interdependence considers the output of a particular user as a requirement for the input to another user. This content is usually uni-directional, and as such, the *retweet functionality* is representative of sequential interdependence. Reciprocal interdependence looks at the outputs of users as they become inputs for others, in a bi-directional flow among actors over time [16]. This can be represented by actors using the *"@" mention* feature, which allows for a user to directly reply or respond to the actor.

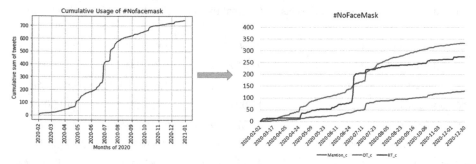

Fig. 3. Breakout of S-Curve into affordance-interdependence relationships by examining retweets, mentions, and original tweets.

Thus, we must look at the relationships between users to characterize what actions take place and how information flows through users and propagates in the connective action network. Figure 3 breaks the s-curve into Retweets, Mentions, and Original tweets to examine how information is spread in these information campaigns.

The #Nofacemask campaign is initially driven by original tweets. About halfway through the campaign, we see that Retweets experience a rapid adoption. By understanding this relationship and knowing that Retweets are part of shared affordances and sequential dependence, we can begin to understand the type of campaign this is and how it is organized. Further, we could remove an early node responsible for this Tweet or early Retweets.

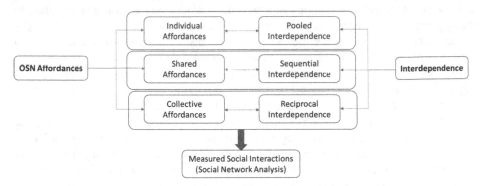

Fig. 4. Relationship between OSN affordances and interdependence

Table 1. Tweet counts for various COVID-19 hashtags from January 1 to December 31, 2020

Hashtag	Number of tweets	Original tweets	Number of RTs	Number of mentions
#Lockdownskill	11,630	3,260 (28%)	5,976 (51%)	2,394 (21%)
#Lockdownswork	444	93 (2%)	125 (28%)	226 (51%)
#Nofacemask	740	334 (45%)	276 (37%)	130 (18%)
#Wearafacemask	2,200	1,300 (59%)	760 (35%)	140 (.06%)
#Novaccineforme	12,225	3,827 (31%)	5,816 (48%)	2,582 (21%)
#Vaccinesavelives	305	79 (26%)	193 (63%)	33 (11%)
#Covidscam	13,395	4,165 (31%)	6,482 (48%)	2,748 (21%)
#BillGatesVirus	20,862	2,993 (14%)	10,826 (52%)	7,043 (34%)
#Coronascam	0,592	3,810 (36%)	4,893 (46%)	1,889 (18%)

Table 1 above shows the affordance-interdependence relationships between 9 different connective action campaigns, while Fig. 5 shows the curve signatures for these events.

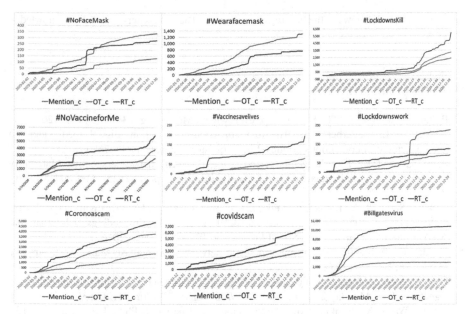

Fig. 5. Affordance-interdependence relationships for Covid-19 connective action events

5   Conclusions and Future Work

Using the s-shaped event signatures from connective action events offers insights into past social science theories that support characterizing these events through a quantitative approach. The observational approach taken in this analysis builds on the diffusion of innovations theory, critical mass theory, and previous s-shaped production function research to provide ideas for modeling future campaigns. A key benefit to this approach is that these curves are platform independent and occur across a range of industries, technologies, and social movements. Our next steps are to parameterize the mathematical equation so that we can predict the probability of the event as it emerges. Once we have established this model, we will track the interdependencies (Retweets, Mentions, Original Tweets) as the event unfolds and remove any deviant nodes, bridges, or communities capable of spreading discourse in the network.

Acknowledgments. This research is funded in part by the U.S. National Science Foundation (OIA-1946391, OIA-1920920, IIS-1636933, ACI-1429160, and IIS-1110868), U.S. Office of Naval Research (N00014-10-1-0091, N00014-14-1-0489, N00014-15-P-1187, N00014-16-1-2016, N00014-16-1-2412, N00014-17-1–2675, N00014-17-1-2605, N68335-19-C-0359, N00014-19-1-2336, N68335-20-C-0540), U.S. Air Force Research Lab, U.S. Army Research Office (W911NF-20-1-0262, W911NF-16-1-0189), U.S. Defense Advanced Research Projects Agency (W31P4Q-17-C-0059), Arkansas Research Alliance, the Jerry L. Maulden/Entergy Endowment at the University of Arkansas at Little Rock, and the Australian Department of Defense Strategic Policy Grants Program (SPGP) (award number: 2020-106-094). Any opinions, findings, and conclusions or recommendations expressed in this material are those of the authors and do not necessarily reflect the views of the funding organizations. The researchers gratefully acknowledge the support.

References

1. Bejan, A., Lorente, S.: The S-curves are everywhere. Mech. Eng. **134**(05), 44–47 (2012)
2. Bennett, W.L., Segerberg, A.: The logic of connective action: digital media and the personalization of contentious politics. Inf. Commun. Soc. **15**(5), 739–768 (2012)
3. Faraj, S., Azad, B.: The Materiality of Technology: An Affordance Perspective. Materiality and Organizing: Social Interaction in a Technological World, pp. 237–258. Oxford University Press, Oxford (2012)
4. Fink, C., Schmidt, A., Barash, V., Kelly, J., Cameron, C., Macy, M.: Investigating the observability of complex contagion in empirical social networks. In: Proceedings of International AAAI Conference on Web and Social Media, vol. 10, no. 1, March 2016
5. Geddes, C.: Achieving critical mass in social networks. J. Database Mark. Cust. Strategy Manag. **18**(2), 123–128 (2011)
6. Hemsley, J.: Studying the viral growth of a connective action network using information event signatures. First Monday **21**(8) (2016). https://doi.org/10.5210/fm.v21i8.6650
7. Hu, H., Wang, X.: Evolution of a large online social network. Phys. Lett. A **373**(12–13), 1105–1110 (2009)
8. Kucharavy, D., De Guio, R.: Application of S-shaped curves. Procedia Eng. **9**, 559–572 (2011)
9. Leonardi, P.M.: When does technology use enable network change in organizations? A comparative study of feature use and shared affordances. MIS Q **37**, 749–775 (2013)
10. Ma, L., Lee, C.S., Goh, D.H.-L.: Investigating influential factors influencing users to share news in social media: a diffusion of innovations perspective. In: Proceedings of the 13th ACM/IEEE-CS Joint Conference on Digital Libraries, pp.403–404 (2013). https://doi.org/10.1145/2467696.2467749
11. Marwell, G., Oliver, P.E., Prahl, R.: Social networks and collective action: a theory of the critical mass III. Am. J. Sociol. **94**(3), 502–534 (1988)
12. Priestley, M., Sluckin, T.J., Tiropanis, T.: Innovation on the web: the end of the S-curve? Internet Hist. **4**(4), 390–412 (2020)
13. Rogers, E.M.: Diffusion of Innovations. Simon and Schuster , New York (2010)
14. Shimogawa, S., Shinno, M., Saito, H.: Structure of S-shaped growth in innovation diffusion. Phys. Rev. E Stat. Nonlin. Soft. Matter. Phys.**85**, 056121 (2012)
15. Trott, V.: Connected feminists: foregrounding the interpersonal in connective action. Aust. J. Polit. Sci. **53**(1), 116–129 (2018). https://doi.org/10.1080/10361146.2017.1416583
16. Vaast, E., Safadi, H., Lapointe, L., Negoita, B.: Social media affordances for connective action: an examination of microblogging use during the Gulf of Mexico Oil Spil. MIS Q. **41**(4), 1179–1205 (2017)

Reinforcement Learning for Data Poisoning on Graph Neural Networks

Jacob Dineen[(✉)], A. S. M. Ahsan-Ul Haque, and Matthew Bielskas

Department of Computer Science, University of Virginia, Charlottesville, USA
jd5ed@virginia.edu

Abstract. Adversarial Machine Learning has emerged as a substantial subfield of Computer Science due to a lack of robustness in the models we train along with crowdsourcing practices that enable attackers to tamper with data. In the last two years, interest has surged in adversarial attacks on graphs yet the Graph Classification setting remains nearly untouched. Since a Graph Classification dataset consists of discrete graphs with class labels, related work has forgone direct gradient optimization in favor of an indirect Reinforcement Learning approach. We will study the novel problem of Data Poisoning (training-time) attacks on Neural Networks for Graph Classification using Reinforcement Learning Agents.

1 Introduction

The success of Deep Learning algorithms in applications to tasks in computer vision, natural language processing, and reinforcement learning has induced cautious deployment behavior, particularly in safety-critical applications, due to their ability to be influenced by adversarial examples, or attacks [1,2]. For real-world deployment, a level of trust in the robustness of algorithms to adversarial attacks is not only desired but required. Research by Goodfellow et al. in the areas of Generative Adversarial Networks (GANs) and Adversarial Training [3], has influenced a sub-field of Deep Learning focused on exactly the above, where the synthesis of viable attacks on neural network architectures, under White-Box, Black-Box or Grey-Box Attacks, can provide meaningful insight into both the models themselves and the preventative measures to be taken to ensure correctness (Adversarial Defenses [4]).

Here, we study a less saturated version of the adversarial setting: The effect of inducing a Poison attack on a Graph Neural Network under the task of graph classification [5,6]. The agent is trained to find an optimal policy, under Reinforcement Learning (RL) principles, such that they can inject, modify, or delete graph structure, or features under Black-Box Attacks. Such attacks assume that the agent has little to no underlying knowledge of the information regarding the target neural network (architecture, parameters, etc.), as is most often the case in real-world applications of machine learning algorithms, although ML-as-a-service systems provide infiltrators with alternative methods to attack [7]. We structure this problem as a Markov Decision Process, in which the agent

© Springer Nature Switzerland AG 2021

R. Thomson et al. (Eds.): SBP-BRiMS 2021, LNCS 12720, pp. 141–150, 2021.
https://doi.org/10.1007/978-3-030-80387-2_14

must interact with their environment(s) (a set of graphs) to fool the underlying learning algorithm into graph misclassification.

2   Background

2.1   Preliminaries

Adversarial Attacks. In this paper, we consider an adversarial example to be an intentionally perturbed instance of data, whose purpose is to fool the learning algorithm [8]. The attacker's main goal is to induce model misspecification given a particular task such that imposed modifications to the original data are unbeknownst to the model or a human observer. Perturbations in the context of Graph Machine Learning are generally of the form of structure modification, i.e. node, edge, or sub-graph modification, or feature modifications concerning particular nodes in the graph. Such attacks fall under a taxonomy noted in [2]: Evasion attacks and Poisoning attacks. Our focus here is on Poison attacks, which occur before or during the model training phase. In this way, we induce model misspecification during the parameter estimation phase of our learning paradigm.

Reinforcement Learning. The purpose of our work is to solve a Markov Decision Process (MDP) in which the agent(s) of the system is to solve sequential decision-making processes. MDPs are popular frameworks for superposing agent-environment interaction and simulation. Formally, an MDP can be represented by the tuple (S, A, P_a, R_a), where S is the state space, A is the action space, P_a, or T defines the dynamics of the state space (the probability that an action at a given state will lead to a subsequent state, the transition probability function), and R_a is the reward function. Given the introduced context, attacker(s) of our learning algorithm is considered as agent(s).

Graph Classification. Graph classification is the task of taking an instance of a graph and predicting its label(s). Graph Convolutional Neural Networks (GNN) in this setting need to map node outputs to a graph-level output ("read-out layer"), and then typically a Multi-Layer Perceptron (MLP) is employed to assign graph embeddings to class labels. [9] have used 2D CNN for Reddit, Collab and IMDB datasets in a supervised setting. [10] have used GNN in a semi-supervised setting for Citeseer, Cora, Pubmed and NELL datasets. Formally, Graph Classification is employed over a set of attributed graphs, $\mathcal{D} = \{(G_1, \ell_1), (G_2, \ell_2), \cdots, (G_n, \ell_n)\}$, each containing a graph structure and a label, and learning a function $f : G \to \mathcal{L}$ where G is the input space of graphs and $\mathcal{L}$ is the set of graph labels. Each graph $G_i \in G$ is comprised of V vertices, and E edges. The cardinality of V and E corresponds to the size and the connectedness (edges) of the graph, respectively. We let $d^{in}(G, i)$ represent the in-degree of the ith node of G.

2.2   Related Work

Attacks (on GNNs) in the Graph Classification setting are unique in that so far, they do not take the form of a gradient or meta-learning optimization problem directly on the model hyperparameters. The first attack introduced by [11] is RL-S2V, a Hierarchical Reinforcement Learning (HRL) approach for both Node and Graph Classification at test-time. They propose this because HRL agents are better suited for attacking graph data which is both discrete and combinatorial. The authors formulate a Hierarchical Q Learning algorithm that relies on GNN parameterization, and they successfully train it to attack by adding or removing edges in test graphs. Ultimately this paper treats Graph Classification as an afterthought and does not provide an insightful way to alter RL-S2V in this setting.

An evasion attack designed for Graph Classification is Rewatt [12], which relies on standard Actor-Critic to perturb individual graphs. This is done via "rewiring" operations that simultaneously delete and add edges between nodes near each other. Rewiring is meant to be subtle so that graph instances maintain similar metrics such as degree centrality. The authors theoretically justify Rewatt by showing that rewiring preserves top eigenvalues in the graph's Laplacian matrix. While Rewatt is useful at test time, it is ill-suited for a Poisoning attack. This is because it is designed to attack a single graph with minimal perturbation. Meanwhile, we have the burden of selecting training graphs that would be best for poisoning, but we can get away with significantly altering graphs if we limit ourselves to a small portion of the training data.

The newest entry to the Graph Classification literature details a backdoor attack on Graph Neural Networks [13]. The purpose of a backdoor attack is to perturb data points in a particular manner (e.g. wearing an accessory in security camera images) so that they are mislabeled as a class of your choice. For Graph Classification, this is done by generating a random subgraph as a trigger and injecting it a portion of training graphs. The reasoning is that random subgraph generation (with good parameters) can produce graphs that aren't "too anomalous" yet are assigned to the attacker's class of choice by the tuned GNN. This paper is interesting because they manipulate subgraphs instead of edges; an RL agent that perturbs subgraphs has potential to learn quickly due to the reduction in its action space.

3 Motivation

Attacks on algorithms designed for classification/regression, detection, generative models, recurrent neural networks and deep reinforcement learning, and even graph neural networks, are well covered in literature [14,15]. The generation of adversarial attacks in these spaces generally leads to corresponding defenses, such as Adversarial Training, Perturbation Detection [16], and Graph Attention Mechanisms [17] among others. Uncovering the root cause of the model malfunction is an equally important part of the process towards widespread industrial adoption of Deep Neural Networks (DNNs). As a practical example, we consider Wikipedia as a graphical representation - a network. Each node in the overarching network, composed of sub-graphs, could represent an article. Under the guise of a graph machine learning problem, a certain task for the learning algorithm could be to classify nodes as being 'real' or 'fake', e.g. "Is this article, that is linked to these other authentic articles, malicious?". A viable Graph Machine Learning algorithm would be able to detect or classify each observation to its correct class. A successful adversarial attack on the algorithm would be one that causes it, through perturbation of its underlying structure, to misclassify the instance. To the best of our knowledge, we present a novel framework to the setting of adversarial data poisoning in graph classification tasks [18].

4   Methodology

The framework for initiating a Poison attack on a graph neural network classifier (GNN) can be decomposed into a multi-step process involving 1) Data Procurement, 2) Graph Classification, and 3) Reinforcement Learning.

4.1   Data Procurement

We require a graph dataset composed of a set of graph/label pairs. Each graph in the dataset is represented by a set of nodes and edges - $G(V, E)$. This dataset is instantiated or partitioned into two distinct sets, forming training and testing data. The training set has two purposes: 1) to train a GNN for the task of label classification over a set of graphs, and 2) to represent the state of the environment, whereby the reinforcement learning agent can act to perturb the state via a finite set of actions A. As noted in Sect. 2.1, this is a Poison attack where we perturb the training set, rather than an evasion attack where the training data and GNN are untouched.

Experimenting with Graph Neural Networks has become more accessible with the release of Deep Graph Library [19]. It includes benchmark datasets, popular GNN algorithms across several settings, and easy compatibility with Pytorch and other Deep Learning libraries. Thus we will rely on DGL for graph models and data.

4.2   Graph Classification

The GNN, f, is a mapping from a raw graph g to a discrete label associated with the respective class of the graph. As a comprehensive survey of GNNs is beyond the scope of this work, we use a simple multi-layer Graph Convolutional Neural Network (GNN) for all experiments which are described below.

4.3   Graph Neural Networks

Here, f features two convolutional layers, Relu activation functions, a readout function, and a fully connected linear layer. We use the node in degrees of $G \in G$, $x_G = d^{in}(i, G)$, as input into the GNN and fix the readout function to $h_g = \frac{1}{|\mathcal{V}|} \sum_{v \in \mathcal{V}} h_v$, averaging over node features for each graph in the batch. A batch of graphs $G \in V$, where $V \subset G$ are fed through the GNN, where a graph representation h_g is learned through message passing and graph convolutions over all nodes before being passed through to the linear layer of the network for classification [19]. As most graph classification tasks that we explore with are multi-class, the outputs of the linear layer are passed through a softmax activation function: $\sigma(\mathbf{z})_i = \frac{e^{z_i}}{\sum_{j=1}^{K} e^{z_j}}$.

We employ categorical cross-entropy as our loss function J throughout, and iteratively learn θ via backpropagation, seen in Eq. 1.

$$\Delta w_{ij} = -\eta \frac{\partial J}{\partial w_{ij}}$$
$$w_{ij} \leftarrow w_{ij} - \eta \frac{\partial J}{\partial w_{ij}}$$

(1)

We make note that we treat the hyper-parameterization of the GNN as fixed to reduce the elasticity of our experiments. Finely-tuned, and extensively trained neural networks may be less prone to poison attacks.

4.4   Reinforcement Learning

Algorithm. We use a Monte Carlo variant of a Deep Policy Gradient algorithm: REINFORCE [20], with additional algorithmic validation saved for future work and use a simple random policy $\pi(s) = \frac{1}{|A|}$ as our baseline. Intuitively, we conceptualize this as: How much better does a Deep Reinforcement Learning Algorithm choose a sequence of poison points (graphs), such that testing accuracy is degraded, against a fully random search of the action space?

Reward. We craft a Reward function, Eq. 2, to pass through p (poison points) rewards during a single episode. Let θ be the original, learned parameters from training on the unperturbed set, and θ^p be the perturbed variant. X^{test} is the original testing set generated a priori. For simplicity, we assume that the application of the actual or perturbed parameters to the testing set yields an accuracy. The Reward function R is a signal that measures the difference between the unperturbed model's application to X^{test} against the perturbed model's application to X^{test}. In other words, if the reward is positive, it means that the Poison attack was successful.

$$R = \theta(X) - \theta^p(X)$$

(2)

Environment. Utilizing a form of Deep Reinforcement Learning, a set of inputs, representing the state of the system, are required at t to pass through to a function approximator. We represent this state $S(G_t)$ as the mean, max, min over all $G_i \in G$ at t. Intuitively, the RL algorithm sees as input a tensor containing each graph's representation (defined by summary statistics) in the training set, chooses an action, and transitions into a new state based on perturbations to the training set.

At any t, the agent can alter the state by performing any $a \in A$. To discretize the action space, we specify that the action of adding a subgraph into an existing graph's structure has a fixed set of parameters n and p, corresponding to the number of nodes, and the probability of an edge between nodes, respectively. We also constrain the action space by permitting node or edge modifications to be random processes, e.g., the agent has no control over which node or edge to perturb, only that they wish to perform that action.

An optimal policy would learn which sequence of actions to take, and their applications to a subset of G, to maximize the episodic degradation in the GNN's ability to generalize over the testing set.

4.5 Poison Attack

The culmination of the multi-step framework results in a poisoning attack on the original graph training set. Given a set of graph, label pairs, we first compute a benchmark accuracy via the application of θ to the training set and perform inference on the testing set. Given a scalar value for p representing the number of poison points that are selected by the RL algorithm at each episode, we begin. After each poison point, decided by the RL algorithm, we perform retraining of θ on the perturbed training set, generating θ^p, and applying it to the testing set. The retraining stage is conducted by transfer learning: we take the original set of parameters and warm-start on them, and then retrain the network for a single additional epoch.

The end-to-end system is designed to take as input a full dataset composed of graphs and have an agent learn to manipulate those graphs, using available actions in A, such that they maximize their episodic reward. The reward is measured as the difference between the baseline accuracy (the GNN's accuracy on the testing set before perturbation on the training set) and the perturbed accuracy (the GNN's accuracy on the testing set after the training set has been perturbed).

5 Experiments

5.1 MiniGCDataset

We first create partitioned sets from the MiniGCDataset, with the size of the training set $= 150$, and the testing set $= 30$. We run the original GNN for 70 epochs to generate θ and our benchmark accuracy on the testing set. During the Poison attack, we allow the RL algorithm to perturb $p = 10$ poison points during each episode. The reward is passed back intra-episode, between poison point selection, to reduce vanishing gradient issues and to encourage learning. This process is conducted 175 times as we fix $NEpisodes = 175$. We also fix $NRuns = 10$. The Graph Neural Network architecture is defined in Sect. 4.2. The RL algorithm procedure is released in our source code. Again, we use $\pi(s) = \frac{1}{|A|}$ as a baseline to measure RL's efficacy at this specific task. Figure 1 shows reward over the episode, averaged over $NRuns$ to reduce stochasticity.

We see a modest improvement over random search, as well as lesser extremes in the incorrect direction, meaning we rarely see cases where our RL agent perturbs a set of graphs during an episode and increases the testing accuracy.

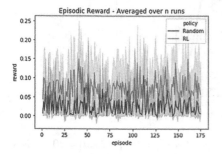

Fig. 1. Simulation results on MiniGC-Dataset: $|Train| = 180$, $|Test| = 30$, GNN epochs $= 70$, RL epochs $= 175$. The blue line represents the accuracy degradation over a random policy, averaged out over 10 runs. The orange line is the accuracy degradation over REINFORCE, averaged out over $NRuns = 10$ (Color figure online)

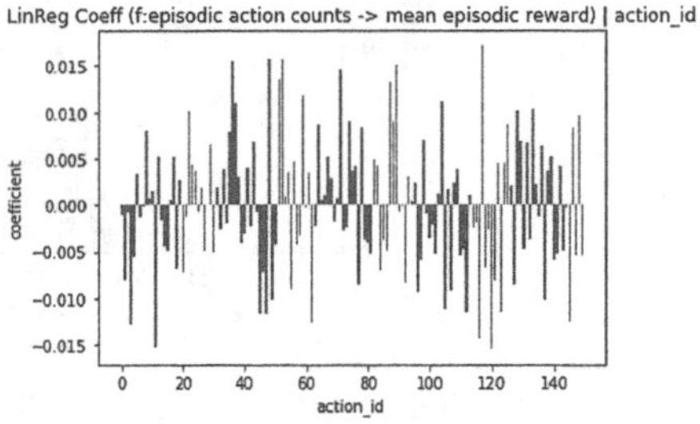

Fig. 2. Linear regression: f: summed episodic actions counts → mean episodic reward. Visualizing the coefficients to show which action selections have a higher impact on accuracy degradation over the testing set. the x-axis represents the graph id of the graph being perturbed, and the y-axis is the linear regression coefficient. *The linear regression model was fit with an r^2 of 0.85.

We also create a way to visualize the efficacy of individual graphs perturbations in Fig. 2. There, the x-axis represents the action id, i.e., the graph id within the training set. The y-axis is the linear regression coefficient found when mapping the summed episodic actions counts to the mean episodic reward over $Nruns$. We see that star graphs appear most often amongst the top ten high valued coefficients and perceive that they are the most likely class within this dataset to have an attack result in net positive reward on the system.

5.2 MiniGC - Larger

Next, we increase the size of the graphs by an order of magnitude. All other parameters are kept the same, including the size of the subgraph inserted into the original graph. In the GNN initial training stage, we see that the size of the graphs has a profound effect on test-time accuracy, meaning higher baselines for our RL algorithms. We visualize these results in Fig. 3, noting that the distribution plots of episodic reward comparing REINFORCE and random selection show that REINFORCE has a laterally shifted distribution and a higher peak. What we do notice is that central tendency measures are dramatically shifted downward, meaning that the increase in network sizes allowed for the GNN to draw more clear decision boundaries and consequently increase its own robustness.

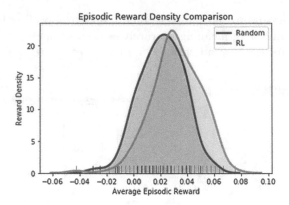

Fig. 3. Simulation results on MiniGCDataset: $|Train| = 180$, $|Test| = 30$, GNN epochs = 70, RL epochs = 175. We increase the size of the networks by an order of magnitude, such that $MinNodes = 1500$ and $MaxNodes = 2000$. These are normalized density plots. The x-axis represents the average episodic reward over each policy, while the y-axis is the density of observing that statistic. Blue corresponds to a random policy, while orange is the utilized REINFORCE algorithm. (Color figure online)

6   Conclusion

In this study, we present a novel end-to-end poisoning attack on a GNN Classifier. The results from our analysis show that our method presents a lift over a random, brute force search of the graph space, and provides additional insight into the vulnerabilities of specific graph structures, e.g., which graph classes have a higher propensity to affect test-time performance given a perturbation. We also experiment with various ways of representing the state of our graph dynamic system. Such research could be important in practice, particularly in the healthcare industry, and has unexplored theoretical ties to the robustness and expressive power of Graph Deep Learning.

7   Future Work

State Representation: We intend to explore further methods for encoding the state of the graphical system at t. Here, we use a flattened representation of summary statistics over each graph in the training set, but we believe that there may be other descriptive methods from the area of network science, such as centrality measures, that may better represent a graph's characteristics in a low-dimensional space.

Expanded Action Space: One of the main things that we intend to explore in the future is the action space. We reduced the action space from containing granular level actions, such as node or edge edits, but believe that translating this to a Hierarchical Reinforcement Learning problem opens the door for us to operate on a larger action space without sacrificing much in the way of compute. In this way, we could subject the RL algorithm to a further customized reward function that rewards more heavily when they choose to alter the graphs in a small way, such that perturbations would be obfuscated from a human detector. Initially, our action space resembled Table 1.

Table 1. A hypothesized action space expansion.

Action	Name	Definition
a_1	Subgraph add	A random gnp graph is inserted into the existing graph structure
a_2	Node delete	A random node with g is removed
a_3	Node add	A node is added to g
a_4	Edge delete	A random edge is deleted from g
a_5	Edge add	A random edge is added to g

Datasets: Most of our experimental process was centered around using the MiniGCDataset from DeepGraphLibrary. Other datasets available feature vastly different APIs. For future work, we intend to expand the testing suite to include real-world datasets e.g. PROTEINS (with known consequences faced by a successful attack), and to make our overall pipeline more extensible.

References

1. Jin, W., Li, Y., Xu, H., Wang, Y., Tang, J.: Adversarial attacks and defenses on graphs: a review and empirical study. arXiv preprint arXiv:2003.00653 (2020)
2. Xu, H., et al.: Adversarial attacks and defenses in images, graphs and text: a review. Int. J. Autom. Comput. **17**, 151–178 (2020). https://doi.org/10.1007/s11633-019-1211-x
3. Goodfellow, I.J., et al.: Generative adversarial networks (2014)
4. Ren, K., Zheng, T., Qin, Z., Liu, X.: Adversarial attacks and defenses in deep learning. Engineering **6**(3), 346–360 (2020)
5. Knyazev, B., Lin, X., Amer, M.R., Taylor, G.W.: Spectral multigraph networks for discovering and fusing relationships in molecules. arXiv:1811.09595 [cs, stat], November 2018
6. Ying, R., You, J., Morris, C., Ren, X., Hamilton, W.L., Leskovec, J.: Hierarchical graph representation learning with differentiable pooling. arXiv:1806.08804 [cs, stat], February 2019
7. Tramèr, F., Zhang, F., Juels, A., Reiter, M.K., Ristenpart, T.: Stealing machine learning models via prediction APIs. In: 25th USENIX Security Symposium (USENIX Security 2016), pp. 601–618 (2016)
8. Zhang, W.E., Sheng, Q.Z., Alhazmi, A., Li, C.: Adversarial attacks on deep-learning models in natural language processing: a survey. ACM Trans. Intell. Syst. Technol. **11**(3) (2020). https://doi.org/10.1145/3374217

9. Tixier, A.J.-P., Nikolentzos, G., Meladianos, P., Vazirgiannis, M.: Graph classification with 2D convolutional neural networks. In: Tetko, I.V., Kůrková, V., Karpov, P., Theis, F. (eds.) ICANN 2019. LNCS, vol. 11731, pp. 578–593. Springer, Cham (2019). https://doi.org/10.1007/978-3-030-30493-5_54

10. Kipf, T.N., Welling, M.: Semi-supervised classification with graph convolutional networks. arXiv preprint arXiv:1609.02907 (2016)

11. Dai, H., et al.: Adversarial attack on graph structured data. arXiv:1806.02371 [cs, stat], June 2018

12. Ma, Y., Wang, S., Derr, T., Wu, L., Tang, J.: Attacking graph convolutional networks via rewiring. arXiv:1906.03750 [cs, stat], September 2019

13. Zhang, Z., Jia, J., Wang, B., Gong, N.Z.: Backdoor attacks to graph neural networks. arXiv:2006.11165 [cs], June 2020

14. Papernot, N., McDaniel, P., Swami, A., Harang, R.: Crafting adversarial input sequences for recurrent neural networks (2016)

15. Akhtar, N., Mian, A.: Threat of adversarial attacks on deep learning in computer vision: a survey. arXiv:1801.00553 [cs], February 2018

16. Xu, X., Yu, Y., Li, B., Song, L., Liu, C., Gunter, C.: Characterizing malicious edges targeting on graph neural networks (2019). https://openreview.net/forum?id=HJxdAoCcYX

17. Zhu, D., Zhang, Z., Cui, P., Zhu, W.: Robust graph convolutional networks against adversarial attacks. In: Proceedings of the 25th ACM SIGKDD International Conference on Knowledge Discovery & Data Mining, KDD 2019, pp. 1399–1407. Association for Computing Machinery, New York (2019). https://doi.org/10.1145/3292500.3330851

18. Xu, K., Hu, W., Leskovec, J., Jegelka, S.: How powerful are graph neural networks? arXiv:1810.00826 [cs, stat], February 2019

19. Wang, M., et al.: Deep graph library: a graph-centric, highly-performant package for graph neural networks. arXiv preprint arXiv:1909.01315 (2019)

20. Williams, R.J.: Simple statistical gradient-following algorithms for connectionist reinforcement learning. Mach. Learn. 8(3), 229–256 (1992). https://doi.org/10.1007/BF00992696

Social Cybersecurity and Social Networks

Simulating Social-Cyber Maneuvers
to Deter Disinformation Campaigns

Janice T. Blane$^{(\boxtimes)}$ (iD), J. D. Moffitt$^{(\boxtimes)}$ (iD), and Kathleen M. Carley$^{(\boxtimes)}$ (iD)

School of Computer Science, Carnegie Mellon University, 5000 Forbes Avenue,
Pittsburgh, PA 15213, USA
`jblane@andrew.cmu.edu`, {`jdmoffit,kathleen.carley`}`@cs.cmu.edu`

Abstract. We develop an agent-based model of a Twitter environment to simulate using social-cyber (BEND) maneuvers to deter a disinformation campaign. We explore the use of the network maneuvers of *back*, *build*, and *neutralize* to manipulate the network and the information maneuvers of *excite*, *dismay*, *explain*, and *dismiss* to control the narrative. Using belief as a measure of effectiveness, we explore the changes in user behavior and the resulting network. We demonstrate that *build* is the most effective network maneuver countermeasure for deterrence. The results also show that affecting a tweet's emotional and logical values through information maneuvers effectively controls the overall network belief.

Keywords: Social cybersecurity · Social network analysis · Social media analytics · Disinformation · Agent based model

1 Introduction

The lines between real and digital life have become indistinguishable. Online platforms have gained distinction as the primary source for news, discussion, sharing ideas, and community building. Maligned actors continuously develop ways to exploit this space to conduct cyber-mediated attacks on civil society. The emerging, transdisciplinary field of social cybersecurity provides theories and tools to help study and counter such cyber-mediated attacks [6].

While recent research in the field of social cybersecurity focuses on establishing a framework to discuss information maneuver [2], social cyber-forensics [3], and diffusion [14–16], the current pandemic and US election have demonstrated the need for greater understanding of social-cyber maneuvers and how to mitigate them. However, studying social-cyber maneuvers poses two problems: there is no exact way to measure the effects of information operations (open research problem). Two, it is not ethically tractable to manipulate human test subjects' beliefs thoughts. Agent-based modeling (ABM) is a tool that can help researchers overcome the latter of the two problems. ABM provides a powerful method for representing complex and dynamic real-world environments. Until recently, modeling efforts in this domain focused on information diffusion and rumor propagation. In most cases, studies abstract away platform-specific

© Springer Nature Switzerland AG 2021
R. Thomson et al. (Eds.): SBP-BRiMS 2021, LNCS 12720, pp. 153–163, 2021.
https://doi.org/10.1007/978-3-030-80387-2_15

mechanics; `twitter_sim` is the first model our team is aware of that evaluates forms of social-cyber maneuver [4] through simulation.

Our principal research question is: can anti-disinformation maneuvers mitigate disinformation campaigns? This paper explores a social network experiencing a disinformation campaign leveraging the *back* disinformation maneuver to sway network belief. We then implement information and network maneuvers to counter the *back* maneuver using belief to measure the effect of disinformation and mitigating strategies. Our study sets out to improve `twitter_sim` and incorporate more realistic mechanics to expand its ability to model social-cyber maneuvers to counter the spread of disinformation.

2   Related Works

2.1   Modeling Information and Beliefs

Many models have been used to simulate the spread of information as well as the dynamics on social networks; e.g., [1,8,12,18]. Construct, an agent-based social-network model, simulates information and belief diffusion using a turn-based social influence approach [5]. Agents have bounded rationality where they use homophily and expertise to help decide with whom they will have interactions. Additionally, agents have general memory of task knowledge, social memory of with whom they are interacting, and a transactive memory of who knows what and whom [7]. This paper builds from Beskow and Carley's `twitter_sim` model [4] and Construct. We further limit the agent's attention by limiting the number of read tweets and exploring more BEND maneuvers. These models are compared in Table 1.

Table 1. Docking lite comparison of `twitter_sim2.0`, `twitter_sim`, and construct.

Features	twitter_sim2.0	twitter_sim	Construct
General population	✔	✔	
Media agents	✔	✔	✔
Opinion leaders	✔	✔	✔
Information access	✔	✔	✔
General memory	✔	✔	✔
Transactive memory			✔
Homophily	✔	✔	✔
Limited attention	✔	✔	
Dynamic network	✔	✔	✔
Emotional response	✔		

2.2   BEND Framework

The BEND framework provides a way of understanding information/influence operations [6]. The framework provides two types of social-cyber maneuvers for manipulating information, ideas, and beliefs: information (or narrative) maneuvers and network maneuvers. BEND describes 16 maneuvers (which contribute to the acronym for "BEND"); we incorporate seven to extend the `twitter_sim` model to explore both information and network forms of maneuver.

2.3 Modeling Emotions and Reason

Several theories discuss how emotions and cognition affect an individual's behavior. In Heise's Affect Control Theory, people create events or conduct actions that confirm their fundamental sentiments of themselves in a particular situation. Emotions then reflect a person's sentiment about themselves and the validations or invalidations of self-created by the situation at the moment [10]. Behavior is thus related to a combination of the state of the person and the situation.

3 Model Description

In `twitter_sim2.0`, we made the discrete event distribution better represent Twitter, modified agent behavior to facilitate BEND maneuvers, and introduced four additional BEND maneuvers for testing. The model is Python 3.7 based. We used the NetworkX package [9] to generate scale-free networks. `Twitter_sim2.0` is a discrete event agent-based model that attempts to replicate individual user behavior and Twitter platform mechanics. The simulation consists of three types of agents- normal users, spreaders (facilitate disinformation operations), and beacons (trusted community leaders or subject matter experts) [15]. Spreaders and beacons both conduct social-cyber maneuvers to change the belief of the simulated network.

3.1 Review of `twitter_sim` Features

Homophily (Similarity): Homophily is the tendency for people to seek out connections with those that are similar to themselves [13]. In this simulation, homophily (similarity) is represented by an adjacency matrix of all nodes in the network; the link values are the Jaccard similarity coefficient between nodes ($similarity = \frac{successors_A \cap successors_B}{successors_A \cup successors_B}$).

Influence: In graph theory, the indegree of a node (vertex) in a directed graph is the number of edges directed into the node. In this model, influence is represented by a normalized value of an agent's indegree.

Tweet: We simulate the Twitter's content sharing functions for tweeting, retweeting, replies, and mentions ($Tweet_{value} = type \times similarity_{ij} \times influence_i$). A tweet's value is a function of tweet type (0: normal, 1: disinformation, -1: anti-disinformation), similarity, and influence [4].

Limited Attention and Changing Beliefs: In our model, agents are influenced only by the content they see. Limited attention [17] constrains how many tweets a user reads each active period.

We measure the effectiveness of disinformation operations by the change in belief using a continuous value between zero and one. An agent with a belief value less than 0.5 believes disinformation, and a value greater than 0.5 does not believe disinformation. Belief is calculated as $belief_t = belief_{t-1} + (mean(tweets_{read}) + global_{perc})) \times (1 - belief_{t-1})$.

3.2   Model Changes Introduced for `twitter_sim2.0`

Discrete Event Simulation: As mentioned, a user is influenced only by seen content. A key component to the content a user sees is their activity level or how often they engage on Twitter's platform. The original `twitter_sim` model simulates user activity ranging from every two months up to eighteen times per day. Based on recent research from Pew's Internet & Technology center, we adjusted the distribution of activity in our model [11]. In the simulation, agents are assigned a daily, weekly, or less often activity attribute in the appropriate percentages. Each time an agent wakes, their next wake time is stored by randomly sampling awake time from 0–24 h daily, 25–168 h weekly, and 169–1081 hours for less than weekly.

Model Agents and their Twitter Actions: In `twitter_sim2.0`, normal users, spreaders, and beacons tweet, send mentions, and retweet. Beacons use replies to counter disinformation. All agents retweet with a given probability a portion of their read tweets. Normal users do not create disinformation but do spread it through retweeting. Beacons will not retweet messages that contain disinformation, but spreaders will.

In `twitter_sim2.0`, agents generate mentions with a given probability instead of every time step as in `twitter_sim`. If an agent-i mentions agent-j, and an edge does not exist between the two agents, agent-i will be added to a recommendation queue for agent-j to form an edge between the two. BEND network maneuvers can exploit the mechanics of mentions [4].

Beacons counter disinformation by replying with anti-disinformation but can only counter disinformation they see. Additionally, beacons conduct BEND maneuvers to counter disinformation. We set the percentage of beacons in the simulation as an independent variable ranging from 0–15%.

Spreaders send messages with a mix of noise and disinformation. Spreaders start the simulation with one link and gradually build connections and work their way into the network. Spreaders use the *back* BEND maneuver and link to influential agents aligned with their disinformation campaign.

Impacts of Emotion and Logic: To facilitate impacts of emotion and logic in our simulations, we adjust the original tweet value calculation to include a multiplier for emotion and a multiplier for logic ($Tweet_{value} = type \times similarity_{ij} \times influence_i \times emotion \times logic$).

Emotion maneuvers represent both the *excite* and *dismay* BEND maneuvers, where *excite* is discussion that brings joy or happiness, and *dismay* is discussion that brings sadness or anger. A very emotional tweet can be either very exciting or very dismaying. In our model, both the tweet and the user have an emotional state. The closer the tweet's and user's emotional state, the higher the emotional strength and impact on the user's actions. Users reinforce their beliefs based on the matched sentiment.

The logic maneuvers also represent two similar and opposite BEND maneuvers. *Explain* are actions that provide more details on a topic. *Dismiss* are actions that provide details on the unimportance of a topic. We model this by assigning tweets a value for how logically persuasive it is. This can range from innocuous cat videos to a scientific paper. Likewise, users have a reasoning level, which is a measure of how open they are to new ideas or their ability to listen and comprehend the logic of a tweet. Our model compares the two values, and the resulting logical strength of the tweet is a function of combining the two values. Therefore, a highly logical tweet sent to a user with a high reasoning level is more affected by the tweet, resulting in a high logical strength for the tweet value.

4 Experiments

This section explores emergent behavior when beacons use BEND maneuvers to counter a spreader disinformation campaign. See Table 2 for a detailed view of the independent and control variables for the experiments. The percentages of spreaders and beacons were selected to ensure three ratios: more spreaders than beacons, parity, and more beacons than spreaders. As a baseline, spreaders conduct disinformation maneuvers (*back*), and beacons do not conduct anti-disinformation maneuvers.

Table 2. Experimental design for `twitter_sim2.0`

Variable	Number variants	Variant values
Spreaders (% of Network)	2	5, 10
Beacons (% of Network)	3	0, 10, 15
Type of network	1	Normal
Spreader BEND	1	BACK
Beacon BEND (Network maneuver)	3	BACK, BUILD, NEUTRALIZE
Beacon BEND (Information maneuver)	3	EMOTION, LOGIC, COMBINED

4.1 Network Based Maneuvers

We selected *back*, *build*, and *neutralize* to provide a diversified representation of network-based BEND maneuvers. These maneuvers affect who is talking and listening to whom in Twitter [2]. Each combination of variables was run 100 times on a 100 node network, and we averaged the results for reporting. Each run simulated 1,680 h of activity or approximately 2.5 months.

Maneuvers. A *back* maneuver increases the importance of opinion leaders [2]. On Twitter, *back* is conducted through following and retweeting. The greater the scale, the greater the impact. Beacons execute *back* by finding opinion leaders with beliefs between 0 and 0.49 (representing unbelief in disinformation) and following them.

The *build* maneuver involves creating a group, or the appearance of a group [2]. This maneuver focuses on building a group around a common bond and then injecting a directed narrative into the group. Beacon agents conduct *build* by co-mentioning other agents in the network. When a beacon co-mentions (*Agent X* and *Agent Y*), the beacon appears in both of the agents' suggested follow queues, and *Agent X* will appear in *Agent Y*'s suggested follow queue and vice versa.

A *neutralize* maneuver involves actions that limit the effectiveness of opinion leaders [2]. Reducing the number of users that follow an opinion leader achieves a *neutralize* effect. *Neutralize* is similar to *back* but with the opposite desired effect. Beacon agents conduct *neutralize* by finding opinion leaders that propagate disinformation and then follow them. The beacon will send anti-disinformation replies by linking to the opinion leader, possibly directing their followers away from disinformation.

4.2 Information Based Maneuvers

We represented *excite, dismay, explain,* and *dismiss* maneuvers by simulating differences in the emotional and logical aspects of a tweet in conjunction with varying user emotional states and reasoning levels. The emotion simulations represent *excite* and *dismay*, and the logic simulations represent *explain* and *dismiss*. The combined experiment combines both simulations. Each experiment executed 100 runs, resulting in 2400 replications.

Maneuvers. For simulating emotional behavior, the beacons send anti-disinformation tweets and retweets at a minimum of 80% emotional strength. This simulates tweets that attempt to persuade by appealing to the users' emotions to either *excite* or *dismay*, understanding that not all tweets can map to every person's emotional state at 100%. When simulating emotional behavior independent of reason, the logic of the tweet and user reasoning levels result in a 50% average logic strength.

As logical behavior is a function of the logic of the tweet and the user's reasoning level, beacons attempt to deter the spread of disinformation by creating very logical tweets in favor of anti-disinformation. To simulate the logic maneuvers, we set all tweets sent by beacons to the maximum logic level. The beacons simulate *explain* when sending very logical tweets in favor of anti-disinformation or *dismiss* when sending against disinformation. This value is combined with the user reasoning level to determine the logic strength. When simulating logic independent of emotion, we maintain an emotional strength of 50%. In the combined emotion and logic maneuver experiment, we execute both methods simultaneously. Beacons send tweets at 80% emotional strength and tweets that are 100% logical.

5 Results and Discussion

5.1 Baseline

From the results in Table 3, we see that when there are no beacons in the network, the overall belief grows in favor of disinformation (0.003). When we double the number of spreaders, we find that the overall belief almost doubles in favor of disinformation (0.005). This result indicates that an increased mass of accounts conducting *back* increases the effectiveness of the maneuver.

Table 3. BEND Maneuver results

Network maneuver		5% Spreaders	10% Spreaders	Information maneuver		5% Spreaders	10% Spreaders
BEND Maneuver	% of Beacons	Change in Network belief	Change in Network belief	BEND Maneuver	% of Beacons	Change in Network belief	Change in Network belief
None	0	0.00257	0.00477	None	0	0.00257	0.00477
Back	10	−0.00112	0.00079	Emotion	10	−0.00142	−0.00083
	15	−0.00220	−0.00056		15	−0.00243	−0.00167
Build	10	**−0.00780**	**−0.00665**	Logic	10	−0.00182	−0.00084
	15	**−0.00913**	**−0.00969**		15	−0.00270	−0.00202
Neutralize	10	−0.00321	−0.00227	Combined	10	**−0.00294**	**−0.00190**
	15	−0.00472	−0.00376		15	**−0.00392**	**−0.00309**

5.2 Network Maneuver

Maneuvers. We see from the results displayed in Table 3 that the *back* is the least effective BEND maneuver when attempting to counter a *back* disinformation maneuver. Anti-disinformation forces employing *back* to counter *back* must have a greater ratio of agents to succeed because, at parity and under-match, disinformation forces will achieve their desired effects. Using the *build* BEND maneuver to counter a *back* disinformation campaign was the most effective countermeasure across all percentages of spreaders and beacons. These results may be evidence of the importance of creating or reinforcing strong, knowledgeable, and resilient communities in combating disinformation. Further, *neutralize* is an effective countermeasure against the *back* disinformation maneuver for all percentages of beacons and spreaders. However, it is not as effective as the *build* BEND maneuver. In the presence of 5% spreaders, *neutralize* is less than half as effective, and as spreaders increase, its effectiveness decreases compared to *build*.

Sensitivity to Network Size: We examined how network size impacted the most effective maneuver, *build*. We keep the percentage of spreaders at 10% of the network. We find that *back* remains an effective countermeasure against the *back* disinformation maneuver. Further studies should explore whether the maneuver's degree of effectiveness decreases as network size increases.

Overall. We discover that how the network belief is changed varies by network BEND maneuver. *build* and *neutralize* have a more significant impact on changing the belief of users who believe disinformation. Whereas, *back* has more of an effect on users who do not believe disinformation.

The results for *neutralize* compared to *build* may further strengthen the importance of building or reinforcing strong, resilient communities as the best policy for fighting disinformation. In some cases, the use of *neutralize* could have the opposite desired effect on disinformation. An example of this could be the disinformation opinion leader's Twitter account is suspended (neutralized), strengthening the opinion leader's position as evidence for whatever narrative they are pushing.

5.3   Information Maneuver

Maneuvers. When we independently simulated the emotional values for *excite* and *dismay* to beacon-sent tweets, the tweet's belief changed in favor of anti-disinformation by −0.00083 and −0.00167 for beacons at 10% and 15%, respectively.

For *explain* and *dismiss*, when we independently simulated the logic maneuvers without beacons, the total network belief grows linearly to 0.00293. At 10% beacons, the total network belief grows slightly until halfway through the period before it declines in favor of anti-disinformation, ending at −0.00084. Finally, at 15% beacons, the total network belief declines more immediately, ending at −0.000202.

The combined emotion and logic maneuvers simulated the effects of both maneuvers acting simultaneously. The combined maneuver caused an immediate and consistent shift in belief towards anti-disinformation for all beacon levels in the network. Table 3 displays the results for these maneuvers.

Overall. Beacons executing emotion and logic maneuvers changed the direction of the overall total network belief. They performed in the same general manner given varying beacon and spreader values. If the number of spreaders were greater than the number of beacons, then the belief value would grow over time in favor of disinformation. In the opposite case, the belief value would decrease over time in favor of anti-disinformation. The final values varied depending on the beacon to spreader ratio.

Compared to the baseline at 10% for beacons and spreaders, the logic maneuver results in approximately the same belief value as the emotion maneuver. They differ because the emotional information effect is immediate but has a gradual decrease, where the logical reasoning effect is delayed with a steeper decrease. This trend suggests that logical reasoning is less effective than emotional information in the short run, but it accumulates a greater effect over time. Combining both the emotion and logic maneuvers proved that there are compounding effects on the total network belief by manipulating both aspects of the message. Table 3 shows the average total change in network belief for the different information maneuvers experiments.

6 Validation

This model is a simulation of personal beliefs, which are difficult to measure, making it difficult to validate the model. Much of the validation for our model is carried forward validation assumptions from the twitter_sim [4]. The simulation has face validity as it represents a Twitter environment with agents who behave like Twitter users. Additional validation relies on statistical and stylized facts using collected tweet artifacts such as the daily volume of tweets and the distribution of tweets, retweets, replies, mentions, and quotes. User behavior is based on Pew Research to accurately define the percentage of users who check Twitter daily, weekly, and greater weekly. Our analysis of 2.2 million tweets from COVID-19 caused us to adjust the model to reflect that about 8% of tweets contain mentions. Though the model cannot accurately reflect true emotional strength or user reason, the model does apply previous research to simulate user actions in response to emotional and logical tweets.

7 Conclusion

Our results demonstrate that using anti-disinformation BEND maneuvers as countermeasures against BEND disinformation maneuvers may be viable for fighting disinformation campaigns. Our results suggest that building and strengthening communities with truth and trust is a far more effective strategy than offensive actions against disinformation opinion leaders.

We demonstrated how to use emotion and logic within a social network as countermeasures. Beacons play an essential role in sending tweets to negate the direction of the total network belief caused by the *backing* maneuver. We found the ratio between beacons and spreaders and beacon tweet characteristics as two critical factors in reducing disinformation diffusion. When the counter maneuvers were combined, the effects on the overall belief were more pronounced, showing the advantage of a well-devised narrative.

There are several limitations. First, we limit our measure of effectiveness to change in belief. Second, we assume that *excite* and *dismay* as well as *explain* and *dismiss* are complementary to each other. In reality, user narratives may be more complex. Further research is necessary to refine the model to reflect human emotion and logic activity better. Finally, because of the complexity of the network, the robustness of the tweet attributes, and the number of agents and agent rules, we were limited by computing power to simulating networks much smaller than in the real world.

Future work should incorporate edge lists of actual tweet networks to seed the network rather than synthetic scale-free networks allowing the results to be directly compared with real-world events. Future studies must include a complete list of BEND maneuvers and maneuver combinations to simulate influence campaigns better. Finally, incorporating more comprehensive tweet attributes to differentiate between types and intensity of emotions and logic will improve information maneuver simulation leading to better maneuver evaluations.

Acknowledgements. The research for this paper was supported in part by the Office of Naval Research (ONR) under grant N00014182106, the Knight Foundation, the United States Army, and by the center for Informed Democracy and Social-cybersecurity (IDeaS). The views and conclusions are those of the authors and should not be interpreted as representing the official policies, either expressed or implied, of the Knight Foundation, the ONR, the United States Army, or the US Government.

References

1. Bass, F.M.: A new product growth for model consumer durables. Manag. Sci. **15**(5), 215–227 (1969)
2. Carley, K.M., Cervone, G., Agarwal, N., Liu, H.: Social cyber-security. In: Thomson, R., Dancy, C., Hyder, A., Bisgin, H. (eds.) SBP-BRiMS 2018. LNCS, vol. 10899, pp. 389–394. Springer, Cham (2018). https://doi.org/10.1007/978-3-319-93372-6_42
3. Beskow, D., Carley, K.M., Bisgin, H., Hyder, A., Dancy, C., Thomson, R.: Introducing bot-hunter: a tiered approach to detection and characterizing automated activity on Twitter. In: SBP-BRiMS. Springer, Heidelberg (2018)
4. Beskow, D.M., Carley, K.M.: Agent based simulation of bot disinformation maneuvers in twitter. In: 2019 WinterSim, pp. 750–761. IEEE (2019)
5. Carley, K.M.: Group stability: asocio-cognitive approach. Adv. Group Process. **7**(1), 44 (1990)
6. Carley, K.M.: Social cybersecurity: an emerging science. Comput. Math. Organ. Theory **26**, 365–381 (2020)
7. Carley, K.M., Martin, M.K., Hirshman, B.R.: The etiology of social change. Topics Cogn. Sci. **1**(4), 621–650 (2009)
8. Daley, D.J., Kendall, D.G.: Stochastic rumours. IMA J. Appl. Math. **1**(1), 42–55 (1965)
9. Hagberg, A.A., Schult, D.A., Swart, P.J.: Exploring network structure, dynamics, and function using networkx. In: Varoquaux, G., Vaught, T., Millman, J. (eds.) Proceedings of the 7th Python in Science Conference, pp. 11–15. Pasadena, CA USA (2008)
10. Heise, D.R.: Expressive Order Confirming Sentiments in Social Actions, 1st edn. Springer, New York (2007). https://doi.org/10.1007/978-0-387-38179-4
11. Pew Research Center: Internet, Science & Tech, Pew Research Center: Demographics of Social Media Users and Adoption in the United States (2021). http://www.pewresearch.org/internet/fact-sheet/social-media/. Accessed 26 April 2021

12. Maki, D.P., Thompson, M.: Mathematical Models and Applications: With Emphasis on the Social, Life, and Management Sciences, Prentice-Hall, Englewood Cliffs (1973)
13. McPherson, M., Smith-Lovin, L., Cook, J.M.: Birds of a feather: homophily in social networks. Ann. Rev. Sociol. **27**(1), 415–444 (2001)
14. Morstatter, F., Pfeffer, J., Liu, H., Carley, K.M.: Is the sample good enough? Comparing data from twitter's streaming API with twitter's firehose. arXiv preprint arXiv:1306.5204 (2013)
15. Serrano, E., Iglesias, C.Á., Garijo, M.: A novel agent-based rumor spreading model in twitter. In: Proceedings of the 24th International Conference on World Wide Web, pp. 811–814 (2015)
16. Wang, C., Tan, Z.X., Ye, Y., Wang, L., Cheong, K.H., Xie, N.g.: A rumor spreading model based on information entropy. Sci. Rep. **7**(1), 1–14 (2017)
17. Weng, L., Flammini, A., Vespignani, A., Menczer, F.: Competition among memes in a world with limited attention. Sci. Rep. **2**, 335 (2012)
18. Zanette, D.H.: Dynamics of rumor propagation on small-world networks. Phys. Rev. E **65**(4), 041908 (2002)

Studying the Role of Social Bots During Cyber Flash Mobs

Samer Al-khateeb[1]([✉]), Madelyn Anderson[1], and Nitin Agarwal[2]

[1] Creighton University, Omaha, NE 68178, USA
{sameral-khateeb1,madelynanderson}@creighton.edu
[2] COSMOS Research Center, UA – Little Rock, Little Rock, AR 72204, USA
nxagarwal@ualr.edu

Abstract. A Cyber Flash Mob (CFM) is an event that is organized via social media, email, SMS, or other forms of digital communication technologies in which a group of people (who might have an agenda) get together online or offline to collectively conduct an act and then quickly disperse. In addition to the humans participating in these events, non-humans, i.e., artificial agents or social bots - which are computer software programmed to accomplish some tasks on your behalf such as tweeting, retweeting, and liking a tweet - also participate in a CFM. In this research, we study the shared orientations of the CFMs' participants and try to understand the role of social bots in disseminating CFMs' agendas by examining the communication network of these accounts, the toxicity of their posts, and the artifacts, e.g., the URLs they share. The goal is to understand how social bots help CFM organizers advertise, recruit, and share their products (e.g., videos, pictures) on various social media platforms.

Keywords: Cyber Flash Mobs · Social bots · Twitter · Toxicity · Social communication network · Human-machine teaming

1 Introduction

A Cyber Flash Mob (CFM) is an event that is organized via social media, email, SMS, or other forms of digital communication technologies in which a group of people (who might have an agenda) get together online or offline to collectively conduct an act and then quickly disperse [1]. To an outsider, such an event may seem arbitrary. However, a sophisticated amount of coordination is involved. In recent years, cyber flash mobs *"have taken a darker twist as criminals exploit the anonymity of crowds, using social networking to coordinate everything from robberies to fights to general chaos"* [2, 3]. More recently, the term *"mob"* has been increasingly used to remark an electronically orchestrated violence such as the recent attack on the State Capital in Washington by Pro-Trump protesters that lead to property damages, government disruption, and injuries or death for some of the protesters [4, 5]. In a recent incident, an army of small investors from all over the world used Reddit to coordinate *"flashmob investing"* [6] to create stock market frenzy causing GameStop's stock value to rise from $20 to $483 in less than a

© Springer Nature Switzerland AG 2021
R. Thomson et al. (Eds.): SBP-BRiMS 2021, LNCS 12720, pp. 164–173, 2021.
https://doi.org/10.1007/978-3-030-80387-2_16

month [7]. These events show that our systems (security, financial, etc.) are not equipped to handle such highly coordinated and flash actions, underscoring the importance of systematically studying such behaviors.

To study CFMs, it is essential to understand the *motivation* of the individuals that coordinate such events. Many researchers highlight the importance of studying the *shared orientations* such as language, location, religious or political views, among the group members as an indication of group organization. *Shared orientations* among individuals often form the basis for *motivation* resulting in *collective actions* [8, 9]. *Shared orientations* among individuals induce a sense of belongingness to the group, giving rise to the group's collective identity. Individuals may be connected along one or more social dimensions, resulting in multiple shared orientations and thus a stronger collective identity. Hence, in this study we analyze the *languages* and *locations* of the flash mobbers in an attempt to understand their motivation.

In addition to the human participants of such events, *automated actors* or *social bots* are also used during these events to advertise, recruit, and share the products of CFMs, e.g., videos and pictures on various social media platforms. One study estimates that 50% of Twitter accounts are automated or social bots [10] and around 16% of spammers on Twitter nowadays are social bots [11]. Social media moderators are aware of this problem and constantly suspend these accounts. However, there is a lack of systematic investigation of such a human-machine teaming that can affect cyber flash mobs. More specifically, what role do computer agents, or social bots play during CFMs? In this study, we try to answer this research question by examining the following: (1) *in general, where are citizens more interested in CFMs?* (2) *What is the role of social bots in disseminating cyber flash mobs' agenda?* More specifically, (a) *Who is more toxic, bots or humans?* (b) *What are the differences between bots and human's communication networks?* (c) *What resources (e.g., images, videos, URLs) are shared via Cyber Flash Mobs participants (bots and humans)?* Next, we provide a brief literature review of the topics related to this paper, then we discuss our methodology, results, and conclusion with possible future research directions.

2 Literature Review

A flash mob is a phenomenon that has been studied in various disciplines such as communication studies [12], marketing [13], cultural studies [14], and other disciplines. However, there is a lack of a systematic and computational model of its formation and prediction of its occurrence or its success and failure. Our research is one step in this direction. On the other hand, social bots are a known problem that is facing social media sites. It has been studied by various researchers. A study that shares a similar methodology to our research is conducted by Khaund et al. [15]. They focused on the role and coordination strategies of Twitter social bots during four natural disasters that occurred in 2017. Here, we focus on the role of social bots during a CFM event and compare its behavior to humans' behaviors on Twitter. Online toxic content such as toxic posts or toxic comments is another problem that is facing social media sites. Using rude, disrespectful, hateful, and unreasonable language to provoke other users or make them leave a conversation are all examples of toxic behaviors. Toxicity analysis is different from sentiment analysis as the latter usually gives a score ranking the text to be either positive, negative, or neutral [16]. In our study, we leverage Google Perspective API to assess a toxicity score for each post shared by bots and humans.

3 Methodology

In this section, we explain the methodology we followed to conduct our research. We first explain how we collected our data and provide information about it, then we explain our analysis.

3.1 Data Collection

For this research, we collected data using Twitter Restful API for the period July 3, 2018, to June 18, 2020. We used the keyword "flash mob" to pull data from Twitter using Twitter Archiving Google Sheets (available at https://tags.hawksey.info) then used Python with GSpread API (available at https://gspread.readthedocs.io) to upload the data to our MySQL database. This resulted in 766,111 records (including 108,372 Tweets[1], 581,484 Retweets[2], and 76,255 Mentions[3]) written by 508,029 Tweeters. We preprocessed the data, e.g., reformatted the date and time columns to match the MySQL required date format. Then we extracted all the hashtags and URLs included in the records. This resulted in 29,931 unique hashtags and 50,634 unique URLs. We used Google Perspective API (available at https://www.perspectiveapi.com/) to obtain the toxicity score (i.e., how toxic a text is) of the 766,111 records we collected. Finally, we used Botometer Pro API (available at https://botometer.osome.iu.edu/api) to obtain the bot score (i.e., how likely a Twitter account is a bot or a human) of all the Tweeters.

3.2 Data Analysis

In this subsection we highlight the analysis we conducted then we explain our results in the following subsection. **First**, we analyzed the diversity of the data by analyzing the user's languages and locations of the records we collected. This analysis should help in shedding a light on the ethnographical nature of CFMs. **Second**, we ran a Python script that utilizes the *Botometer Pro API* on our 508,029 unique Twitter accounts to find the likelihood of an account being bot or human. This API calculates various scores, so we selected the "universal" score because it is language-independent, and we have multi-language records. The score is returned in the range of 0.0 to 1.0 representing the likelihood of an account being bot or not. The closer the score is to 1.0, the more likely that account is a social bot. We multiplied the returned values by 100 to better visualize the likelihood of an account being a bot. Since the returned values are continuous and we wanted to have two distinct classes to categorize our Tweeters, i.e., human or bot we considered all the users who have a $\leq 10\%$ bot score to be most likely human

---

[1] These are the records that do not start with RT and do not contain @.
[2] These are the records that start with RT.
[3] These are the records that do not start with RT but contain @.

accounts and all the users who have a ≥90% bot score to be most likely social bots [15]. This method should help in accounting for the Botometer Pro API misclassification and making sure that these accounts have distinct features to be considered as human or bot accounts. **Third**, we used the *Google Perspective API* to calculate the toxicity score of each record (tweets, mentions, or retweets) in our database. The score is returned in the range of 0.0 to 1.0 representing the likelihood of a text being toxic or not. The closer the score is to 1.0, the more likely that text is toxic. We multiplied the returned values by 100 to better visualize the likelihood of toxicity. Then, we calculated the user toxicity score by taking the average toxicity score of all the records of a Tweeter. A threshold of 0.5 (50%) was used to determine if a user is toxic or not. **Fourth**, we extracted all the users mentioned or retweeted from each tweet by using a Python script to create two communication networks, namely the *bot's communication network* and the *human communication network*. This resulted in a *bot's communication network* that consists of 1,861 nodes and 1,675 edges. This network is divided into 407 connected components, which consist of 2 Isolates (single nodes), 229 Dyads (two nodes connected), 81 Triads (three nodes connected), and 95 larger than Triads (more than three nodes connected) (See Fig. 1). We filtered the network and removed all the small-connected components (contain less than 100 nodes), as they did not contribute to our analysis and instead focused on the two largest-connected components which have 495 nodes and 658 edges (See Fig. 2). On the other hand, the *human communication network* consists of 59,530 and 70,418 edges. This network is divided into 6,287 connected components which consist of 46 Isolates (single nodes), 3,719 Dyads (two nodes connected), 1,254 Triads (three nodes connected), and 1,268 larger than Triads (more than three nodes connected) (See Fig. 3). We filtered the network and removed all the small-connected components (contain less than 100 nodes) as they did not contribute to our analysis and instead focused on the largest-connected component which contains 39,182 nodes and 54,520 edges (See Fig. 4). **Fifth**, we applied the *Newman Clustering algorithm* to auto locate communities in the filtered version of both networks. The *bot's communication network* has a *global clustering coefficient* (*transitivity*) of 0.0 and a *Newman modularity* of 0.679, while the *human communication network* has a *global clustering coefficient* (*transitivity*) of 0.069 and a *Newman modularity* of 0.667. **Sixth**, we analyzed the number of retweets and mentions shared by bots and humans. Then, we calculated the *Retweets to Mentions Ratio* to find differences and similarities of these accounts' behaviors. The *Retweets to Mentions Ratio* give us an idea about how many retweets a bot or a human account posted for each mention. **Finally**, we extracted all the URLs included in the collected records using another Python script. We investigated the popularity of the twelve most known social media sites. This analysis can help us in identifying potential platforms for CFM activities and the difference in humans and bots sharing activities. Finally, we investigated the URL shortening services used by both account types. We used the list of URL shortening services mentioned in [17] to guide us. This analysis also shed light on the difference between humans and bots' behaviors when it comes to shortening URLs and including them in a tweet.

4 Results and Analysis

In this section, we try to answer the research questions listed in our *introduction* using the analysis mentioned in the *data analysis* subsection. We use the **first** analysis to answer the first research question (**RQ1: In general, where are citizens more interested in CFMs?**) We found that 239,985 (31.3%) records were shared with Tweeter's language information (49,216 tweets, 34,950 mentions, and 155,819 retweets) and only 186,553 (36.7%) users shared their language (1,503 bots and 36,872 humans). There are 47 different languages of users who share these records as shown in Fig. 5. The top 10 languages are English, Italian, Portuguese, Spanish, Indonesian, French, Thai, Japanese, Korean, German.

We also found 941 (0.123%) records were shared with geolocation information (517 tweets and 424 mentions) and only 532 (0.105%) users shared their geo-location (0 bots and 88 humans). Figure 6 shows a map of the users (humans) who shared their location. We note that most of the users who shared their geolocations are located in Europe and the USA with few in other parts of the world. This also aligns with our findings of the user's languages, as six out of the top 10 languages mentioned above are languages usually spoken in Europe and the USA. This analysis also shows that social bots share their location and language less than humans.

We used the rest of the analysis to answer the second research question (**RQ2: What is the role of social bots in disseminating cyber flash mobs' agenda?**). More specifically, to answer **a) Who is more toxic, bots or humans?** Using the **second** and **third** analysis, we found that 44,642 Twitter accounts were humans, 1,647 were social bots, and 387,359 accounts were "unknown" because Botometer Pro API did not return their bot score, which could be due to the account being set to private (most likely human in this case) or suspended (most likely a bot that is caught by Twitter algorithm). We found 20,710 (4.1%) toxic users, i.e., users who have a $\geq 50\%$ toxicity score. We also found that social bots tend to post less toxic tweets than humans, i.e., the average user toxicity score for social bots is 15.63% while the average user toxicity score for humans is 17.58%. This could be due to the fact that humans tend to express their feelings about a specific CFM which might lead to more toxic posts (e.g., if people are against the CFM), while bots tend to share a URL of the event or prompt participation in a CFM.

We used the **fourth, fifth,** and **sixth** analysis to answer, **b) What are the differences between bots and human's communication networks?** We found that the human communication network is more fractured but has stronger within-community cohesion than the bot communication network. There are 6,287 connected components in human communication network (the mean is 38.07) compared to 407 connected components in bot communication network (the mean is 12.18). The *human communication network* has a higher value of transitivity which indicates that the human communities are more cohesive than the *bot communities*. This means the humans are more communicative about the cyber flash mob, while the bots work as disseminators of information about CFMs.

Finally, we found that both bot and human networks have more retweets than mentions. However, the *retweet to mention ratio* is much higher for the bots than the humans, meaning bots have a higher tendency to retweet than humans (see Table 1).

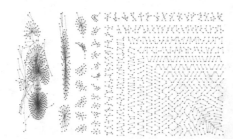

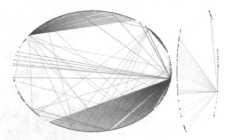

Fig. 1. Bot communication network. Green edges are mentions and blue edges are retweets (Color figure online)

Fig. 2. Filtered bot communication network. Central nodes are high in betweenness centrality (bridges). Green edges are mentions and blue edges are retweets (Color figure online)

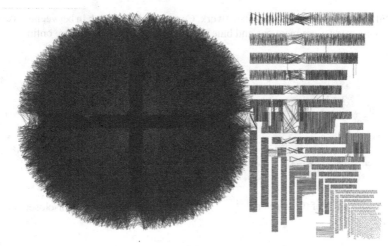

Fig. 3. Human communication network. Green edges are mentions and blue edges are retweets (Color figure online)

We used the **seventh** analysis to answer, **c) What resources (e.g., images, videos, URLs) are shared via Cyber Flash Mobs participants (bots and humans)?** Humans share more URLs than bots regardless of the site type. While humans share more Twitter URLs (e.g., status, picture of the CFM), bots share a disproportionately large number of YouTube URLs. We also found Twitter, YouTube, Facebook, Instagram, and Blogs to be the top 5 social media sites shared by these two account types. Other social media sites seem to be of less interest to both account types. Several researchers note that cyber commentaries on YouTube and Facebook pages are usually written by supporters and/or participants of CFMs, while blogs and online news sources attract *"more heterogeneous readership and offer a glimpse of strong criticism that is also directed at flash mobs"* [18].

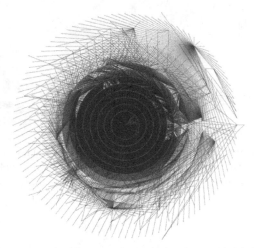

Fig. 4. Filtered human communication network. Central nodes are high in betweenness centrality (bridges). Green edges are mentions and blue edges are retweets (Color figure online)

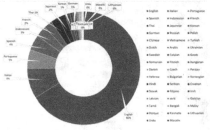

Fig. 5. Distribution of the tweeters' languages

Fig. 6. Tweeters' self-disclosed location

Table 1. The retweets and mentions count for both human and bot communication network. Bots have a higher retweet to mention ratio.

	Retweets	Mentions	(Retweet/Mentions) ratio
Bots	1,345	330	≈4.1
Humans	52,192	18,226	≈2.7

Figure 7 shows these sites and the number of URLs shared by both account types. We also found that humans tend to shorten URLs more than bots, probably due to the number of characters limit imposed by Twitter and so humans can squeeze in more content along with the URL (See Fig. 8).

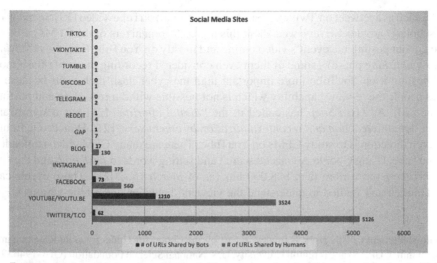

Fig. 7. Top social media sites shared by social bots (orange bars) and humans (blue bars). (Color figure online)

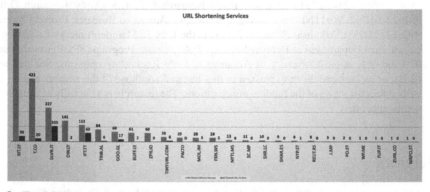

Fig. 8. Top 25 URL shortening services along with the number of URLs shortened and shared by humans (blue bars) and bots (orange bars). (Color figure online)

5 Conclusion and Future Research Directions

In this study, we examined the shared orientations of the CFMs' participants. We focused on the role of social bots and humans in disseminating CFM agenda by examining the communication networks of these two types of accounts, the toxicity of their posts, and the artifacts shared by such accounts, i.e., the URLs in an attempt to understand how social bots help CFM organizers advertise, recruit, and share the products of the CFMs (e.g., videos, pictures) on various social media platforms. We found various differences between the two types of networks highlighted in our analysis subsection. This research sets the foundation for a deeper examination of the role of human-machine teaming on CFM events. The CFM process includes the *planning phase*, the *recruitments phase*, the *execution phase*, and the *replaying and republishing of the products of the flash mob phase* such as a retweet on Twitter, a Facebook post, or a YouTube Video [19]. In many of the scholarly articles we reviewed about this topic, the organizers of the CFMs cared so much about posting the event's video online and mostly on YouTube (i.e., the *replaying and republishing* phase) - some of them even considered recording the cyber flash mob and posting it on YouTube more important than the cyber flash mob itself because it gives people the re-view capability which is not possible without recording and posting online [20]. As *Lissa Soep* has called it, the *"digital afterlife"* [21] has a significant *"potential for a reaction and recontextualization by other users"* [21]. Hence, one future research direction is to study CFMs on YouTube. There are many CFM videos uploaded to YouTube. Using Google Advanced search and setting words to *flash mob* and the site to *YouTube* returns more than 808,000 hits (as of March 15, 2021). These events can be identified and studied to understand the viewers' behavior toward the various CFM forms.

Acknowledgments. This work is funded in part by the Center for Undergraduate Research and Scholarship (CURAS) at Creighton University, U.S. National Science Foundation (OIA-1946391, OIA-1920920, IIS-1636933, ACI-1429160, and IIS-1110868), U.S. Office of Naval Research (N00014-10-1-0091, N00014-14-1-0489, N00014-15-P-1187, N00014-16-1-2016, N00014-16-1-2412, N00014-17-1-2675, N00014-17-1-2605, N68335-19-C-0359, N00014-19-1-2336, N68335-20-C-0540, N00014-21-1-2121), U.S. Air Force Research Lab, U.S. Army Research Office (W911NF-20-1-0262, W911NF-16-1-0189), U.S. Defense Advanced Research Projects Agency (W31P4Q-17-C-0059), Arkansas Research Alliance, the Jerry L. Maulden/Entergy Endowment, and the Australian Department of Defense Strategic Policy Grants Program (SPGP) (award number: 2020-106-094) at the University of Arkansas at Little Rock. Any opinions, findings, and conclusions, or recommendations expressed in this material are those of the authors and do not necessarily reflect the views of the funding organizations. The researchers gratefully acknowledge the generous support.

References

1. Al-khateeb, S., Agarwal, N.: Cyber flash mobs: a multidisciplinary review. Soc. Netw. Anal. Mining (2021)
2. Tucker, E., Watkins, T.: More flash mobs gather with criminal intent. NBC News, 09 August 2011. https://www.nbcnews.com/id/wbna44077826. Accessed 7 Jan 2021
3. Steinblatt, H.: E-incitement: a framework for regulating the incitement of criminal flash mobs. Fordham Intell. Prop. Media Ent. LJ **22**, 753 (2011)
4. Staff, W.P.: Woman dies after shooting in U.S. Capitol; D.C. National Guard activated after mob breaches building. Washington Post, 06 January 2021
5. Barry, D., McIntire, M., Rosenberg, M.: 'Our president wants us here': the mob that stormed the capitol. The New York Times, 09 January 2021
6. Pratley, N.: The Reddit flash mob won't be able to work the GameStop magic on silver. The Guardian, 01 February 2021. http://www.theguardian.com/business/nils-pratley-on-fin ance/2021/feb/01/reddits-flash-mob-wont-be-able-to-work-the-gamestop-magic-on-silver. Accessed 5 Feb 2021
7. Brignall, M.: How GameStop traders fired the first shots in millennials' war on Wall Street. The Guardian, 30 January 2021. http://www.theguardian.com/business/2021/jan/30/how-gam estop-traders-fired-the-first-shots-in-millenials-war-on-wall-street. Accessed 5 Feb 2021

8. Melucci, A.: Nomads of the Present: Social Movements and Individual Needs in Contemporary Society. Temple University Press, Philadelphia (1989)
9. Maheu, L.: Social Movements and Social Classes: The Future of Collective Action, vol. 46. SAGE Publications Ltd., London (1995)
10. Koh, Y.: Only 11% of new twitter users in 2012 are still tweeting. Wall Street Journal, 21 March 2014
11. Grier, C., Thomas, K., Paxson, V., Zhang, M.: @spam: the underground on 140 characters or less. In: Proceedings of the 17th ACM Conference on Computer and Communications Security, pp. 27–37. New York, NY, USA, October 2010. https://doi.org/10.1145/1866307.1866311
12. Nicholson, J.A.: Flash! Mobs in the age of mobile connectivity. Fibreculture J. (6), 1–15 (2005). http://citeseerx.ist.psu.edu/viewdoc/summary?doi=10.1.1.566.6906, http://www.softhook.com/coverage/Fibreculture.pdf
13. Barnes, N.G.: Mob it and sell it: creating marketing opportunity through the replication of flash mobs. Market. Manag. J. **16**(1), 174–180 (2006)
14. Do Vale, S.: Trash mob: Zombie walks and the positivity of monsters in western popular culture. In: The Domination of Fear, pp. 191–202. Brill Rodopi, Amsterdam (2010)
15. Khaund, T., Al-Khateeb, S., Tokdemir, S., Agarwal, N.: Analyzing social bots and their coordination during natural disasters. In: International Conference on Social Computing, Behavioral-Cultural Modeling and Prediction and Behavior Representation in Modeling and Simulation, pp. 207–212 (2018)
16. Obadimu, A., Mead, E., Hussain, M.N., Agarwal, N.: Identifying toxicity within youtube video comment. In: International Conference on Social Computing, Behavioral-Cultural Modeling and Prediction and Behavior Representation in Modeling and Simulation, pp. 214–223 (2019)
17. Yang, K.-C. et al.: The COVID-19 infodemic: twitter versus facebook. arXiv preprint arXiv:2012.09353 (2020)
18. Molnár, V.: Reframing public space through digital mobilization: flash mobs and contemporary urban youth culture. Space Cult. **17**(1), 43–58 (2014). https://doi.org/10.1177/1206331212452368
19. Zellner, A., Sloan, C., Koehler, M.: The educational affordances of the flash mob: from mobs to smart mobs. In: Koehler, M., Mishra, P. (eds.) Proceedings of SITE 2011–Society for Information Technology & Teacher Education International Conference, Nashville, Tennessee, USA, pp. 3042–3047. Association for the Advancement of Computing in Education (AACE) (2011). https://www.learntechlib.org/primary/p/36779/. Accessed 11 June 2021
20. Venning, W.: Funk up your Japanese with a flashmob!. Thomson (ed.) pp. 60–69 (2016)
21. Soep, E.: The digital afterlife of youth-made media: implications for media literacy education. Comun. Media Educ. Res. J. **20**(1) , 93–100 (2012)

The Integrated Game Transformation Framework and Cyberwar: What 2×2 Games Tell Us About Cyberattacks

Alexander L. Fretz[1]($\boxtimes$), Jose J. Padilla[2]($\boxtimes$), and Erika F. Frydenlund[2]($\boxtimes$)

[1] Graduate Program in International Studies, Old Dominion University, Norfolk, USA
afret001@odu.edu
[2] Virginia Modeling, Analysis and Simulation Center, Old Dominion University, Suffolk, USA
{jpadilla,efrydenl}@odu.edu

Abstract. We apply a game theory framework to cyber war in the areas of Cyber Influence Operations (CIOs), Advanced Persistent Threats (APTs), and Traditional Cyber Attacks (TCAs). For greater generalizability, we rely on a set of actor typologies based on posture rather than individual countries. Our findings suggest that cyberwar will remain problematic going forward, especially with respect to APTs and that great powers will be the most likely offenders. They will also be the most stubborn actors when it comes to diminishing cyberwar in general, even when the base game can be transformed into one that favors cooperation over conflict.

Keywords: Game theory · CIO · APT · Cyber security

1 Introduction

Among the many forms of unconventional warfare that have risen to prominence in the 21st century, cyberwarfare has steadily become the most consequential and frequently deployed by the great powers as well as numerous lesser actors [1, 2]. At their most extreme cyberattacks can paralyze a nation's economy and endanger the lives of citizens by targeting critical infrastructure such as banking institutions, stock exchanges, and power grids. These kinds of attacks tend to be reserved as part of a greater conventional war scenario and are less numerous, excluding cases of excessive capability asymmetries. Far more common are less severe cyber operations that are conducted for the purposes of espionage and/or influence campaigns. Thus, two new kinds of cyberattacks have been deployed by states with intensifying frequency over the last decade: Cyber Influence Operations (CIOs) and Advanced Persistent Threats (APTs). Alongside Traditional Cyberattacks (TCAs), these ever-increasing threats pose a serious challenge to States and to the stability of international relations.

CIOs have been roughly defined as the deliberate use of information by one actor on an adversary to confuse, mislead, and ultimately express influence upon the choices and decisions that the adversary makes [3]. Many scholars also include tactics such as

© Springer Nature Switzerland AG 2021
R. Thomson et al. (Eds.): SBP-BRiMS 2021, LNCS 12720, pp. 174–183, 2021.
https://doi.org/10.1007/978-3-030-80387-2_17

subversion, demoralization, and paralysis as well as specifying the use of social media platforms, human mimicry (trolls & bots), and online forms of traditional media [4, 5].

TCAs refer to the broad spectrum of offensive maneuvers in the cyber space that targets computer information systems, infrastructures, or networks. These can be differentiated from CIOs in that they are more technical in nature and target the online infrastructure of websites and services whereas CIOs involve weaponizing information for social engineering in order to influence the decision-making of other actors [6].

APTs and the groups that execute them are stealthy actors that make use of organized and long-term cyber-attacks, designed specifically to covertly access computer systems and/or networks with the goal of extracting intelligence [7]. They are particularly effective because actors can slowly and carefully infiltrate their targets aided by the use of low-impact custom malware that allows them to enter the environment without creating a disturbance [8].

This paper aims to establish a method of utilizing game theory for analyzing and addressing the challenges of CIOs, APTs, and TCAs among rival dyads. We use simplified representations of strategic interactions in the form of 2×2 games. More uniquely, we develop actor typologies based on posture and capabilities in order to capture the behavior of certain kinds of nations rather that any one specific country. Using the initial Game Transformation Framework (GTF), developed by Nandakumar and Padilla [9], we formulate an Integrated GTF model as the primary method for running our games. This approach involves using a base game structure based on macro-level factors and "integrating" micro-level factors based on the actor types and attack types.

2   Game Theory and Cyberwarfare

Game theory has been used to explain how alliances can successfully deter kinetic aggression [10] as well as explaining nuclear deterrence [11] and overcoming the security dilemma [12]. However, cyberwarfare presents challenges that complicate traditional means of deterrence and cooperation. For example, severe cyberattacks that target critical infrastructure are recognized as a breach of sovereign equality and territorial integrity in accordance with the UN Charter [13] similar to kinetic attacks, but softer cyberattacks such as CIOs, ATPs, and certain TCAs (small scale defacements and DDoS) typically do not lead to significant responses [14]. Moreover, it is difficult for states to take legal action with respect to international law because enforcement provisions are nonexistent [15]. There is also the attribution problem because of the prevalence of non-state offenders and the difficulty of establishing state-sponsorship [16]. In this way cyberattacks present a similar challenge in international relations (IR) as did privateering in the 19th century, which was overcome by establishing relevant international law and norms at the Declaration of Paris in 1856 [17]. These complexities also worsen the security dilemma considerably primarily due to the offensive advantage and the obfuscation of posture [18]. According to Jervis, this leads to a "doubly dangerous" world [12].

Ample work has been done utilizing ordinal games to model strategic interactions in the international relations literature [12, 19, 20]. However, when it comes to cyberwar much less work has been done. Jormakka and Mölsä [21] applied game theory to information warfare finding that the game yields similar results to other forms of asymmetric warfare and that when iterated, the players adopt the "bold" or single strategy rather than selecting for equilibrium. Ma et al. [22] develop a game theoretic model for the strategic interactions between private firms and government agencies when these private entities face the threat of cyber espionage. Goel and Hong [23] model cyberwar between countries taking in to account the poor attribution environment. Matusitz [24] applies game theory and social network theory to cyberterrorism. While these examples demonstrate that some work has applied game theory to TCAs and to a lesser extent APTs, there has not yet been any application to CIOs. We aim to establish a model that can do all three.

3   Methodology

Following the thinking of scholars such as Barrett [25], we approach these strategic interactions from the perspective that they can be captured with the simplified 2×2 game framework. The game-theoretic approach we use makes all of the same assumptions typical of this approach, such as rational actors, complete information, and utility maximization. However, our approach is unique in how it ranks preferences. We use the typical payoff structure of 2×2 games using the 1–4 scoring system, which are qualitatively assigned based on macro-level factors. We then modify the payoff structure based on micro-level factors comprised of quantified data taken from the Dyadic Cyber Incidents Database (DCID) [26].

Originally categorized by state dyads, we reorganize the DCID data into Actor Typologies which we call the Hegemon, Revanchist, Rising Power, Rogue State, and Minor State. The preferences of the first two typologies are derived from the US and Russia respectively. The second two are synthesized from the data from two separate actors, China and India as rising powers and Iran and North Korea as rogue states. Lastly, the minor state typology is a composite of 10 different actors from across Eastern Europe, The Middle East, South Asia, and Southeast Asia.

Our approach extends the ordinal rankings of preferences by accounting for the 3 Strategic Choices: CIOs, APTs, and TCAs. Each attack type is awarded a score of 0–3 based on their efficacy related to four separate attack vectors: Political, Economic, Social, and Technological (PEST). The data is categorized for both actor types and attack types. We allow for singular cyber incidents to be counted as more than one attack type, if it targets more than one of the PEST spheres. The scoring system for Actor Typologies and Strategic Choice will be explained further in Sect. 3.1.

We lastly focus on how games can be transformed based on payoff swaps. We center this approach on the logic of the "New Periodic Table" (NPT) of 2×2 games introduced by Robinson and Goforth [27] that demonstrates how games can be shifted across a spectrum if preferences and payoffs are altered. We utilize this logic as a secondary level of analysis to control for macro-level factors and the base game.

3.1   The Integrated GTF Model

We utilize a variant of the Game Transformation-Based Framework (GTF) established by Nandakumar and Padilla [10] which we call the Integrated GTF model. It differs primarily in its ability to capture both macro-level and micro-level factors in the payoff matrix. The second and third steps change places to allow for aggregation based on the data for actor types and attack types. Figure 1 illustrates the Integrated GTF model.

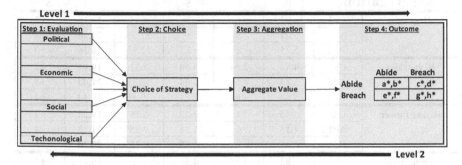

Fig. 1. The integrated GTF model

The model is based on a 4-step process and 2 levels of analysis. Level 1 involves progressing through the 4 steps and producing the initial game while level 2 involves analyzing whether or not changes in the base game (macro-level factors) might produce a different future outcome.

In step 1 values are assigned to the PEST spheres based on actor typology aggregated from the DCID database. Each has its own score based on past cyber activity. The individual cyber incidents are categorized as attacks that target one or multiple spheres. Because the number of incidents varied for each group, we averaged the totals and multiplied by a factor of 10 to get a whole number value. This is represented as aggregate value (AV).

Step 2 involves the strategic choice of the players. There are 3 strategies or attack types that actors can choose from: the CIO, APT, or TCA. Because these attack types have different timelines and nations are either running all three games simultaneously or not playing certain games at all, we assume that in the choice phase both actors choose the same strategy for the purpose of comparison. The value modifiers for three attack types are based on a 0–3 scoring system with regard to their efficacy as it relates to the PEST spheres. The scores are awarded based on usage frequency and success rate.

Step 3 is the aggregation phase in which the AV is calculated for each possible combination of actor type and attack type. We then synthesize a Simplified Value Modifier (SVM) meant for the 2×2 framework. As the unmodified game is based on the 0–4 scoring system, it is necessary to establish parity among values. The SVM is calculated by awarding a "+1" value to the lowest AV whole number (18) and progresses to the highest whole number (21), which receives a "+4." Table 1 demonstrates the process of aggregation.

The final step involves a plug-in of the aggregate values in the form of the SVM into the base game. This can be done for any of the games across the NPT spectrum, be that the "Chicken Game," the "Prisoner's Dilemma Game," or the "Harmony Game." The base game depends on the qualitative assessment of the macro-level factors: The structure of capabilities (offensive/defensive advantage), posture transparency (attribution), and the existence of international agreements or laws. For this study, we select the "Battle Game,"

Table 1. Aggregate value modifiers by actor type and attack type.

Aggregation		P	E	S	T	AV	SVM
Hegemon	CIO	6.21	1	3	8.95	19.16	+2
	APT	5.21	3	1	11.95	21.16	+4
	TCA	5.21	2	1	11.95	20.16	+3
Revanchist	CIO	4.84	2.64	4.64	7.31	19.41	+2
	APT	3.84	4.64	2.64	10.31	21.43	+4
	TCA	3.84	3.64	2.64	10.31	20.43	+3
Rising Power	CIO	5.15	4	3.38	6.75	19.28	+2
	APT	4.15	6	1.38	9.75	21.28	+4
	TCA	4.15	5	1.38	9.75	20.28	+3
Rogue State	CIO	4.83	2.54	4.54	5.3	18.46	+1
	APT	3.83	6.5	1.83	8.3	20.46	+3
	TCA	3.83	5.5	1.83	8.3	19.46	+2
Minor State	CIO	5.59	2.54	4.79	5.38	18.3	+1
	APT	4.59	4.54	2.79	8.38	20.3	+3
	TCA	4.59	3.54	2.79	8.38	19.3	+2

a highly uncooperative game, based on the qualitative assessment of the macro-level factors we discussed earlier. Figure 2 illustrates the base game payoff structure.

"Battle"		Nation 1	
		Abide	Breach
Nation 2	Abide	2*,2*	4*,3*
	Breach	3*,4*	1*,1*

Fig. 2. Base game structure

4 Findings

After running every combination of games, 3 different variations arose that are significant. The first variation represents games played by the Hegemon, the Revanchist, and the Rising Power. Essentially, this represents the strategic interactions between great powers. Any combination of these three players produces the same 2×2 outcome in terms of payoff differentials. Figure 3 (Rev vs. Heg.) is a representation of this outcome. We observe that there is no significant change from the "Battle Game" base model in that the differences in the payoff structure are roughly the same across games with respect to the in-game variation across possible outcomes. The most favored game is the APT because of the stealth factor or the ability to go undetected for long periods, allowing for large scale espionage. In contrast, TCAs and CIOs seem to provoke more of a response. For example, the 2016 election interference has had lasting effects on US-Russia relations.

The comparative parity among outcome differentials reflects the relative balance of power among the "big three" actor types in the cyber space with respect to capabilities and posture. However, because the base game structure favors conflict, we can expect a continued rise in cyberwarfare going forward. One the other hand, a level 2 analysis suggests that a change in the base game can lead to a more cooperative future. Therefore, if these nations focus on altering the macro-level factors, they can push the base game towards a "Harmony Game" or a "No Conflict," which would result in cooperation rather than conflict.

HEG/REV	"CIO"	REV/RSP Abide	REV/RSP Breach
	Abide	4,4	6,5
	Breach	5,6	3,3

HEG/REV	"APT"	REV/RSP Abide	REV/RSP Breach
	Abide	6,6	8,7
	Breach	7,8	5,5

HEG/REV	"TCA"	REV/RSP Abide	REV/RSP Breach
	Abide	5,5	7,6
	Breach	6,7	4,4

Fig. 3. Revanchist/rising power vs. Hegemon 2×2 games

The second variation represents games played by the Hegemon, Revanchist, or the Rising Power and the Minor State or the Rogue State. This essentially reflects a game between a great power and a lesser power. Prominent examples would include Russia and Eastern European minors, China and regional minors such as Taiwan or Japan, or the US and Iran or North Korea. Here it becomes clear that the strategic postures and choices of the great powers push the game into a more perilous state than of the base game (See Fig. 4). In terms of outcomes or payoffs, the great power considers both players abiding as a win, both players breaching as a win, and strongly prefers a breach/abide outcome in its favor while a breach/abide outcome in the minor's favor is neutral. Similar to the previous variation the most favored attack type is the APT while the least is the CIO.

The payoff preferences are representative of a belligerent posture towards the rival and strongly favors opportunism, which is the result of the power imbalance. Even the case of abiding while the other breaches is considered a neutral outcome. A greater power does not mind this outcome because it offers a justification for reprisal in a future interaction, serving to satisfy the desire for maintaining the status quo in the case of the Hegemon, recapturing a lost sphere of influence in the case of the Revanchist, or capturing a new one in the case of the Rising Power. Moreover, it provides the rationale for further aggrandizement at the expense of the rival or further hardening its posture towards the rival.

A level 2 analysis suggests that game transformation will be challenging. Macro-level changes that could alter the base game structure are not alone enough. Even if the base game was more cooperative, the greater powers would still have strong incentives to pursue conflict. While a more cooperative base game raises the costs of engaging in conflict by closing the payoff gap, a greater power still has the incentive to pursue the national interest despite a more marginal return.

The third variation represents games played between the Rogue State and the Minor State. This reflects the strategic interaction between rivalrous lesser powers, one having a more innate belligerence than the other. Examples with be that of Iran and Saudi Arabia

MNS/RGS	"CIO"	HEG/REV/RSP	
		Abide	Breach
	Abide	4,3	6,4
	Breach	5,5	3,2

MNS/RGS	"APT"	HEG/REV/RSP	
		Abide	Breach
	Abide	6,5	8,6
	Breach	7,7	5,4

MNS/RGS	"TCA"	HEG/REV/RSP	
		Abide	Breach
	Abide	5,4	7,5
	Breach	6,6	4,3

Fig. 4. Great power vs. lesser power 2×2 games

or Israel as well as North Korea and South Korea or Japan. Similar to the games played among great powers, the variance among payoffs for the separate in-game outcomes are static (see Fig. 5). Additionally, the most favored game is the APT, followed by the TCA and the CIO. However, these games are less favored across the across the board when compared to the set of games played by the great powers. This is due to the imbalance of capabilities as lesser powers are not able to execute attacks on the scale that greater powers do. That being said, there is still strong incentives for conflict.

Similar to the great powers, the comparative parity among outcome differentials points toward a relative balance of power among the lesser powers with respect to capabilities and posture. In addition, the base game is ultimately conflict prone which presents the primary issue at hand. Thus, the level 2 analysis reveals that the macro-level indicators are the best means of game transformation.

MNS	"CIO"	RGS	
		Abide	Breach
	Abide	3,3	5,4
	Breach	4,5	2,2

MNS	"APT"	RGS	
		Abide	Breach
	Abide	5,5	7,6
	Breach	6,7	4,4

MNS	"TCA"	RGS	
		Abide	Breach
	Abide	4,4	6,5
	Breach	5,6	3,3

Fig. 5. Rogue state vs. minor state 2×2 games

5   Discussion and Conclusion

When it comes to cyberwar several issues stand out as acutely problematic for the future. Particularly, our models reveal three distinct issues that will plague the cyberspace moving forward: the growing threat of APT, the great power problem, and dealing with macro-level factors. First, the APT threat has made a dramatic entrance into cyberwar in the last five years or so and has quickly become the most effective means of conducting political and economic espionage. This explains why it is the preferred attack type in every model, but also suggests that it will persist into the future and continue to grow in both frequency and severity. Nations seem to be drawn to APTs because they can go undetected for very longer periods of time when compared to other attacks, allowing for successful espionage while dampening the risk of more immediate retaliation. There is also the flexibility with which APTs can be applied. For example, China is interested in economic dominance and will therefore seek access to intellectual property that can erase an edge potential rivals might have in various markets. Conversely, Russia tends to deploy APTs more in favor of political espionage so that it can better influence the political sphere. So long as APTs remain useful for espionage, this type of cyberwarfare is unlikely to diminish.

Second, the issue of great powers maintaining overwhelming capability over lesser powers and the likelihood of using that power imbalance to dominate them. The games point toward a continuation of conflict regardless of base game transformation particularly because of this fact, which presents a serious challenge for the future. This finding agrees with scholars such as Gartze [28]. The greater powers can also tinker with micro-level factors in an effort to extend their payoffs should a game transformation shrink the benefits of conflict. For example, they can invest far more economic power into expanding their capabilities. Additionally, traditional balancing is more difficult to achieve for the minors because an alliance still faces the attribution problem. It would also require levels of military cooperation on par with NATO to develop a cyber force on par with that of a great power. This makes appeasement the only realistic option for the foreseeable future. As a result, lesser powers will continue to be at the mercy of greater powers going forward.

Third, dealing with macro-level factors that structure the base game being will be the most effective means of achieving cooperation in the future. This means solving the problems of offensive advantage, attribution, and international law. Scholars such as Healy and Jervis have already suggested that the best way to establish deterrence against cyberattacks is to ensure cyber defenders take the advantage [29]. Attribution is another major problem area. So far, cybersecurity firms loosely attribute the cyberattacks of nonstate actors to particular countries based on whose interests are being served and on some of the typical behavior and tactics that certain groups tend to display. But establishing state-sponsorship remains problematic and countries typically maintain plausible deniability. Perhaps an international consensus that attributes blame to nations whose territory attacks emanate from could help, but there is no guarantee that countries would not begin to sponsor groups from outside their borders as a counter. This will likely remain the most challenging factor to change. Lastly, the development of international law would do a great deal to stabilize the system. However, even if there was legal recourse on the international level, attribution will continue to disrupt opportunities for reprisal whether that be from individual states, alliances, or the international community. Furthermore, self-binding agreements require the cooperation of the great powers, who have the most to lose and lack any incentive to relinquish their advantage in this sphere. These macro-level factors, if solved, would lead to a more stable system.

Future work could expand on the inputs for greater comparison among model variations. As an example, actor types whose posture is based on the DCID data could be substituted for hypothetical or antithetical postures to test for different outcomes. Another interesting approach would be to incorporate models that account for discovery chance or the length of time attacks may go undiscovered for. Further work could also focus on how attack types are used together to enhance their efficacy rather than modeling them independent of one another. For example, APTs can be used to enhance CIOs. These would be valuable for expanding the robustness of our model and confirming the justification for our assumptions.

Acknowledgments. This material, in part, is based on research sponsored by the Office of the Assis-tant Secretary of Defense for Research and Engineering (OASD(R&E)) under agreement number FAB750-15-2-0120. The U.S. Government is authorized to reproduce and distribute reprints for Governmental purposes notwithstanding any copyright notation thereon. The views and conclusions contained herein are those of the authors and should not be interpreted as necessarily representing the official policies or endorsements, either expressed or implied, of the Office of the Assistant Secretary of Defense for Research and Engineering (OASD(R&E)) or the U.S. Government.

References

1. Liff, A.P.: Cyberwar: a new 'absolute weapon'? The proliferation of cyberwarfare capabilities and interstate war. J. Strat. Stud. **35**, 401–428 (2012)
2. Mazanec, B.M.: Why international order in cyberspace is not inevitable. Strat. Stud. Q. **9**(2), 78–98 (2015)
3. Lin, H.S., Kerr, J.: On cyber-enabled information/influence warfare and manipulation. SSRN, pp. 4–5, 13 August 2017
4. Cohen, D., Bar'cl, O.: The use of cyberwarfare in influence operations. In: Yuval Ne'eman Workshop for Science, Technology and Security. Tel-Aviv University, October 2017
5. Brangetto, P., Veenendaal, M.A.: Influence cyber operation: the use of cyberattacks in support of influence operations. In: 8th International Conference on Cyber Conflict, pp. 113–126. IEEE, June 2016
6. INSA Cyber Council: Getting Ahead of Foreign Influence Operations. Intelligence and National Security Alliance, p. 2, May 2018
7. Andress, J., Winterfeld, S.: Cyber Warfare: Techniques, Tactics and Tools for Security Practitioners. 2nd edn. Elsevier, Amsterdam (2014). Ch. 9
8. Ghafir, I., et al.: Detection of advanced persistent threat using machine-learning correlation analysis. Futur. Gener. Comput. Syst. **89**, 349–359 (2018)
9. Nandakumar, G.S., Padilla, J.: A game-transformation-based framework to understand initial conditions and outcomes in the context of cyber-enabled influence operations (CIOs). In: 13th International Conference SBP-BRiMS. Washington, DC, 18–21 October 2020
10. Snyder, G.H.: Alliance Politics. Cornell University Press (2007).
11. Powell, R.: Nuclear Deterrence Theory: The Search for Credibility. Cambridge University Press, Cambridge (1990)
12. Jervis, R.: Cooperation under the security dilemma. World Polit. **30**(2), 167–214 (1978)
13. Green, JA.: The regulation of cyber warfare under the jus ad bellum. In: Cyber Warfare: A Multidisciplinary Analysis, pp. 96–124 (2015)
14. Sklerov, M.J.: Solving the dilemma of state responses to cyberattacks: a justification for the use of active defenses against states who neglect their duty to prevent. Military Law Rev. **201**, 1–85 (2009)
15. Shackelford, S.: From nuclear war to net war: analogizing cyber attacks in international law. Berkley J. Int. Law (BJIL) **25**(3), 193–224 (2009)
16. Rowe, N.C.: The attribution of cyber warfare. In: Cyber Warfare: A Ultidisciplinary Analysis, pp. 75–86. Routledge (2015)
17. Lemnitzer, J.: Power, Law, and the End of Privateering. Springer, London (2014). https://doi.org/10.1057/9781137318633
18. Buchanan, B.: The Cybersecurity Dilemma: Hacking, Trust, and Fear between Nations. Oxford University Press, New York (2016)
19. Glaser, C.L.: The security dilemma revisited. World Politics **50**(1), 171–201 (1997)
20. Axelrod, R., Keohane, R.O.: Achieving cooperation under anarchy: strategies and institutions. World Politics **38**(1), 226–254 (1985)
21. Jormakka, J., Mölsä, J.V.E.: Modelling information warfare as a game. J. Inf. Warfare **4**(2), 12–25 (2005)
22. Ma, Z., Chen, H., Zhang, J., Krings, A.W.: Has the cyber warfare threat been overstated?: A cheap talk game-theoretic analysis on the google-hacking claim? J. Inf. Warfare **11**(1), 21–29 (2012)

23. Goel, S., Hong, Y.: Cyber war games: strategic jostling among traditional adversaries. In: Jajodia, S., Shakarian, P., Subrahmanian, V.S., Swarup, V., Wang, C. (eds.) Cyber Warfare. AIS, vol. 56, pp. 1–13. Springer, Cham (2015). https://doi.org/10.1007/978-3-319-14039-1_1
24. Matusitz, J.: A postmodern theory of cyberterrorism: game theory. Inf. Secur. J. Global Persp. **18**(6), 273–281 (2009)
25. Barrett, S.: Environment & Statecraft: The Strategy of Environmental Treaty-Making. Oxford University Press, Oxford (2003)
26. Maness, R.C., Valeriano, B., Jensen, B.: Dyadic Cyber Incidents Database. Version 1.5, 1 June 2019
27. Robinson, D., Goforth, D.: The Topology of the 2 × 2 Games: A New Periodic Table. Routledge, New York (2005)
28. Gartze, E.: The myth of cyberwar: bringing war in cyberspace back down to earth. Int. Secur. **38**(2), 41–73 (2013)
29. Healy, J., Jervis, R.: The escalation inversion and other oddities of situational cyber stability. Texas Natl. Secur. Rev. **3**, 30–53 (2020)

Bot-Based Emotion Behavior Differences in Images During Kashmir Black Day Event

Lynnette Hui Xian Ng$^{(\boxtimes)}$ (ID) and Kathleen M. Carley (ID)

CASOS, Institute for Software Research, Carnegie Mellon University,
Pittsburgh, PA 15213, USA
{huixiann,carley}@andrew.cmu.edu

Abstract. A picture speaks a thousand words. Images are extremely effective at evoking emotions and presents a potentially damaging force to the health of digital discourse. While text-based emotion analysis has been studied, little work has examined the emotions images invoke on social media platforms. This work analyzes bot-based emotion behavior differences in the images surrounding the 2020 Kashmir Black Day event. Through Twitter data, we observed at least half the agents in the conversation are bots, which dominate image conversations calling for action, e.g. "Be The Voice of Kashmir". Sadness and trust dominates the emotions in images. We further analyze a sub-dataset as a case study and discern the role of digital media in heightening online conflicts.

Keywords: Social cybersecurity · Emotion analysis · Bot analysis

1 Introduction

Social cybersecurity examines large-scale efforts to sway public opinion on digital platforms. This involves the development of computational tools to characterize online information operations, which are a national security concern, given their ability to incite offline violence [3].

Images can be extremely effective at invoking emotions, which enhances the threat of offline violence. The use of visual images increases message engagement [13], even result in polarized communities [9]. While significant efforts have tackled text emotion analysis in social media content [5,8], image emotion analysis

The research for this paper was supported in part by the Knight Foundation and the Office of Naval Research grant N000141812106 and by the center for Informed Democracy and Social-cybersecurity (IDeaS) and the center for Computational Analysis of Social and Organizational Systems (CASOS) at Carnegie Mellon University. The views and conclusions are those of the authors and should not be interpreted as representing the official policies, either expressed or implied, of the Knight Foundation, Office of Naval Research or the US Government.

© Springer Nature Switzerland AG 2021
R. Thomson et al. (Eds.): SBP-BRiMS 2021, LNCS 12720, pp. 184–194, 2021.
https://doi.org/10.1007/978-3-030-80387-2_18

is less studied. This paper contributes to the image emotion literature by examining how images play a part in invoking emotions, particularly in conjunction with bot-driven information operations.

Kashmir has been under a state of lockdown since 5th August 2019. It is a disputed territory between India and Pakistan, where both countries lay claim but rule it in part. The lockdown sparked widespread street protests and Indian government deployed troops to maintain peace.

We analyze bot-based emotion behavior differences in image conversations on Twitter surrounding the Kashmir Black Day. Specifically, we aim to answer the following research questions: (1) What were the themes of images shared during the Black Day event? (2) What were the emotions invoked in the shared images? (3) What were the difference in image sharing behavior between bots and non-bots? We collect data surrounding the event and annotate the agents in the dataset on their bot probability. For our analysis, we first identify image clusters, where each cluster represents an image visual theme. We then annotate each image with one of the eight core emotions from an image emotion classifier we constructed. Using the image clusters and emotions, we analyze the differences in the type of images shared between two agent groups: bots and non-bots. Finally, we present a case study based on a filtered sub-dataset which shows a small information operation.

2 Data and Methodology

2.1 Data Collection and Processing

We collected Twitter data with the Twitter REST API on the hashtag #5thAugustBlackDay from 5th August to 13th August 2020, covering a week of the curfew implementation of the one-year anniversary of the lockdown. To retrieve as many tweets related to the event, we opted for only one specific hashtag, reducing the risk of data dilution that tends to occur with multiple hashtags. Images were downloaded from tweets if they were present. The final dataset contained 269k tweets and 35k images.

Bot Annotation. We annotated the data by performing bot-probability annotation using the BotHunter algorithm with a 90% accuracy [1]. BotHunter has previously been successfully applied in previous information studies [2,11]. It extracts account-level metadata and classifies agents using a supervised random forest method through a multi-tiered approach, each tier making use of more features. For each user agent, BotHunter provided a probability that the account is inorganic. A probability over 70% indicates a bot.

2.2 Bot-Based Emotion Behavior Differences

In this section, we describe our methodology to analyze bot-based behavior differences in terms of image clusters. Image clusters are thematic groups of images in the dataset. After grouping images by themes, we analyze the dominant image emotion expressed in each theme, and observe the differences between the emotions expressed by the bots and non-bots. We also identify the themes with the highest bot percentage, which may be a weak signal of information operation.

Image Cluster Identification. In identifying image clusters, we represented each image as a vector using the pre-trained ResNet50 image model implemented with Tensorflow. The image feature dimensions were further reduced using Principal Component Analysis (PCA) to the most salient 100 principal components used for the next step.

Using the vector representations generated in the last step, we clustered the images through pairwise comparison of the vectors with Euclidean distances using the DBScan (Density-Based Spatial Clustering) algorithm. The algorithm performs well over large and noisy datasets like our Twitter image dataset. Ten clusters were chosen for their distinct separation of image clusters. Each cluster was represented by a cluster centroid, the image vector with the minimum Euclidean distance from all images in the cluster. Each image was annotated with an image cluster number based on its Euclidean distance to the cluster centroid in the projected space. We manually interpret each image cluster.

Image Emotion Detection. Plutchik defined eight core emotions in his psychoevolutionary classification approach for general emotional responses: anger, anticipation, disgust, fear, joy, sad, surprise and trust [6]. We constructed an image emotion classification pipeline by adapting the pipeline from Gajarla and Gupta [4] whose classifier yielded a 0.40 accuracy.

We trained a Flickr image emotion classifier based on the VGG16 model, which uses convolutional neural network. We constructed our model training dataset from Flickr. When users upload an image to the image sharing site Flickr, they tag keywords. We downloaded the top 1000 images for each of the emotional categories using the Flickr's API service, searching by emotion names as image tags. This provided us a stream of human-annotated tagged photos for model training. Based on crowd-sourced user engagement., the images retrieved are most relevant to the tag.

We fine-tuned the VGG16 image emotion classifier using the Keras python library and implemented model training on the 8000 collected images with a 80-20 train-test split across 200 epochs with five-fold cross-validation. Our classifier yields a 0.47 accuracy, outperforming chance selection (accuracy = 0.125).

Using our image classifier, we performed image emotion inference on the Twitter images. For each image, we predicted a vector $E = [e_1, e_2, ..., e_8]$ where e_i is the probability of the image being classified into one of the eight emotions. We determined the image emotion with $argmax(E)$.

2.3   Community Case Study

Through manual inspection of the dataset, we identified a group of agents where their reported location and their Tweet language do not match. These agents Tweet in Urdu, one of the main languages spoken in Pakistan, but indicate their location elsewhere, i.e. China. We extracted a sub-dataset comprising of these 173 agents and profile their activity through image and network analysis. We profiled the distribution of images and dominant emotions across bots and non-bot groups across the image clusters. For network analysis, we analyzed the communication network between bots and non-bots to understand the interactions between both groups. The influence of each agent within the network was profiled using the eigenvector centrality network measure.

3 Results

3.1 Image Clusters in Kashmir's Black Day

Image Cluster Analysis. We present representative images of each cluster in Table 1. A summary of the joint analysis of the bot percentage and image emotions given the image clusters is presented in Table 2.

Image Cluster 1. Accounting for 9.5% of images, images in this cluster call out the injustice to Kashmir. Mixed in this cluster are images of hope. Sad, trust, joy and fear are dominant emotions, with bot percentage at 65%.

Image Cluster 2. Accounting for 8.8% of images, images in this cluster call out to "Free Kashmir". Sad, trust, joy and fear are dominant emotions, with bot percentage at 63%.

Image Cluster 3. Accounting for 9.0% of images, this cluster contains general images of the wounded during the military occupation. Dominant image emotions are sad, disgust and trust, with bot percentage at 38%.

Image Cluster 4. Accounting for 10.0% of images, this cluster contains images bearing the words "Black Day", "Kashmir bleeds" and "1 year Kashmir lockdown", referring to the military occupation. Dominant image emotions are sad, trust and fear, with bot percentage at 60%.

Image Cluster 5. Accounting for 11.3% of images, this cluster calls out to stop the genocide in Kashmir together with many images containing the Kashmir flag. Sad and fear are dominant emotions, with bot percentage at 44%.

Image Cluster 6. Accounting for 9.7% of images, there are two main image themes: calling for the end of the military occupation and a bloody hand sign. Sad and trust are dominant emotions, with bot percentage at 59%.

Image Cluster 7. Accounting for 12.3% of images with a 49% bot percentage, this cluster is characterized by bloody images and images of Kashmir's women. Dominant image emotions are sad, trust, and disgust.

Image Cluster 8. Accounting for 8.4% of images and having 64% bots, this cluster is characterized by a two-finger hand signifying "free kashmir". Sad and trust are dominant emotions.

Image Cluster 9. Accounting for 12.2% of images, images in this cluster show the Pakistan Map and explain how it has changed since the military occupation. Dominant image emotions are sad and trust, with bot percentage at 52%.

Table 1. Examples of images in each cluster. The less violent images are selected as examples, so that the readers may have a less nightmarish time reading.

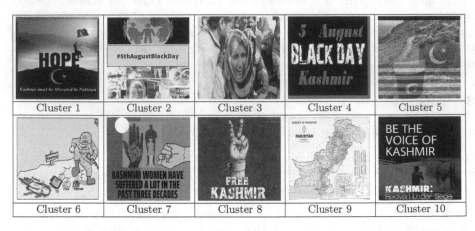

Table 2. Summary of image clusters, together with the dominant emotions expressed in the images and the bot prediction percentage.

	Image cluster description	Dominant image emotion	Bots (%)
1	United Nations' injustice to Kashmires, Images of hope	Non-bots: sad, trust, disgust Bots: sad, trust, joy	65%
2	Images calling to "Free Kashmir"	Non-bots: sad, trust, joy, fear Bots: sad, trust, joy	63%
3	General images of the wounded	Non-bots & Bots: sad, trust, disgust	38%
4	Images of Black Day, Kashmir Bleeds, 1 year Kashmir lockdown	Non-bots & bots: sad, trust, fear	60%
5	Stop genocide in Kashmir, Kashmir flag	Non-bots & bots: sad, trust, joy	44%
6	Images calling the end of the military occupation Images of a bloody hand sign	Non-bots & bots: sad, trust, disgust	59%
7	Bloody images Images of injured women	Non-bots & bots: sad, trust, disgust	49%
8	Images of a two-finger hand sign to signify "free kashmir"	Non-bots & Bots: sad, trust	64%
9	Images of the Pakistan Map	Non-bots & Bots: sad, trust	52%
10	Images calling to Be the Voice of Kashmir	Non-bots & Bots: sad, trust, disgust	61%

Image Cluster 10. Accounting for 8.7% of images, images in this cluster calls out to Be the Voice of Kashmir. Dominant image emotions are sad, trust, and disgust, with bot percentage at 61%.

Bot Activity. Overall, 56% of the agents in the entire dataset are predicted as bots. Bot activity is notable and fairly consistent across the clusters, ranging approximately from 38% to 65%. Notably, a larger proportion of the image conversation for Clusters 1, 2, 8 and 10 are dominated by bots.

Emotions in Images. All eight core Plutchik emotions are represented in the full dataset, with sadness being the most dominant emotion expressed in 31% of the images. Anticipation and surprise are the least dominant emotion, expressed in only 2% and 8% of images respectively. Examples of the less violent images that are classified into each of the eight emotions are presented in Table 3. The distribution of emotions across the dataset is shown in Fig. 1a. Sadness is expressed in all clusters. Anticipation is found in clusters 8, 3, 1; and surprise is found in clusters 6, 4 and 8. The rest of the emotions contain an even distribution of images (~12% each).

For most image clusters, bots and non-bots share images of similar emotions and did not have significantly different image sharing behaviors. However, for Image Clusters 1 and 2, the dominant image emotions are different for the two types of agents. These clusters also contain the highest bot percentage. Specifically, non-bots share images related to disgust and fear in addition to the common image emotions of sad/trust/joy.

3.2 Community Case Study

In our community case study, we studied a smaller group of agents. These agents have declared their location in China, yet Twitter's language detection algorithm identifies their tweets to be in Urdu. Using the same image clusters as the main dataset, we summarize the activity of these 173 agents in terms of the dominant emotions expressed in the images and bot activity in Table 4.

Bot Activity. 66% of the agents in this case study are classified as bots by the BotHunter algorithm. This is a higher proportion as compared to the full dataset (56% bots). Bot activity is the highest in image cluster 2 at ~70% which contains images calling for freedom of Kashmir. When inspecting a communication network between the agents (Fig. 2), the bots constantly communicate with each other, shown by their high interconnectivity at the network's center, yet they are able to attract the attention of non-bots to disseminate their messages. Sizing nodes in the network figure by eigenvector centrality values, we observe that bots show the highest centrality values, reflecting their high influence in this network.

Images and Emotions. In this sub-dataset, the images are not equally distributed across all the image clusters. Image cluster 8, which presents images of a two-finger hand sign, has no agents present. Bots are present in only two of the image clusters: images calling to "Free Kashmir" and general images of the wounded.

Table 3. Examples of less violent images that are classified into each of the eight emotions defined by Plutchik.

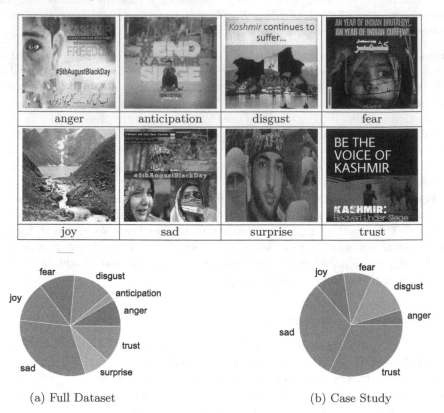

| anger | anticipation | disgust | fear |
| joy | sad | surprise | trust |

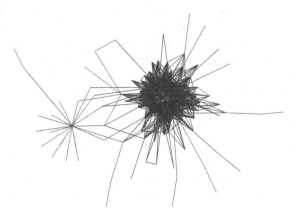

(a) Full Dataset   (b) Case Study

Fig. 1. Emotion distribution of images in the full dataset and case study

Fig. 2. Communication network of the agents in the case study. Red nodes are bots, while green nodes are non-bots. Agents are sized by eigenvector centrality. Bots have the largest node sizes, reflecting their high influence.

On the emotions that the images expressed, sadness and trust are the two most dominant emotions, covering 31.8% of images each. Fear and joy are expressed the least, covering 9.1% each. Unlike the main dataset, these agents do not dominate the conversation with sad images. For image clusters that have no bots present, we infer that the emotions represented are that of non-bot actors, which include humans.

Table 4. Summary of the activity of agents in the case study.

Image cluster	Dominant image emotion	Bot percentage
1	Sad, trust	0
2	Sad, trust, joy	70%
3	Disgust, trust, joy	50%
4	Fear	0
5	Sad, fear	0
6	Sad	0
7	Trust	0
8	Anger, trust	0
9	(No agents)	(No agents)
10	Sad, disgust	0

4 Discussion

Characterization of Images and Their Emotions. A large proportion of images call out the military occupation and its effect to land and life (Clusters 6 and 9) as well as bloody images. All image clusters are characterized by sad and fearful images, which is not surprising given the violent nature of the event. While the dominant emotion across all image clusters are sadness, trust and joy is another dominant emotion which depicts hope in a change of situation. Disgust is a highly dominant emotion in clusters that call for action, given that images of disgust hold one's attention longer [12]. The least dominant emotion, anticipation, is found in clusters that call for the stop of killing and look to a larger authority like the United Nations for hope of the end of the situation.

Bot-Behavior in Image Sharing. Overall, 56% of the agents in the entire dataset are predicted as bots, higher than previous studies with BotHunter on Asian datasets [10,11]. Bot and non-bot agents have similar image sharing behavior, hence we postulate that the bots mimic the emotions of non-bot agents to blend in and avoid detection.

Bots are found in high prevalence in image clusters calling out for freedom and drawing awareness to injustice to the region and women. They are found in low proportions in image clusters that describe the situation or the wounded.

Through these observations, bot accounts are perpetuating the message of freedom using ideas of gender and fairness. They are most dominant in image clusters that have calls for actions, e.g. "Be the Voice of Kashmir". In addition, the emotion of sadness stimulated through the loss of valued members of the Kashmir community provides opportunities for integration into the community, which bot agents may captialize on [7]. These are worrying trends as propagation of messages on Twitter calling for social justice warriors can result in social unrest.

Community Case Study. In our case study, we focus on a group of agents that call out for the fight for freedom in Kashmir. In our analysis, bot agents are consolidated in image clusters that represent the fight for freedom, possibly indicating their stance towards the situation. A network analysis perspective shows that bots are successful in catching the attention of non-bot agents who then propagate their messages. Network centrality metrics point to bots having the largest values, reflecting their high influence position in the network. This is concerning as bots in this case study could have the intention to invoke physical freedom fighting, destabilizing the region.

Despite the sadness of the situation, represented by the dominance of sad images, the dataset presents a high proportion of images of trust, urging the people of Kashmir to trust that freedom is near and hence fight for it. Bots are prevalent in image clusters that invoke emotions of sadness and trust, building behavioral effects of affiliation and integration [7], possibly leading to their successful penetration of the communication network. Images with themes that call for freedom have the highest percentage of bots, suggesting themes to watch for future information operation campaigns.

On closer manual inspection of the accounts surfaced in this case study, many accounts are purporting the Pakistan War Industry. We further observe that several accounts were originally food accounts that evolved into accounts calling for freedom. Given the large percentage of bots and the topics the bots converse about, this may be a small information operation.

Limitations and Future Work. Several limitations nuance our conclusions from this work. Sampling Twitter data remains limited by API generalizability issues, suggesting caution in extrapolating findings. Though the Flickr image dataset had been manually annotated by Flickr users through tags during image upload, it may not be the gold standard in image emotion annotation. The images can be very varied, ranging from natural to abstract images. Additionally, the top images from Flickr for each emotion may not be relevant to social movements and thus affect classifier accuracy. The classifier accuracy is important in the analysis pipeline as it affects downstream tasks of group characterizations. Nonetheless, we hope our work provides insights into characterizing emotions through images on Twitter during an Asian protest. Future work on image emotion classification involves a more targeted image training dataset, curated for social cybersecurity events; and training a multimodal image classifier that captures image text content and the image features to identify emotions.

5   Conclusion

This work characterized bot-based emotion behavior differences in the Kashmir Black Day event. We trained an image emotion classifier based on a dataset collected from Flickr, pushing the value of image analysis from a mere prediction to a more comprehensive engagement in a concrete setting. We identified major themes and emotions in images that were salient in online discourse on Kashmir's Black Day and how the community-level prevalence may be driven by bot activity. Our most significant insights point to: (a) a large proportion of images call out the Kashmir military occupation and changes in life since then; (b) bots dominate in image clusters calling for action - freedom and justice. We also presented a case study of a small information operation, finding that bots are dominant the conversations in image clusters that call for action, emphasizing how interoperable frameworks of machine learning and network science tools can surface potential information operations that rely on visual resources. These techniques are key in social cybersecurity, and may readily be adapted in a wide variety of analytical contexts in studying the multimodal nature of online discourse.

References

1. Beskow, D.M., Carley, K.M.: Bot-hunter: a tiered approach to detecting & characterizing automated activity on twitter. In: Conference paper. SBP-BRiMS: International Conference on Social Computing, Behavioral-Cultural Modeling and Prediction and Behavior Representation in Modeling and Simulation, vol. 3, p. 3 (2018)
2. Beskow, D.M., Kumar, S., Carley, K.M.: The evolution of political memes: detecting and characterizing internet memes with multi-modal deep learning. Inf. Process. Manage. **57**(2), 102170 (2020). https://doi.org/10.1016/j.ipm.2019.102170
3. Carley, K.M.: Social cybersecurity: an emerging science. Comput. Math. Organ. Theory (2020). https://doi.org/10.1007/s10588-020-09322-9
4. Gajarla, V., Gupta, A.: Emotion detection and sentiment analysis of images
5. Ng, H.X.L., Lee, R.K.W., Awal, M.R.: I miss you babe: analyzing emotion dynamics during COVID-19 pandemic. In: Proceedings of the Fourth Workshop on Natural Language Processing and Computational Social Science, pp. 41–49. Association for Computational Linguistics, November 2020. https://doi.org/10.18653/v1/2020.nlpcss-1.5
6. Plutchik, R.: A general psychoevolutionary theory of emotion. In: Theories of Emotion, pp. 3–33. Elsevier (1980)
7. Plutchik, R.: Emotions and life: perspectives from psychology, biology, and evolution. American Psychological Association (2003)
8. Sailunaz, K., Dhaliwal, M., Rokne, J., Alhajj, R.: Emotion detection from text and speech: a survey. Soc. Netw. Anal. Min. **8**(1), 28 (2018)
9. Schill, D.: The visual image and the political image: a review of visual communication research in the field of political communication. Rev. Commun. **12**(2), 118–142 (2012). https://doi.org/10.1080/15358593.2011.653504
10. Uyheng, J., Carley, K.M.: Bots and online hate during the COVID-19 pandemic: case studies in the united states and the Philippines. J. Comput. Soc. Sci. **3**(2), 445–468 (2020). https://doi.org/10.1007/s42001-020-00087-4

11. Uyheng, J., Ng, L.H.X., Carley, K.M.: Active, aggressive, but to little avail: characterizing bot activity during the 2020 Singaporean elections. Comput. Math. Org. Theory, May 2021. https://doi.org/10.1007/s10588-021-09332-1
12. van Hooff, J.C., Devue, C., Vieweg, P.E., Theeuwes, J.: Disgust- and not fear-evoking images hold our attention. Acta Physiol. (Oxf) **143**(1), 1–6 (2013). https://doi.org/10.1016/j.actpsy.2013.02.001
13. Wang, Y., et al.: Understanding the use of fauxtography on social media. Proc. Int. AAAI Conf. Web Soc. Media **15**(1), 776–786 (2021). https://ojs.aaai.org/index.php/ICWSM/article/view/18102

Using Social Network Analysis to Analyze Development Priorities of Moroccan Institutions

Imane Khaouja[1,2](✉), Ibtissam Makdoun[1,2](✉), and Ghita Mezzour[1,3](✉)

[1] International University of Rabat, Rabat, Morocco
{imane.khaouja,ibtissam.makdoun}@uir.ac.ma
[2] University Mohammed V in Rabat, Rabat, Morocco
[3] DASEC, Rabat, Morocco
ghita.mezzour@dasec.ma

Abstract. Encompassing development priorities of different categories in the national development plan provides the engagement of different categories while responding to their development expectations. A special commission for the development model was formed in Morocco by royal instructions in order to identify the development priorities of different institutions.

In this paper, we use social network analysis to analyze the development priorities of Moroccan Institutions. We identify, using dictionary-based content analysis, top concerns from the written contributions received by the commission from different entities. Then using social network analysis, we rank the top priorities of the different entities.

Our results reveal that the top three concerns converge to social protection, education and employability. However, the top concerns differ from one category to another when taken exclusively.

Keywords: Development priorities · Dictionary-based content analysis · Social network analysis

1 Introduction

A special commission for the development model was formed in Morocco by royal instructions in 2019. The commission should proceed to an inventory of the achievements of the Kingdom, of the reforms undertaken, taking into account the expectations of different institutions and citizens. To this end, the commission carried out different consultation processes such as working sessions, including listening and consultation sessions, field visits and debate with experts, in addition to working sessions by members of the Committee, which allowed the commission to interact with around 10 000 people directly. In particular, more

This work was partly funded by CSMD and was carried out within the framework of a partnership agreement between CSMD and the International University of Rabat.

© Springer Nature Switzerland AG 2021
R. Thomson et al. (Eds.): SBP-BRiMS 2021, LNCS 12720, pp. 195–203, 2021.
https://doi.org/10.1007/978-3-030-80387-2_19

than 200 written contributions were sent to the commission by political parties, economic operators, unions, civil society, public administration, universities and international organizations, as well as, the commission organized listening sessions to consult with these institutions. The Covid19 pandemic and the restrictions measures put in place in the context of the health crisis have not interrupted this process of consultations, but made it more necessary to understand the impact of the epidemic, specifying that this process of broad consultations continued online and by video-conference.

In this paper, we focus on the written contributions sent by different institutions to the commission. We examine the importance given to the concerns and development priorities of different institutions to clarify their concerns. More specifically, we analyze the cited concerns in the written contributions, using dictionary-based content analysis. Then we generate networks of top concerns for the different institutions.

Our results reveal that overall the top three concerns converge to social protection, education and employability. The term social protection refers particularly to the health system more than to social protection in its broad sense which includes social prevention such as unemployment insurance and retirement. Social prevention is not very present in the expressed concerns in the written contributions. However, the top concerns differ from one category to another when taken exclusively. Such divergence can be explained by the focus of different institutions on their interests.

2   Related Work

Content analysis is a research technique for making valid inferences from text and useful for examining trends and patterns in textual documents. It was widely used to infer topics and concerns from textual documents [4,8]. In an attempt to automate content analysis, dictionary-based text analysis was first introduced by [6] and is considered as the most exhaustive content analysis method that can be automated with the use of computers. This method was employed in different studies to analyze big social data [5,7]. However, dictionary-based text analysis requires several subjective steps to adapt and tune the content. Like manual content analysis, the researcher needs to develop a predetermined list of categories as well as word lists to indicate the categories. It is important to assess whether the predetermined list can adequately reflect the entire big dataset [3].

3   Proposed Approach

An important number of reports[1] were sent to the commission from different institutions (see Table 1), to express their concerns and priorities concerning the development plan in Morocco. These reports were written in Arabic and in French and varied in page length from one page to more than 10 000 pages. These

---

[1] The contributions can be accessed online on https://www.csmd.ma/.

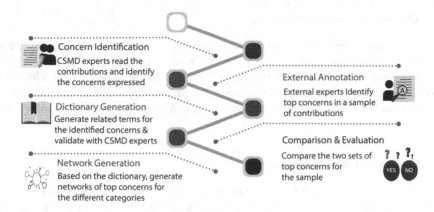

Fig. 1. Proposed approach for concern identification

contributors belong to civil society, political parties, public institutions, unions, professional chambers and international organizations. The proposed approach to analyze and identify those concerns is detailed in Fig. 1. CSMD experts start by identifying the main reported concerns in the received documents. Then, we generate a list of terms for the identified concerns using DBpedia [2], DBpedia is a structured knowledge base generated from Wikipedia pages. For each concern, we extracted the direct page redirects and hyperlinks. The final list of terms is, then validated and enriched by CSMD experts, the output of this process is considered as our final dictionary. Once the dictionary is set, we identify the cited concerns for each document. Moreover, we generate networks based on these concerns.

In order to validate our results, we give a sample of the received documents (10%) to external experts to identify and rank the observed concerns in the documents. Once the concerns are identified from the sample of the documents, we compare the resulted concerns using our dictionary-based method.

Table 1. Number of contributors in each category

Category	Number of contributors
Associations	50
Political parties	25
Think-Tanks & Academic sector	20
Institutions & Public organisations	15
Professional chambers	6
Unions	7
International organizations	2

3.1   Dictionary Generation

The commission identified a list of concerns from their preliminary reading of the reports. Based on these concerns, we develop a dictionary of terms to identify those concerns in the written contributions. From these initial identified concerns, we generate related terms using DBpedia [2]. Then, CSMD experts validated this list of related terms and added Moroccan-specific terms to the dictionary.

3.2   Concern Identification

Once the dictionary was defined, we identified the concerns of each contributor. We scale the weights of the extracted concerns where the sum of the weights is 1. Each concern has a weight that varies from 0 to 1. The more terms are cited for a specific concern, the more its weight increases. Such scaling is performed to give the contributors the same importance as the contributions vary significantly in page length.

3.3   Network Generation

Once the weight is computed for each concern in the contributors' documents. We generate networks where nodes are the contributors categories and the concerns. The edges represent the weight given by the category to the concern.

3.4   Evaluation

In order to validate our results, we give a sample of the received documents (10%) to external experts to identify and rank the observed concerns in the documents. Once the concerns are identified from the sample of the documents, we compare the resulted concerns using our dictionary-based method. When compared, the concerns identified from the documents were similar to our automated dictionary-based annotation. Moreover, the final results concorded with the listening session results.

4   Results

The networks in the figures depict the concerns of the different contributors of different categories. In Fig. 2, we can see that social protection, employability & employment, education and regionalization are the top overall concerns of the different categories in terms of in-degree centrality. When we only keep contributors of the professional world such as professional associations, socio-professional associations, unions and professional chambers, we can observe from Fig. 3 that the concern ranking changes slightly. The employment & employability concern become the most important concern in terms of in-degree centrality.

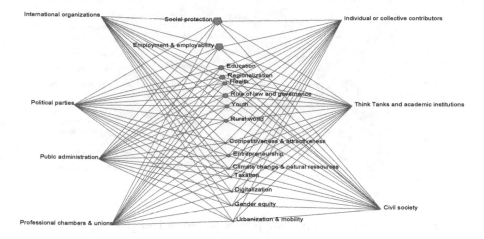

Fig. 2. The concerns of the contributors ranked by in-degree centrality

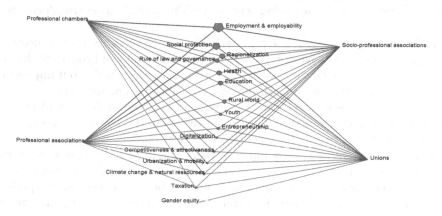

Fig. 3. The concerns of the professional contributors ranked by in-degree centrality

Figure 4 depicts concerns of political parties in Morocco. Regionalization, rule of law and governance, social protection are their top concerns, which can be explained by the political nature of such contributors. However, regionalization is also among the top concerns as Morocco has adopted a regionalization strategy [1] to reduce disparities between regions and contribute to regional economic development. The represented political parties (Fig. 5) and non-represented political parties (Fig. 6) share the top three concerns social protection, rule of laws and governance and regionalization. Aside from shared top concerns, professional associations give more importance to entrepreneurship, digitalization and climate change & natural resources as depicted in Fig. 7. Such concerns show that civil society is aware of the challenges facing Morocco and urges the development model to encompass measures for facing global warning.

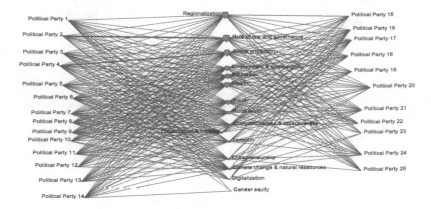

Fig. 4. The concerns of political parties ranked by in-degree centrality

Moreover, professional associations encourage the digitalization transformation of Morocco.

In Fig. 8, we can see that socio-professional associations are more concerned by employability & employment, education, regionalization and youth. Such concerns indicate that socio-professional associations are concerned about improving employability and education in all regions, in particular among youth.

The top concerns of unions are employment & employability, social protection, rule of law and governance, health and regionalization as depicted in Fig. 9. Also, unions are concerned about youth and education. Taxation appears among the top concerns. Unions are a group of workers who unite to improve their working conditions such as wages, benefits and safety through collective negotiation with employers. Decreasing income tax is among their top concerns. For professional chambers, taxation and the rural world are among the top concerns as

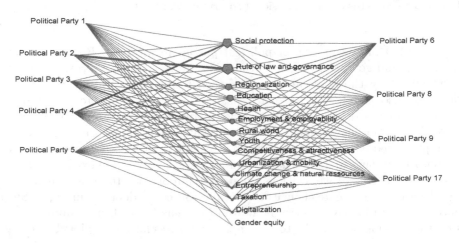

Fig. 5. The concerns of represented parties ranked by in-degree centrality

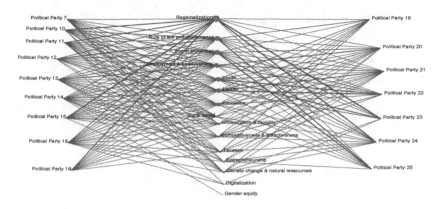

Fig. 6. The concerns of non-represented political parties ranked by centrality

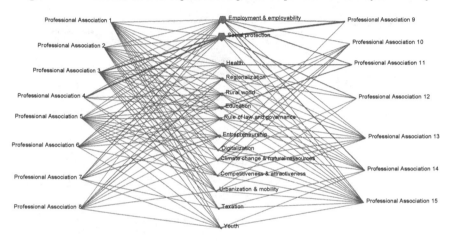

Fig. 7. The concerns of professional associations ranked by centrality

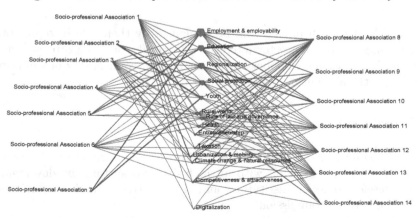

Fig. 8. The concerns of socio-professional associations ranked by in-degree centrality

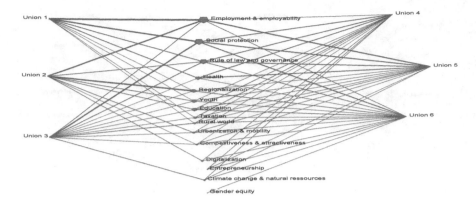

Fig. 9. The concerns of unions ranked by in-degree centrality

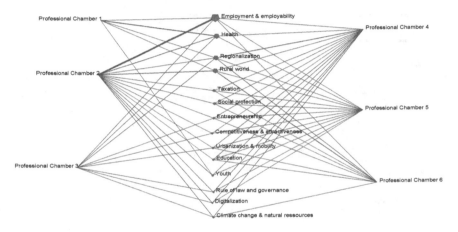

Fig. 10. The concerns of professional chambers ranked by in-degree centrality

shown in Fig. 10. Such interest can be explained by their role in representing and defending the private sector interests. The rural world is a top concern for agricultural professional chambers as enhancing the connectivity of rural areas will improve the agriculture sector in Morocco.

5   Conclusion

A special commission for the development model was formed in Morocco by royal instructions to come up with an inclusive and sustainable development model. Therefore different institutions and citizens were invited to submit their contributions to the commission.

In this paper, we examine the top concerns of institutions in their contributions to a new development model for Morocco. With the help of CSMD experts, we define a dictionary for different concerns and their related terms. Then, we use semantic networks to represent the top concerns of different institutions in Morocco. Our study reveals that the top three concerns converge to social protection, education and employability. Social protection in the contributions refers more to the health system more than to social protection social prevention which is given little importance. However, the top concerns differ from one category to another when taken exclusively. Such divergence can be explained by the stakeholder's focus on their interests. The methodology helps to analyze and summarize top concerns for different categories. The methodology identifies the urgent concerns and development priorities that the new development model should encompass.

Acknowledgements. The authors would like to thank CSMD experts: Nadia Hachimi and Marwane Fachane for their valuable feedback. The authors would also like to thank Jean-Noël Ferrié and Zineb Omary for their external validation and annotation of the results reported in this paper.

References

1. Benkada, A., Belouchi, M., Iallouchen, A., Essarsar, M.: Regional financial governance: a lever for change for advanced regionalization in Morocco. Int. J. Sci. Eng. Res. **9** (2018)
2. De Mauro, A., Greco, M., Grimaldi, M., Ritala, P.: Human resources for big data professions: a systematic classification of job roles and required skill sets. Inf. Process. Manage. **54**(5), 807–817 (2018). https://doi.org/10.1016/j.ipm.2017.05.004. http://www.sciencedirect.com/science/article/pii/S0306457317300018
3. Guo, L., Vargo, C.J., Pan, Z., Ding, W., Ishwar, P.: Big social data analytics in journalism and mass communication: comparing dictionary-based text analysis and unsupervised topic modeling. J. Mass Commun. Q. **93**(2), 332–359 (2016). Publisher: SAGE Publications Sage CA: Los Angeles, CA
4. Riffe, D., Lacy, S., Fico, F., Watson, B.: Analyzing Media Messages: Using Quantitative Content Analysis in Research. Routledge, New York (2019)
5. Russell Neuman, W., Guggenheim, L., Mo Jang, S., Bae, S.Y.: The dynamics of public attention: agenda-setting theory meets big data. J. Commun. **64**(2), 193–214 (2014). publisher: Oxford University Press
6. Stone, P.J., Dunphy, D.C., Smith, M.S.: The General Inquirer: A Computer Approach to Content Analysis. MIT Press, Cambridge (1966)
7. Vargo, C.J., Guo, L., McCombs, M., Shaw, D.L.: Network issue agendas on Twitter during the 2012 US presidential election. J. Commun. **64**(2), 296–316 (2014). Publisher: Oxford University Press
8. Wester, F.P.J., Krippendorff, K.: Content Analysis. An Introduction to Its Methodology: 2005 9780761915447 (2005)

Influence Dynamics Among Narratives
A Case Study of the Venezuelan Presidential Crisis

Akshay Aravamudan[1]([✉]) [iD], Xi Zhang[1] [iD], Jihye Song[2] [iD], Stephen M. Fiore[2] [iD],
and Georgios C. Anagnostopoulos[1] [iD]

[1] Department of Computer Engineering and Sciences,
Florida Institute of Technology, Melbourne, FL, USA
{aaravamudan2014,zhang2012}@my.fit.edu, georgio@fit.edu
[2] University of Central Florida, Orlando, FL, USA
chsong@knights.ucf.edu, sfiore@ist.ucf.edu

Abstract. It is widely understood that diffusion of and simultaneous interactions between narratives—defined here as persistent point-of-view messaging—significantly contributes to the shaping of political discourse and public opinion. In this work, we propose a methodology based on Multi-Variate Hawkes Processes and our newly-introduced Process Influence Measures for quantifying and assessing how such narratives influence (Granger-cause) each other. Such an approach may aid social scientists enhance their understanding of socio-geopolitical phenomena as they manifest themselves and evolve in the realm of social media. In order to show its merits, we apply our methodology on Twitter narratives during the 2019 Venezuelan presidential crisis. Our analysis indicates a nuanced, evolving influence structure between 8 distinct narratives, part of which could be explained by landmark historical events.

Keywords: Influence dynamics · Social media narratives ·
Multi-Variate Hawkes Processes · Granger causality · Venezuelan
presidential crisis

1 Introduction

Discourse on social media platforms has become increasingly relevant in political contexts throughout the world [11,19]. There have been quite a few studies that have illustrated the power that social media yields in shaping public opinions and influencing political dialogues [10,13]. While it has been argued that the Internet has not revolutionized how politics is conducted [9], it does provide an efficient medium for rhetoric and discussion born out of existing cultural, economic, and political situations [4]. As such, understanding the nature of how ideas, opinions, and calls to action diffuse over social media provides insights into how real-world events may unfold, especially in the context of areas experiencing political and economic turmoil. Such insights may help interested parties in developing a more comprehensive portrait of developing socio-geopolitical contentions.

© Springer Nature Switzerland AG 2021
R. Thomson et al. (Eds.): SBP-BRiMS 2021, LNCS 12720, pp. 204–213, 2021.
https://doi.org/10.1007/978-3-030-80387-2_20

The Venezuelan presidential crisis is an indicative example of significant political events. It began on January 10th, 2019, when Nicolás Maduro was sworn in for a second presidential term following a disputed election in May of 2018. This development prompted international reactions with nations siding with either Nicolás Maduro or Juan Guaidó, then president of Venezuela's National Assembly. The situation escalated after Guaidó declared himself as the interim president on January 23rd, 2019. All the while, there was significant participation on social media, which drove conversations globally. During this period, Twitter was heavily employed, resulting in competing opinions and, thus, rendering it as an effective tool for political communication [16].

We study this in the context of narratives associated with this event. We develop an interdisciplinary approach that unites social science theorizing on narratives with computational approaches to information evolution. Our narrative approach combines a parsimonious definition of online narratives (see [2]) as "... *recurring statements that express a point of view.*" We use this in conjunction with stances taken on a subset of narratives (*i.e.*, support or disagree). In this way, we are able to model the mutual influence of complementary and competing narratives evolving over time. As such, we put forward an approach for quantifying and assessing the co-evolving influences among narratives under the prism of Granger causality as it is defined in the context of Temporal Point Processes (TPPs). This allows one to discern concrete influence motifs between online narratives. Such motifs serve as an additional facet of characterizing online discussions, which may prove useful to social scientists studying online discourse. We next describe work related to our approach. Then in Sect. 3, we provide an overview of our dataset. Section 4 describes our methodology in some detail. As proof of concept, we then apply our methodology to study the influence dynamics of 8 concurrent narratives connected to the Venezuelan presidential crisis using Twitter data as outlined in Sect. 3. Finally, we provide and comment on our findings in Sect. 5. In particular, we are able to identify influence patterns that can be correlated to milestones of the crisis.

2   Related Works

Our work on narrative modelling falls under a broader umbrella of dynamic topic modelling for temporally sequenced documents, where topics correspond to narratives and documents to tweets. A first example of such an approach is temporal LDA [17], which extends traditional Latent Dirichlet Allocation (LDA) and allows for learning topic transitions over time. A couple of other examples that we mention here are based on TPPs, which are used to model topic dynamics. Lai et al. [12] propose using a marked Hawkes process to uncover inter-topic relationships. Similarly, Mohler et al. [14] propose a Hawkes-Binomial topic model to detect mutual influence between online Twitter activity and real-world events. While our work also falls under this general strand, it focuses solely on learning topic dynamics (see Sect. 4), since the topics themselves are ultimately provided by Subject Matter Experts (SMEs) as described in Sect. 3.

3   Data

The Twitter dataset used for this study is a subset of a larger social media dataset curated by a data provider as part of a larger grant program and is described in [2,3]. In accordance with the requirements of the organization funding the research, the data were anonymised prior to sharing it with researchers to protect user privacy. Overall, it encompasses over 7 million tweets, overwhelmingly in Spanish, from December 25th, 2018, to February 1st, 2019, a period which coincides with the commencement of the presidential crisis.

Each tweet has been annotated with potentially multiple narrative labels. The narratives were derived by applying topic modelling on the tweets' textual content via Non-negative Matrix Factorization combined with tf-idf statistics [18]. These topics were subsequently refined and formalized by SMEs. A subset of the tweets was then manually annotated by the same SMEs and were then used to train a BERT-based multilingual cased multi-label classification model [15] that was then used to annotate the remaining tweets. A similar procedure was followed to label tweets as pro- or anti-Maduro.[1]

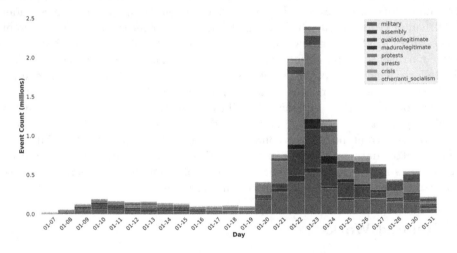

Fig. 1. Histogram of Twitter event (tweet) counts per narrative in 2019. One observes a burst of activity between January 19th and 21st, during which there was a small-scale coup initiated by 27 soldiers. (https://www.theguardian.com/world/2019/jan/21/venezuela-claims-foiled-attempted-military-uprising) The peak activity occurring on January 23rd appears strongly related to massive protests, which demanded Maduro to step down. (https://www.theguardian.com/world/2019/jan/23/venezuela-protests-thousands-march-against-maduro-as-opposition-sees-chance-for-change) During the same day, we also witness a significant increase in anti-Maduro tweets.

---

[1] Narrative and stance labelling was carried out by the data provider and was provided to us as is with very limited description of the process followed.

Table 1. Stance distribution per narrative of the Venezuela Twitter data.

Narrative	Total tweets	% anti-Maduro	% pro-Maduro
military	1,534,242	67.38	21.87
assembly	252,448	95.79	1.82
guaido/legitimate	1,014,726	95.43	3.02
maduro/legitimate	304,127	2.92	96.72
protests	1,746,615	85.87	2.42
arrests	570,574	97.73	0.74
crisis	305,291	73.25	3.58
anti-socialism	101,716	78.04	14.77

We only considered narratives that were present in at least 100,000 tweets. As a result, we analyzed a total of 8 narratives. The daily event counts of these narratives in the time period of interest is shown in Fig. 1, while Table 1 shows the distribution of stances per narrative. The *military* narrative includes discussions about the Venezuelan army, security services, or other organizations that reported to Maduro's government. *Assembly* includes any mentions of Venezuela's National Assembly. *Guaido/legitimate* and *maduro/legitimate* consist of tweets that expressly support the legitimacy of Guaidó and Maduro, respectively. *Protests* includes tweets that mention anti-Maduro demonstrations, public gatherings, or rallies. *Arrests* includes tweets that refer to people who had been imprisoned at the time. Moreover, the *crisis* narrative label refers to the Venezuelan humanitarian crisis[2] and finally, *anti-socialism* includes tweets that mention socialism, communism, or leftism as the primary cause of the humanitarian crisis.

4 Methodology

4.1 Multi-Variate Hawkes Processes

In this work, we employed Multi-Variate Hawkes Processs (MVHPs) [8]—a particular system of TPPs—to characterize the temporal dynamics of tweets comprising the narratives of interest. MVHPs have been widely used for modelling social media events (*e.g.*, [7,20]) as they are capable of describing self- and mutually-exciting modes of event generation. TPPs are completely characterized by their conditional intensity $\lambda(t \mid \mathcal{H}_{t-})$ at time t given their past observations $\mathcal{H}_{t-}$, which is referred to as the history of the process. The quantity $\lambda(t \mid \mathcal{H}_{t-})dt$ yields the expected number of events such a process will generate in the interval

---

[2] https://www.reuters.com/article/us-venezuela-politics-un/venezuelans-facing-unprecedented-challenges-many-need-aid-internal-u-n-report-idUSKCN1R92AG.

$(t, t + dt)$. Assume an MVHP comprising P processes. Its ith process features a conditional intensity of the form

$$\lambda_i(t \mid \mathcal{H}_{t-}) = b_i(t) + a_{i,i} \sum_{t_k^i \in \mathcal{H}_{t-}^i} \phi_{i,i}(t - t_k^i) + \sum_{\substack{j \in \mathcal{P} \\ j \neq i}} \alpha_{i,j} \sum_{t_k^j \in \mathcal{H}_{t-}^j} \phi_{i,j}(t - t_k^j) \quad (1)$$

where $i \in \mathcal{P} \triangleq \{1, 2, \ldots, P\}$, $\mathcal{H}_{t-}^i$ is the history of the ith process and $\mathcal{H}_{t-} \triangleq \cup_{i \in \mathcal{P}} \mathcal{H}_{t-}^i$ is the history of the multi-variate process. In Eq. 1, $b_i(\cdot) \geq 0$ is a history-independent intensity component, which we will refer to as base or background intensity. The quantity $\phi_{i,j}(t - t_k) \geq 0$ reflects the (typically, momentary and, subsequently, diminishing) increase in intensity that the ith process will experience at time t due to an event of the jth process, which occurred at time t_k. The functions $\phi_{i,j}(\cdot)$ are referred to as memory kernels. Also, for fixed base intensity and memory kernels, the non-negative $a_{i,j}$'s constitute the MVHP's model parameters. Finally, apart from the background intensity, one can distinguish two additional terms in Eq. 1: **(i)** a second, self-exciting term, through which past events of the process may generate further future events and **(ii)** a third, cross-exciting term, through which events of the other processes may do the same. An MVHP's base intensity and memory kernel functions are typically inferred non-parametrically, while the $a_{i,j}$ parameters are learned via (sometimes, penalized) maximum likelihood estimation.

For the purposes of studying influences among narratives, we used MVHPs that featured one process per narrative. Each process utilised a constant background intensity b_i and process-independent memory kernels $\phi_{i,j}(t) = e^{-t}$, which is a typical choice for Hawkes processes. Moreover, we opted to fit such MVHPs using overlapping time frames within the period of interest for two reasons: **(i)** initial experimentation gave strong indications—via probability-probability plots—that fitting a single MVHP to the entire period would result in a bad-fitting model. It appears that the tweet dynamics under consideration are not well approximated by MVHPs at long time ranges, but only at shorter ones. And, **(ii)** modeling tweet dynamics on overlapping (instead of disjoint) time frames has a beneficial smoothing effect on the estimated model parameters, when comparing time-adjacent models.

4.2   Quantifying and Comparing Influence Effects

Based on the notion of causality for continuous stochastic processes, Eichler et al. [6] extend the definition of Granger (predictive) causality to MVHPs and show that, if $a_{i,j} > 0$, then the jth process Granger-causes the ith process. This allows one to discern influences within and between processes by mere inspection or, more formally, by testing against $a_{i,j} = 0$. Moreover, the base intensity of each process can be naturally interpreted as a nonspecific cause, which is external to the system of processes. One can reject the absence of such unaccounted-for cause by testing against $b_i = 0$. Of course, the entirety of this causal reasoning is predicated on the absence of other confounding processes (or factors, in general), which is an assumption that we will also adopt for our studies. Thence, in our exposition we will make use of the notion of apparent influences among processes as a stand-in for Granger-causal effects sans confounding.

While testing for $a_{i,j} = 0$ determines a potential influence via a binary decision, quantifying the magnitude of the influence effect based on the $a_{i,j}$ parameters in order to draw comparisons is a much more nuanced issue. Some prior work, such as [1] and [12], for example, attempt to directly interpret the parameters' magnitudes as corresponding magnitudes of influence for the purpose of comparisons. This approach, though, is fraught with problems: (i) the first one is semantic in nature: the physical meaning of the $a_{i,j}$'s is intrinsically linked to the specific form of the memory kernels they multiply and, hence, is often far from straightforward to describe. And, (ii) solely relying on the $a_{i,j}$'s values to quantify the influence effect completely ignores memory kernels, which are equal contributors in shaping event dynamics; based on this approach, this renders comparing influence effects problematic.

In order to circumvent these shortcomings, we introduce and advocate the use of Process Influence Measures (PIMs) for quantifying the magnitude of process-to-process influences. In particular, we define the j-to-i PIM $\pi_{i,j} \in [0,1]$ as the probability that an event of the jth process is the most likely cause for an event of the ith process. These probabilities can be easily estimated from the results provided by a fitted model. This is due to the following fact that readily stems from the theory of TPPs, whose conditional intensity is represented additively: if E_k^i is the kth event of process i, which occurred at time t_k^i, and $\mathcal{E}^j$ denotes the set of all events of process j, then

$$\mathbb{P}\left\{E_k^i \text{ was caused by any earlier event in } \mathcal{E}^j\right\} = \frac{a_{i,j} \sum\limits_{t_\ell^j \in \mathcal{H}_{t_k^i-}^j} \phi_{i,j}(t_k^i - t_\ell^j)}{\lambda_i(t_k^i \mid \mathcal{H}_{t_k^i-}^i)} \quad (2)$$

for $i,j \in \mathcal{P}$. It is important to note that these probabilities jointly depend on model parameters, memory kernels as well on relative event timings and, thus, capture multiple facets of process dynamics. One can iterate over all events of process i and, each time, record the process j, which exhibits the largest probability given by Eq. 2. Then, the PIM $\pi_{i,j}$ is estimated via the frequency with which an event of process j was the likeliest cause for an event of process i.

5   Results and Discussion

We used the setup described in Sect. 4 to infer intra/inter-narrative influences from the available data. We split the observed time period into 35 two-day time frames that were overlapping by one day.[3] For each such time frame, we trained

---

[3] In particular, we chose two-day time frames to reduce the computational burden of training. Also, we used an hourly timescale to represent event time stamps to maintain the numerical stability of our training algorithm.

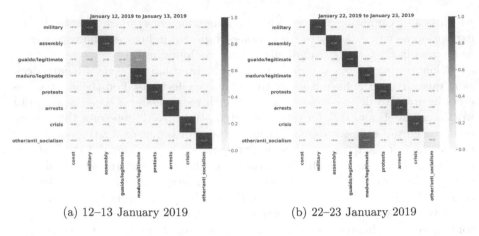

(a) 12–13 January 2019   (b) 22–23 January 2019

Fig. 2. PIM heat maps for two time frames. Columns show influence sources, while rows depict the narratives we studied. These maps illustrate self-driving narratives (prominent diagonal entries), as well as inter-narrative influences (sizeable off-diagonal entries).

a MVHP and ensured that it was exhibiting (at least) a reasonably good fit as judged by probability-probability plots. For investigating influences among narratives, each MVHP encompassed a process per narrative. Furthermore, we computed and tabulated PIMs as described in Sect. 4.2. Figure 2 provides PIMs as heat maps for two indicative time frames: one before Guaidó's self-proclamation as legitimate president and one after. These particular examples illustrate self-reinforcing narratives ($\pi_{i,i} \approx 1$ diagonal values), as well as narratives being influenced by one or more of the remaining narratives (appreciable $\pi_{i,j}$ off-diagonal values). We characterize influences in terms of PIMs as follows: "significant" influences, when $\pi_{i,j} \in (0.2, 0.6]$, "strong" influences, when $\pi_{i,j} \in (0.6, 0.99]$ and "decisive," when $\pi_{i,j} > 0.99$. Figure 3 highlights such influences over the time frame that our study considered.

Judging from the results in Fig. 3, we notice that *anti-socialism* strongly influences *guaido/legitimate* between December 25th and 26th, which may reflect Guaidó's support from Western governments and the growing dissatisfaction of the Venezuelan populace with existing conditions under Maduro's government, which were prevalent even before the crisis. Moreover, after Maduro's inauguration on January 10th, we notice that *maduro/legitimate* significantly influences *military*. The tweets during these days were posted after the first open cabildo— a political action convention—held by then president of the National Assembly, Guaidó, where he was voted in as acting president. This could have motivated Maduro supporters to favor a strengthening of his claim to the presidency and, by extension, favor the military's response to the unrest.[4] Between January 12th and 13th, *maduro/legitimate* significantly affects *guaido/legitimate* showing a

---

[4] https://www.bbc.com/news/world-latin-america-47036129.

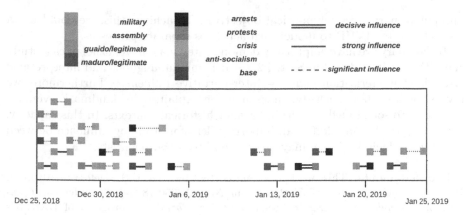

Fig. 3. Timeline indicating noteworthy influences based on our estimated PIMs. Weaker cross-narrative influences with PIM values $\pi_{i,j} \in [0.0, 0.2]$ and self-influences of any kind have been omitted for clarity. Note that, during the period of January 6th through the 11th, narratives only influence themselves ($\pi_{i,i} \approx 1$).

reactionary response to the mobilisation of Maduro's military in response to rising support for Guaidó. This also coincides with Guaidó's brief arrest[5] by the Bolivarian Intelligence Service on January 13th. Finally, we notice that between January 20th and 21st, *military* significantly influences *protests*, which overlaps with when some National Guardsmen rose against Maduro.[6] This was followed by a widespread and vociferous protests against Maduro. Between January 24th and 25th, we observe a significant influence from *anti-socialism* to *military*, which may be attributed to the wake of anti-Maduro protests[7] on January 23rd that called for the military to relinquish their allegiance to Maduro.

Finally, a potentially interesting extension of our work could specifically address strategic platform manipulation that may drive narratives during political crises. For instance, an operation involving Facebook and Instagram accounts attributed to a U.S. communications firm was flagged and removed for coordinated inauthentic behavior targeting Venezuela[8] and was found to have primarily posted anti-Maduro content [5]. Additionally, Twitter announced the removal of a large Venezuelan state-backed operation from its platform.[9] Given the role of the Venezuelan political crisis in international geopolitics, future work could examine how domestic and foreign actors are leveraging social media to

---

[5] https://www.nytimes.com/2019/01/13/world/americas/venezeula-juan-guaido-arrest.html.

[6] https://www.nytimes.com/2019/01/21/world/americas/venezuela-maduro-national-guard.html.

[7] https://www.theguardian.com/world/2019/jan/23/venezuela-protests-thousands-march-against-maduro-as-opposition-sees-chance-for-change.

[8] https://about.fb.com/news/2020/09/august-2020-cib-report/.

[9] https://blog.twitter.com/en_us/topics/company/2019/further_research_information_operations.html.

manipulate narratives for political objectives. To achieve this, we could use a more elaborate MVHP to model such actors as separate processes.

In summary, we have empirically demonstrated via our Venezuelan case study that PIMs, as introduced in Sect. 4.2, furnish an unambiguous and interpretable dynamic characterisation of intra-/inter-narrative influences. Furthermore, we have showcased how such dynamics may be explained by landmark events— exogenous to social media—within broader historical contexts. In this capacity, our work provides an additional, important lens for studying influence between narratives, which, we hope, may prove useful to social scientists.

Acknowledgments. This work was supported by the U.S. Defense Advanced Research Projects Agency (DARPA) Grant No. FA8650-18-C-7823 under the *Computational Simulation of Online Social Behavior (SocialSim)* program of DARPA's Information Innovation Office. Any opinions, findings, conclusions, or recommendations contained herein are those of the authors and do not necessarily represent the official policies or endorsements, either expressed or implied, of DARPA, or the U.S. Government. Finally, the authors would like to thank the manuscript's anonymous reviewers for their helpful comments and suggestions.

References

1. Alvari, H., Shakarian, P.: Hawkes process for understanding the influence of pathogenic social media accounts. In: 2019 2nd International Conference on Data Intelligence and Security (ICDIS), pp. 36–42 (2019). https://doi.org/10.1109/ICDIS.2019.00013
2. Blackburn, M., et al.: Corpus development for studying online disinformation campaign: a narrative + stance approach. In: STOC@LREC (2020)
3. Blackburn, M., Yu, N., Memory, A., Mueller, W.G.: Detecting and Annotating Narratives in Social Media: A Vision Paper. ICWSM, US, June 2020. https://doi.org/10.36190/2020.23
4. Calderaro, A.: Social media and politics. In: Outhwaite, W., Turner, S. (eds.) The SAGE Handbook of Political Sociology, 1st edn., 2v, chap. 44, pp. 781–796. Sage Publications Ltd., December 2017
5. Cryst, E., Ponce de León, E., Suárez Pérez, D., Perkins, S.: Bolivarian factions: Facebook takes down inauthentic assets. Technical report, Stanford's Freeman Spogli Institute for International Studies, September 2020
6. Eichler, M., Dahlhaus, R., Dueck, J.: Graphical modeling for multivariate hawkes processes with nonparametric link functions. J. Time Ser. Anal. **38**(2), 225–242 (2017). https://doi.org/10.1111/jtsa.12213
7. Farajtabar, M., Gomez-Rodriguez, M., Wang, Y., Li, S., Zha, H., Song, L.: COEVOLVE: a joint point process model for information diffusion and network co-evolution. In: Companion Proceedings of the The Web Conference 2018, WWW 2018, pp. 473–477. International World Wide Web Conferences Steering Committee, Republic and Canton of Geneva, CHE (2018). https://doi.org/10.1145/3184558.3186236
8. Hawkes, A.G.: Spectra of some self-exciting and mutually exciting point processes. Biometrika **58**(1), 83–90 (1971). https://doi.org/10.2307/2334319
9. Hindman, M.: The Myth of Digital Democracy. Princeton University Press, Princeton (2009)

10. Jost, J.T., et al.: How social media facilitates political protest: information, motivation, and social networks. Polit. Psychol. **39**(S1), 85–118 (2018). https://doi.org/10.1111/pops.12478
11. Jungherr, A.: Twitter use in election campaigns: a systematic literature review. J. Inf. Technol. Polit. **13**(1), 72–91 (2016). https://doi.org/10.1080/19331681.2015.1132401
12. Lai, E., et al.: Topic time series analysis of microblogs. IMA J. Appl. Math. **81**, hxw025 (2016). https://doi.org/10.1093/imamat/hxw025
13. Margetts, H., John, P., Hale, S., Yasseri, T.: Political Turbulence: How Social Media Shape Collective Action. Princeton University Press, Princeton (2015)
14. Mohler, G., McGrath, E., Buntain, C., LaFree, G.: Hawkes binomial topic model with applications to coupled conflict-Twitter data. Ann. Appl. Stat. **14**(4), 1984–2002 (2020). https://doi.org/10.1214/20-AOAS1352
15. Pires, T., Schlinger, E., Garrette, D.: How multilingual is multilingual BERT? In: Proceedings of the 57th Annual Meeting of the Association for Computational Linguistics, pp. 4996–5001. Association for Computational Linguistics, Florence, July 2019. https://doi.org/10.18653/v1/P19-1493
16. Sytnik, A.: Digitalization of diplomacy in global politics on the example of 2019 Venezuelan presidential crisis. In: Alexandrov, D.A., Boukhanovsky, A.V., Chugunov, A.V., Kabanov, Y., Koltsova, O., Musabirov, I. (eds.) DTGS 2019. CCIS, vol. 1038, pp. 187–196. Springer, Cham (2019). https://doi.org/10.1007/978-3-030-37858-5_15
17. Wang, Y., Agichtein, E., Benzi, M.: TM-LDA: efficient online modeling of latent topic transitions in social media. In: Proceedings of the 18th ACM SIGKDD International Conference on Knowledge Discovery and Data Mining, KDD 2012, pp. 123–131. Association for Computing Machinery, New York (2012). https://doi.org/10.1145/2339530.2339552
18. Xu, W., Liu, X., Gong, Y.: Document clustering based on non-negative matrix factorization. In: Proceedings of the 26th Annual International ACM SIGIR Conference on Research and Development in Information Retrieval, SIGIR 2003, pp. 267–273. Association for Computing Machinery, New York (2003). https://doi.org/10.1145/860435.860485
19. Zhang, W., Johnson, T.J., Seltzer, T., Bichard, S.L.: The revolution will be networked: the influence of social networking sites on political attitudes and behavior. Social Sci. Comput. Rev. **28**(1), 75–92 (2010). https://doi.org/10.1177/0894439309335162
20. Zhou, K., Zha, H., Song, L.: Learning social infectivity in sparse low-rank networks using multi-dimensional Hawkes processes. In: Carvalho, C.M., Ravikumar, P. (eds.) Proceedings of the Sixteenth International Conference on Artificial Intelligence and Statistics. Proceedings of Machine Learning Research, vol. 31, pp. 641–649. PMLR, Scottsdale, 29 April–01 May 2013. http://proceedings.mlr.press/v31/zhou13a.html

Network Structures and Humanitarian Need

Mackenzie Clark, Erika Frydenlund$^{(\boxtimes)}$, and Jose J. Padilla

Virginia Modeling, Analysis, and Simulation Center, Suffolk, VA, USA
efrydenl@odu.edu

Abstract. Financial transactions between humanitarian aid donors and recipients form a network that can inform our understanding of how resilient these efforts are collectively. This is particularly critical considering that certain humanitarian "crises" can span years or even decades, leading to donor fatigue in already chronically underfunded situations. We focus on refugee and migrant situations in Greece and Colombia to highlight the disparity between the humanitarian aid networks in responses that have each stretched beyond five years. We observe these financial networks over time to show that they grow through preferential attachment, with large nodes/actors remaining prominent as the years go on. Drawing on the insights from network resilience and foreign aid literature, we highlight the role that private donors play in Greece in providing some robustness to long-term humanitarian responses with no end in sight. Donor diversity appears to play some role in closing the funding gaps, even as need and funding requests for humanitarian response increase over time.

Keywords: Social network analysis · Network resilience · Humanitarian aid

1 Introduction

Early 2020 marked the beginning of the COVID-19 pandemic for most of the world. The lock-down measures and redirection of government and private financing to meet the demands of hospitals, frontline workers, and medical research for vaccines impacted everyone. Refugees and migrants, often already on the margins of society, were disproportionately impacted by these events [1]. And then, in September 2020 amidst lockdown and peaking numbers of COVID-19 cases, the largest refugee camp in Europe burned to the ground displacing over 12 thousand people [2].

At the heart of humanitarian response is funding, connecting donors and recipients in a web of action that allows those on the frontlines to meet emerging needs. Understanding these financial networks helps reveal political motivations and investments of donors, potential longevity and robustness of funding, and gaps in addressing changing needs. In this study, we look at the financial networks of two humanitarian response plans for refugees and migrants in Greece and Colombia. While the scale of migration has been similar, the funding amounts and donor/recipient network structures have been very different. Through these cases, we can infer insights about the role of private donors and the way these actors may be changing the humanitarian landscape.

© Springer Nature Switzerland AG 2021
R. Thomson et al. (Eds.): SBP-BRiMS 2021, LNCS 12720, pp. 214–223, 2021.
https://doi.org/10.1007/978-3-030-80387-2_21

The rise of private donors in the international humanitarian regime has contributed to bridging the funding gap for some of the most chronically underfunded refugee and migrant crises around the world. Over the last decade, private contributions to the United Nations High Commissioner for Refugees (UNHCR) have increased from less than 2% to 10% of the organization's total income [3]. And while studies of general foreign aid typically suggest this fragmentation of donorship from traditional government actors can cause ineffective aid allocation [4], networks of humanitarian aid may differ in this regard.

Historically, major international organizations like the UNHCR and the International Organization for Migration (IOM) govern the humanitarian response to migration: donors give to these organizations which in turn give to smaller organizations operating "on the ground." However, in a situation where need is rapidly changing, private donors and organizations with greater flexibility and fewer bureaucratic barriers may be key in addressing the dynamic nature of these crises [5]. Private donors have demonstrated their ability to contribute to some of the most complex, multifaceted humanitarian situations around the world. In this study, we find that Greece, where there is a higher concentration of private donors, has been more successful at closing the gap between funding requirements and actual funding received over a shorter period of time than in Colombia which relies primarily on governmental contributions.

2 Humanitarian Response as a Network

This study draws on two disciplines to guide the network analysis and understanding of donors and actors within the financial transaction network of humanitarian migration response in Greece and Colombia: network science and resilience, and international relations theory. Using this approach, we quantify characteristics of the networks that allow inferences about how they may adapt to changing needs over time.

2.1 Measuring Network Robustness

In these networks of aid, there are a variety of "shocks" that could disrupt funding and the long-term sustainability of humanitarian response. Examples include changing political contexts in the donor country or donor fatigue leading to funding reduction or withdrawal, and sudden increased need when disaster re-displaces a large number of people, as in the case of the refugee camp fire in Greece [2]. Diverse donors and recipients in an aid network may provide resilience against shocks such as these [6].

Physics and engineering research provide characteristics of network structures that have different strengths and weaknesses when confronted with shocks. Heterogenous networks where nodes/actors have unique and critical roles are resilient to failure of minor nodes, but suffer cascading failures when major nodes fail or are "attacked" [7–9]. While attacks of these magnitudes on very important nodes in the network are rare, it can be risky to rely heavily on a select few nodes or hubs to support the structure of the network [9], leading to collapse of certain parts or all of the network. This is the case in energy grids where major suppliers are attacked and other major nodes then get overloaded by demand, leading to failures across the network [9]. For networks to be resilient against these types of attacks, they would ideally have a higher degree of clustering and modularity [10], and would be more homogenous [7], meaning that all of the nodes perform the same general function and can be interchanged with another. This does not just apply to critical infrastructure or hypothetical networks, but has applications in humanitarian aid where clustering and homogeneity can maintain network structure even after the loss of a major donor [6].

As aid networks evolve over time, however, there is very little control over the structure. These networks tend to evolve through preferential attachment, with large, important nodes becoming more important over time as new nodes in the network connect to these major hubs—visually following a power law distribution. In this paper, we illustrate that in the case of humanitarian networks, another feature, donor diversification, can influence a migration response's ability to close funding gaps and possibly provide stability over the long haul required of humanitarian response.

2.2 The International Aid Regime

Studies on foreign aid echo the findings of network studies: often groups of donors that are too fragmented lead to ineffective aid allocation [4]. In aid operations, it is most effective to have a "lead donor" that coordinates the allocation of funds and prevents interests from colliding [11]. In humanitarian response, these organizations are often established brokers of aid, like UNHCR and IOM, that can align priorities and interests.

Humanitarian aid networks are becoming more complicated than a simple structure of donors led by UNHCR or IOM to allocate funding to organizations on the frontlines attending to human needs. In this context, large portions of the network, especially private donors, contribute directly to on-the-ground initiatives. This makes cooperation and collaboration a bit messier; now the "lead donors," or aid brokers like UNHCR and IOM get circumvented. But, this can, in some situations, promote a more effective, dynamic response [12]. If the organizations on the ground manage to self-organize in the complex crisis environment and adapt to fill gaps, this can create a more coherent, agile response to meet emergent needs, but there are barriers like trust, political and donor interest, and workload that can constrain cooperation [13].

Using network characteristics of humanitarian financial transactions in Greece and Colombia, we highlight their distinctness to infer how these networks would adapt and stand up to increasing needs over time. The next section discusses the network analysis methodology and measurements, followed by comparative analysis of the two countries' aid networks specifically for refugee and migrant humanitarian response.

3   Methodology

In this study, we use the social network analysis platform, Gephi, to analyze the donor-recipient relationships in two prominent refugee and migrant humanitarian situations in Greece and Colombia. Both countries experienced the major onset of their respective migrant arrivals in 2015. While they are alike in that each country struggles to meet the vast humanitarian need, the composition of their donor networks is distinct.

3.1 Data and Limitations

To construct the two networks of humanitarian aid, we utilize data from the Financial Tracking Service (FTS) of the United Nations Office for the Coordination of Humanitarian Affairs. This ensures that all of the financial transactions in the networks are funds that are designated for the specific humanitarian cause, as they must meet certain criteria to be included in the dataset [14]. In each of these cases, the financial transactions that constitute relationships (edges/links) between donors and recipients (nodes) are officially given specifically for refugee and migrant humanitarian response in each country. This can include support to both migrants and host populations.

In this study, we examine the Greece case from 2016–2019 and the Colombia case from 2017–2019, which are the years for each case when the most data are available from both the FTS dataset and UNHCR yearly budget reports (Table 1).

Table 1. Edges/number of transactions per year for Greece (99 nodes) and Colombia (112).

Year	Greece	Colombia
2016	48	N/A
2017	187	7
2018	47	36
2019	14	163
2020	N/A	166

To determine the level of funding that is being met each year, we rely on UNHCR yearly budget reporting since it is the overarching response organization for each country. While this is not representative of the entirety of each situation, it is the most comprehensive, consistent, and reliable data available for funding gaps in the two cases.

There are limitations of the data. First, it does not capture foreign aid more generally, or aid that is given without humanitarian intent, specifically for refugee and migrant support. Second, reporting funding contributions to the FTS is optional, so not all humanitarian aid transactions—especially those of small, privately funded operations—are reflected in this analysis. In the FTS dataset, many of these private donations are simply designated as coming from "Private Individuals and Organizations." For this reason, all of these transactions are represented by one aggregate private donor node in our analysis. We have reason to believe, based on qualitative fieldwork, that this contribution of private donors may be even larger than it appears in the Greece case [15]. Lastly, some years have sparser reporting than others, especially before a more formal response network had developed. While there are ways to expand these networks to create a more granular analysis of all the organizations involved through qualitative analysis of field reports and funding documents, these efforts are outside of the scope of this study and should be addressed in future work.

3.2 Measurements

Modularity. Modularity is a metric used to detect communities, or clusters of nodes, in a network. Because clustering plays a large role in network resilience [10], we use modularity to determine how the communities play a role in maintaining the network's structure as funding needs continue to grow.

Page Rank. Page rank accounts for the links of a node—direction, weight, and connectedness. This allows us to take into consideration the role that direction (donor to recipient) plays in transactions. because humanitarian aid is typically granted without condition of repayment, the flow of aid often only goes in one direction. Here, we weight the edges based on the amount of money in each transaction. Since page rank also accounts for weight, it provides better insight into how these factors contribute to the centrality of nodes in the network.

4 Humanitarian Aid Under Stress

The aid transactions in Greece and Colombia both centralize around a few prominent donors. However, we find that the network typology of the donors and the way this impacts network clustering diverges greatly. The biggest distinction is the types of donors that fall into each community. These distinct structures suggest how each would hold up to increasing humanitarian need.

4.1 Divergent Donor Typologies

The primary divergence in the two cases is the types of donors that are prominent and the communities that form around them. In Colombia, the major communities tend to center on prominent governmental nodes, such as the Government of the United States, the European Commission's Humanitarian Aid and Civil Protection Department (ECHO), and the Government of Germany. This is indicative of a traditional international government response to crisis. In Greece, however, the two major communities form around ECHO and private donors. The aid network in Greece points to the phenomenon that UNHCR is noticing: a rise in the contributions of private donors. What the network shows us beyond simply that they contribute more money, is that they are making significant contributions directly to the organizations on the ground.

In Greece, there are only six communities, but there is a clear distinction between the government (about 50% of the network) and private donor communities (about 37%). This network also has four smaller, less distinctive communities, each constituting between two and six percent of the network. The low clustering and distinct community clustering of these networks is particularly interesting considering that the humanitarian response in Greece consistently supports about six times as much funding as the Colombian network to serve far fewer refugees and migrants. Of the 365 transactions, the node representing private contributions is responsible for about 38%; ECHO is responsible for about 11%, but maintains influence through the amount of funds given in each transaction, rather than by number of transactions.

In Colombia, there are 12 communities of donors and recipient organizations, but these are weakly defined from one another. Eight of the communities each constitute three percent of fewer of the nodes in the network. In fact, most of the network (about 56%) are in the community where the US Government is the most prominent node. The other three communities also center around prominent government nodes (The European Commission, Germany, and Switzerland). Even nodes/organizations outside of the US community still rely primarily on the US for funding. The network structure shows a relatively high degree of connectivity for a network with 112 nodes (density = 0.091), possibly indicating a high degree of coordination between humanitarian response efforts on the ground in Colombia, as our ongoing qualitative research suggests.

Each of these networks exist in real-world humanitarian contexts that are not only complex and evolving, but are also long-lasting with no end in sight. From the data available, the years of response have only solidified these network characteristics, with preferential attachment to the larger nodes simply growing the existing communities (Table 2). Neither of these cases show a high degree of clustering that would theoretically make the network more resilient against a shock like primary donor withdrawal [10]. Their divergent network structures, however, suggest different ways they might respond to stress, such as an increase in humanitarian need.

4.2 Addressing Changing Needs

Network structure isn't the only difference between these two cases. Greece and Colombia look different in the amount of need they address and the amount of funding they have to work with. Based on UNHCR budget reports that show funding needs and actual funding received on a yearly basis and how the composition of donors changed in each of these years, we can infer how donor typology plays a role in addressing need.

Figure 1 shows each country's requested and actual funding for their respective crises over time. Colombia's financial requests are much lower than those in Greece, with a gap between request and actual funding holding steady at about 30–50% of funding needs left unmet each year. The US is consistently a prominent donor in Colombia's humanitarian response (page rank is 0.13 in 2019; 0.2 in 2020; see Table 3). Overall, this network does not change much in structure from year-to-year (Table 2, column 2), and consistently relies heavily on governmental donors. Colombia has been unable to close its funding gap year after year (Fig. 1). This gap exists despite the fact that Colombia hosts more refugees and migrants than Greece and asks for much less money.

Greece, on the other hand, has a donor composition that evolves more over time. When the funding gap is the largest in 2016 (Fig. 1), the network is composed of mostly governmental contributions (Table 2), primarily ECHO (page rank = 0.16, Table 3). Private donors are there, but have very little reach over the network (page rank = 0.03). When the funding gap starts to shrink in 2017, private donors (page rank = 0.11) are growing in prominence within the network (Table 2, row 2). ECHO maintains its position (page rank = 0.15), but now the two communities have their distinct presences in the network. In 2018, when the gap is the smallest, private donors have maintained a page rank of 0.11 and ECHO declined to 0.12. By this year, both private donors and governmental donors have become stable presences within the network with established

Table 2. Greece and Colombia humanitarian aid networks year-by-year, based on years where data were available. Colors represent communities/clusters.

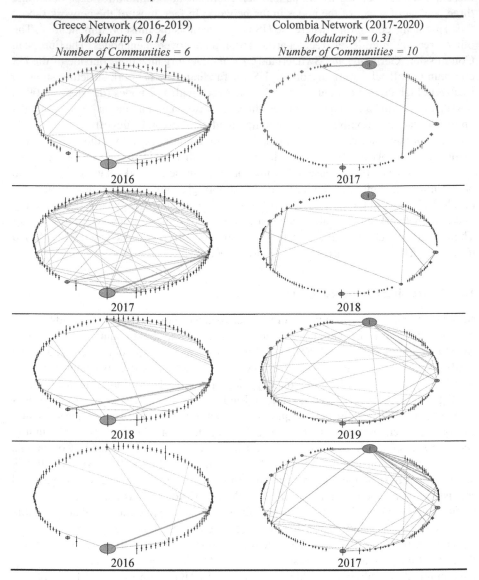

communities. Over these three years, the funding gap shrinks from over 35% of required funding unmet in 2016, to only about 3% in 2018.

Notice in Fig. 1 that requests and funding both go up over time, reflecting the increasing support required in each country to serve the growing needs of refugees and migrants. Greece, the more diverse network of private and governmental donors, was able to close much of the funding gap in that time; Colombia, on the other hand, has not. In reality,

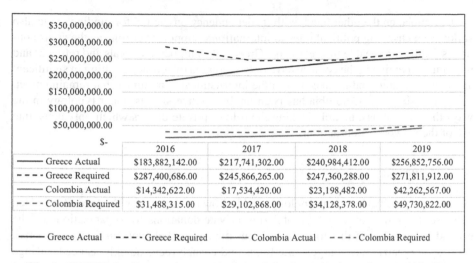

	2016	2017	2018	2019
—— Greece Actual	$183,882,142.00	$217,741,302.00	$240,984,412.00	$256,852,756.00
- - - - Greece Required	$287,400,686.00	$245,866,265.00	$247,360,288.00	$271,811,912.00
—— Colombia Actual	$14,342,622.00	$17,534,420.00	$23,198,482.00	$42,262,567.00
- - - - Colombia Required	$31,488,315.00	$29,102,868.00	$34,128,378.00	$49,730,822.00

—— Greece Actual   - - - - Greece Required   —— Colombia Actual   - - - - Colombia Required

Fig. 1. UNHCR funding requirements and actual funds received year-by-year [16, 17].

Table 3. Page rank statistics by year for top three donors

Greece					Colombia				
Node	2016	2017	2018	2019	Node	2017	2018	2019	2020
ECHO	0.16	0.15	0.12	0.16	United States	0.006	0.05	0.13	0.2
UNHCR	0.09	0.13	0.08	0.14	UNHCR	0.006	0.04	0.07	0.06
Private donors	0.03	0.11	0.11	0.01	Germany	0.006	0.05	0.04	0.01

the Greece network is even more diverse than it at first appears, since its major donor, ECHO, is actually a collection of governmental donors through the European Union.

The large role of private donors may protect against the effects of long-term donor fatigue that can result from long-lasting humanitarian response. Much of the private funding goes directly to organizations working on the ground in humanitarian response, where government money is often routed through international organizations like UNHCR or IOM. This allows the funding and these organizations to respond quickly to address emergent gaps in meeting needs that governments and other bureaucratic organizations struggle to do [5, 12]. The fact that these two major nodes in the Greece network are aggregate actors also suggests that they may be resilient to small changes in their own structure; that is, when private donors or EU member states choose to withdraw funding, but others may join, the overall funding contributions remain relatively stable. In this way, the network structure—both of two distinct communities and of aggregate actors— may create a robustness to donor fatigue over time and at least partially explain Greece's ability to close its funding gap despite the increasing budget requests over time.

Colombia, on the other hand, with its dependency on the US for funding could also suffer from changing political tides in international support over time as the crisis continues for the foreseeable future. Unlike Greece, the number of arrivals and refugees and migrants remaining in the country have only increased over time, resulting in significant and compounding funding needs from the international community that remain unmet. In the last six years, Colombia has been unable to diversify its donor typology in the ways that Greece has, notably lacking the role of private donors which only constitute 4% of the network's transactions.

5   Discussion

Greece's more diverse donor typology could allow needs, even in the face of a sudden increase, to be quickly met by mobilizing private donations for organizations on the ground. We have seen this to be true in our qualitative fieldwork in Greece on small "pop-up" organizations that generate funds and change focus quickly as needs change [12, 15]. On the other hand, the organizations are subject to the whims of the global public consciousness about certain crises and can lose momentum after the situation subsides from media or public view, even though they may last for years longer.

Colombia's donor and recipient network looks more "traditional," where a lead actor, in this case the US, is at the center of much of the humanitarian response. While this network is still very robust to small changes (e.g., private donors leaving the network, on-the-ground organizations that stop working in certain sectors), a change in the political or financial situation of the US could dramatically alter humanitarian response in Colombia. This brings into question the long-term stability of this financing network to meet the ongoing needs of migrants and the host country to manage displacement that has not stopped, even as COVID-19 led to border closures.

Future research will include looking into extended versions of these networks by including even smaller implementing organizations that work on-the-ground to provide even more granularity to these results, and will include the use of these additional sources of data to create a more comprehensive network analysis. Additional data of humanitarian funding trends over time is required to truly understand how these networks will hold up, which may become apparent as data during 2020 and 2021 become available, reflecting the stress of pandemic response in addition to humanitarian operations for migrants and refugees.

Acknowledgments. This research is funded by grant number N000141912624 by the Office of Naval Research through the Minerva Research Initiative; none of the views reported in the study are those of the funding organization.

References

1. OECD: What is the impact of the COVID-19 pandemic on immigrants and their children? OECD (2020)
2. Kingsley, P.: Fire Destroys Most of Europe's Largest Refugee Camp, on Greek Island of Lesbos. The New York Times. The New York Times Company (2020). www.nytimes.com

3. Executive Committee of the High Commissioner's Program: Private sector fundraiding and partnerships. UNHCR (2018)
4. Annen, K., Kosempel, S.: Foreign aid, donor fragmentation, and economic growth. B.E. J. Macroecon. **9**, 33–33 (2009)
5. Desai, R., Kharas, H.: The California consensus: can private aid end global poverty? Survival **50**, 155–168 (2008)
6. Clark, M.: Shifting sources of humanitarian aid: the importance of network resiliency and donor diversification. Master's Thesis. Old Dominion University (2020)
7. Motter, A.E., Lai, Y.C.: Cascade-based attacks on complex networks. Phys. Rev. E Stat. Nonlin. Soft Matter. Phys. **66**, 065102 (2002)
8. Crucitti, P., Latora, V., Marchiori, M.: Model for cascading failures in complex networks. Phys. Rev. E Stat. Nonl. Soft Matter Phys. **69**, 045–104 (2004)
9. Albert, R., Jeong, H., Barabási, A.L.: Error and attack tolerance of complex networks. Nature **406**, 378–382 (2000)
10. Ash, J., Newth, D.: Optimizing complex networks for resilience against cascading failure. Physica A **380**, 673–683 (2007)
11. Steinwand, M.C.: Compete or coordinate? Aid fragmentation and lead donorship. Int. Organ. **69**, 443–472 (2015)
12. Haaland, H., Wallevik, H.: Beyond crisis management? The role of citizen initiatives for global solidarity in humanitarian aid: the case of Lesvos. Third World Q. **40**, 1869–1883 (2019)
13. Seybolt, T.B.: Harmonizing the humanitarian aid network: adaptive change in a complex system. Int. Stud. Quart. **53**, 1027–1050 (2009)
14. Financial Tracking Service: Criteria for inclusion of reported humanitarian contributions into the Financial Tracking Service database, and for donor/appealing agency reporting to FTS (2004)
15. Frydenlund, E., Padilla, J.J., Haaland, H., Wallewik, H.: The rise and fall of humanitarian citizen initiatives: a simulation-based approach. In: SBP BRiMS 2020 (2020)
16. UNHCR Global Focus: Colombia (2021)
17. UNHCR Global Focus: Greece (2021)

Identifying Shifts in Collective Attention to Topics on Social Media

Yuzi He, Ashwin Rao, Keith Burghardt[(✉)], and Kristina Lerman

USC Information Sciences Institute, Marina del Rey, CA 90292, USA
keithab@isi.edu

Abstract. The complex, ever-shifting landscape of social media can obscure important changes in conversations involving smaller groups. Discovering these subtle shifts in attention to topics can be challenging for algorithms attuned to global topic popularity. We present a novel unsupervised method to identify shifts in high-dimensional textual data. By utilizing a random selection of date-time instances as inflection points in discourse, the method automatically labels the data as before or after a change point and trains a classifier to predict these labels. Next, it fits a mathematical model of classification accuracy to all trial change points to infer the true change points, as well as the fraction of data affected (a proxy for detection confidence). Finally, it splits the data at the detected change and repeats recursively until a stopping criterion is reached. The method beats state-of-the-art change detection algorithms in accuracy, and often has lower time and space complexity. The method identifies meaningful changes in real-world settings, including Twitter conversations about the Covid-19 pandemic and stories posted on Reddit. The method opens new avenues for data-driven discovery due to its flexibility, accuracy and robustness in identifying changes in high dimensional data.

Keywords: Change point detection · Confusion · Reddit · Twitter · Attention shift detection

1 Introduction

The ever-growing volume of information in online social media creates a competition among content producers for the limited attention of content consumers [11]. When coupled with the news cycle of media reporting, the attention competition creates a highly dynamic information environment, where some topics are widely discussed and frequently re-shared, while others languish in obscurity. Changes in these patterns suggest important events, which have been detected with generative models, such as dynamic topic models [6]. These approaches, however, focus on the most popular topics and may miss the subtler changes and shifts in attention that occur in the heterogeneous and dynamic information ecosystem of social media. These shifting patterns of attention hold a clue to what disparate

© Springer Nature Switzerland AG 2021
R. Thomson et al. (Eds.): SBP-BRiMS 2021, LNCS 12720, pp. 224–234, 2021.
https://doi.org/10.1007/978-3-030-80387-2_22

communities find important and can help better explain the complex dynamics of the information ecosystem.

We propose a change detection method called *Meta Change Detection* (MtChD), to detect change points, such as shifts of attention to topics in online conversations. This method is in contrast to previous papers on emerging topics (c.f., [2]) by finding subtle changes in the words and sentences discussed. While a number of change detection methods have been developed (c.f., [21]), they do not work well with high-dimensional data, such as text, due to large number of fitting parameters or high memory usage, and few provide confidence about the quality of segmentation of data before and after the change. Another challenge is that social media data is both massive and sparse: many people participate, but most contribute relatively little text. This sparsity introduces noise into change point estimation because changes could be due to shifts in active users rather than topics, and we know rather little about each person individually.

To address these challenges, MtChD uses "confusion" [22] and a novel model to estimate both when the change occurs and the fraction of data affected by change for any number of changes in the dataset. Confusion attempts to *confuse* a model by labeling the same state (before or after the change) with respective labels, even when these labels may be wrong [22]. Significant changes in the accuracy of predicting these labels indicate a potential change. In contrast to previous work, however, we detect changes as differences in accuracy compared to a null model (predicting the majority class). The novel mathematical model for accuracy also estimates the amount of data changed. This acts as a confidence proxy, something not often used in previous methods: we are more confident in a change point if a large fraction of subsequent data changes. Also, in contrast to Nieuwenburg et al. [22], data is then split at this change point and the method is repeated recursively until an arbitrary stopping point, allowing us to detect any number of change points, as well as the degree of confidence we have in each change.

We validate MtChD by applying it to high-dimensional data sets, both synthetic and real-world, to demonstrate that it can detect changes with higher accuracy than [22] and other state-of-the-art baselines, even on sparse and noisy data with missing values. We apply the proposed method to large public datasets of tweets about the Covid-19 pandemic [7] and scary stories posted in the nosleep subreddit (www.reddit.com/r/nosleep). We show that MtChD accurately infers meaningful shifts within Twitter conversations and Reddit stories.

The rest of the paper is organized as follows. We first review related work, and then present details of the confusion-based change detection method including a mathematical model that quantifies these changes. Finally, we present results for changes in synthetic and real datasets and discuss implications.

2   Related Works

Change detection, especially when the number of change points is unknown, usually involves three steps: choosing a cost function on how homogenous data is between change points, determining the change search method, and constraining the number of change points [21]. There is a wide variety of costs functions, from parametric (e.g., maximum likelihood of parametric distributions) to non-parametric (e.g., non-parametric maximum likelihood or kernel-based methods). Parametric methods started with CUMSUM [16], which is used to detect changes in the mean of data with a normal distribution, and have been advanced on with methods such as generalized likelihood ratio (GLR) test [20] and extensions of this technique (c.f., [4,21]). The GLR test seeks to reject a null assumption which states that observations before and after a proposed change point t_0 follow the same distribution, therefore when the null assumption is rejected, a change probably occurs. Parametric cost functions, however, often suffer from the curse of dimensionality and require data to follow a particular distribution, which does not necessarily hold in general. While semi-parametric [5] and non-parametric methods, such as non-parametric maximum likelihood estimation, have been developed, they can be computationally expensive on large or high-dimensional data [21]. Kernel methods have instead been introduced to describe distributions that are difficult to parameterize [3]. Finally, there are some methods that do not easily fit into these categories, such as hidden Markov models (HMM) [18] and Bayesian change detection methods [1,23].

A method that is similar to our change detection method is by Nieuwenburg et al. [22], in which a *confusion scheme* is used to distinguish phases in a physical system. More specifically, the classifier identifies the critical point $c \in (a, b)$ by testing candidate critical points $c' \in C'$. When c' is between (a, c), the classifier correctly labels data between (a, c') and but incorrectly labels data between (c', c) with a different label. We take advantage of this unexpected similarity between changes and phases to create MtChD, which also uses confusion, but is more robust to change points occurring near the edges of data streams. Furthermore, Nieuwenburg et al., [22] do not attempt to discover multiple phase changes (i.e., change points), while the present method can. Another method similar to ours is called Unsupervised Change Analysis [10] which also creates change labels. The paper, however, focuses on explaining changes and not on finding changes.

We improve on these previous methods in several critical ways. First, we offer an agnostic framework to detect change points through the accuracy variation of an arbitrary classifier. This can significantly reduce the time and space complexity of the method compared to using cost functions like kernels or search methods like dynamic programming [21], and the method is flexible to the type of data that it analyzes. By fitting classifier accuracy to a model, we also avoid significant costs of analyzing high-dimensional data. Due to memory complexity, MtChD can easily handle data with several thousand dimensions, while GLR or Kernel based methods can struggle with a few dozen. Moreover, the model can estimate how much of the data was modified after a change, which acts as a change confidence proxy. Finally, our method can outperform baseline methods with near-linear time complexity and near-constant memory complexity, which is a significant improvement over the competing methods.

3   Methods and Materials

Problem Statement. Assume we have data of the form $(X_i, t_i), i = 1, \ldots, n$, where X is an arbitrarily high dimensional vector and t is an external control parameter such as time. We refer to t as the *indicator* and look for a change point in indicator t. Assume there is a change at t_0 such that some data before the change and some after the change have different distributions. In many datasets, however, only a fraction of data, $0 \leq \alpha \leq 1$, may show observable changes. *Our goal is to infer the change point, t_0, and the fraction of data that undergoes a change, α, given the observations (X_i, t_i).* For clarity, just as t_0 varies with each change point, so does α.

Step 1: Confusion-Based Training. Similar to [22], we assume a trial change point $t = t_a$ and label the observed data before t_a as belonging to class $\tilde{y}_i = 0$ (no change), and the data after t_a as class $\tilde{y}_i = 1$ (change). We then train a classifier to predict the labels $\tilde{y}_i$ from the features X_i, of an arbitrary number of dimensions. We plot the accuracy of the classifier as a function of trial change point t_a. In case a true change point exists in the observed range of t, the accuracy vs. t_a curve will significantly increase over the baseline accuracy, which is the majority class ratio of labels $\tilde{y}_i$. The shape of the curve will be affected both by the actual change point, t_0, and the fraction of data points affected by change, α. The classifier one can use could be anything – we use random forest and multi-layer perceptrons as examples in this paper. For each candidate change point, t_a, classifiers are trained on a random subsample of 50% of data, validated on 30%, and tested on 20%. The test set is used to judge the accuracy of the learned models for each t_a. This step is known as confusion-based training.

Accuracy as a function of t_a varies significantly. Near the beginning and end of the dataset, accuracy is nearly 1 because we can almost certainly say that data is before (if $t_a \ll t_0$) or after (if $t_a \gg t_0$) a change point. More specifically, if we have a candidate critical point near the beginning or end of the data, almost all of the data would be after or before the critical point respectively. Accuracy predictably decreases away from the extremes, but, peaks around the true change point, thus forming a W shape [22].

Step 2: Modeling Acc. vs. t_a Curve. The novelty of our work is to model this accuracy curve in order to infer t_0 and α. This natural extension of the previous work provides substantial improvements in change point estimation. We first define the cumulative distribution of t, $F(t) = 1/T \sum_i t_i < t$. Data can fall into three categories (or three distinguishable distributions), a distribution that does not change, S_u, which comprises $1 - \alpha$ of all data, a distribution before a change $(t \leq t_0)$, S_0, and a distribution after this change $(t > t_0)$, S_1. We do not know these distributions *a priori* but we assume the trained classifier will be able to distinguish these distributions using data X.

Assume that the distribution of t is independent of the discrete distribution that a data point belongs to a distribution that does not change after the change point, S_u, or changes from a distribution S_0 to a distribution S_1.

For clarity, we never have to know what distribution each data point lies in, nor even what the distribution looks like. For confusion, recall we label data as 0 if $t' \leq t$ and 1 otherwise. Given candidate change point t_a, $P_{S_{u,0}} = (1 - \alpha)F(t_a)$ of data in S_u is labeled 0 and $P_{S_{u,1}} = (1 - \alpha) - P_{S_{u,0}}$ is labeled 1. On top of this, for a data point in S_u, the expected predicting accuracy should be $\frac{1}{1-\alpha} \max(P_{S_{u,0}}, P_{S_{u,1}})$. Similarly, we can calculate the ratio of data labeled as 0 or 1 in S_0 and S_1. With real change point locate at t_0, given any t, we assume that among α fraction of data affected by change, $\theta(t - t_0)$ fraction of data belongs to S_1 and $1 - \theta(t - t_0)$ fraction of data belongs to S_0. Here $\theta(\cdot)$ is the Heaviside step function. We can calculate for S_1, which has fraction $\alpha(1-F(t_0))$, $P_{S_{1,1}} = \max\left(\alpha(F(t_a) - F(t_0)), 0\right)$ and $P_{S_{1,0}} = \alpha(1-F(t_0)) - P_{S_{0,1}}$. The expected predicting accuracy for S_1 is thus $\frac{1}{\alpha(1-F(t_0))} \max(P_{S_{1,0}}, P_{S_{1,1}})$. Finally, S_0 has a fraction of $\alpha F(t_0)$. The total fraction of data labeled "0" in S_0 and S_1 is $\alpha F(t_a)$, $P_{S_{0,0}} = \alpha F(t_a) - P_{S_{1,0}}$, therefore the proportion of S_0 incorrectly labeled "1" is $P_{S_{0,1}} = \alpha F(t_0) - P_{S_{0,0}}$. The expected predicting accuracy for data point in S_0 is then $\frac{1}{\alpha F(t_0)} \max\left(P_{S_{0,0}}, P_{S_{0,1}}\right)$.

We then leverage the results above to estimate the accuracy curve as $\tilde{A}cc(t_a) = \max\left(P_{S_{u,0}}, P_{S_{u,1}}\right) + \max\left(P_{S_{0,0}}, P_{S_{0,1}}\right) + \max\left(P_{S_{1,0}}, P_{S_{1,1}}\right)$. These variables only depend on $F(t)$, which can be directly estimated from data, and the free parameters t_0 and α. We therefore do not need explicit knowledge of distributions S_0, S_1 and S_u. To estimate t_0 and α, we can do a fast grid search such that the squared difference between the estimated and actual accuracy is minimized, and find data closely aligns with this model.

Multiple Changes. To identify multiple changes, we use recursive binary splitting. We first use the change detection method to find a change point, and split at this point. This, in turn, creates two subsets of data from which we can find additional changes, and split this data, in recursion. We stop splitting a node when we hit the minimum length of range t_c or maximum depth of the binary tree D. The time complexity is $O(TD)$ for binary segmentation depth, D, and number of data points, T. Because D is fixed to a small value, such as 3, the splitting process is almost linear in time. The space complexity only depends on the classifier used, so it can be efficient even in high-dimensional datasets. Relevant code pertaining to our analysis has been made publicly available through GitHub[1].

4 Data

Online Discussions About Covid-19. We apply our method to a large dataset of Covid-19 tweets [7]. This dataset consists of 115M tweets from users across the globe, collected since January 21, 2020. These tweets contain at least one of a predetermined set of Covid-19-related keywords (e.g., coronavirus, pandemic, Wuhan, etc.). Since this dataset provides geolocation data for only 1% of the users, we leverage a fuzzy matching approach [12] to geolocate users within

---

[1] https://github.com/yuziheusc/confusion_multi_change.

the US. We want to understand the significant shifts in attention during the earliest era of Covid-19, from January 21 until March 31, 2020, of which 7.6 million tweets are geo-located to within the US using methods by Chen et al., [7]. We then subsample 200K tweets at random each month, for a total of 600K tweets, to simplify our analysis. Text is pre-processed through removal of stopwords, links, account names, and special characters (e.g., !?%#). Only English language tweets are considered. We then use the tf-idf vectorizer (with 2.2K terms) from Python's *scikit-learn* library [17] in order to generate the tf-idf vectors.

Reddit Stories. We extract Reddit posts from a popular horror story writing subreddit called *nosleep* using the Python Reddit API Wrapper (PRAW). We focus on posts created between January 1, 2019 and June, 2020 to understand both seasonal changes in stories (e.g., Halloween and Christmas), as well as changes in stories since the Covid-19 pandemic, creating 35.4K stories. Data pre-processing includes removing posts labeled "[removed]" and "[deleted]". Text cleaning and tf-idf vectorization (with 25K terms) follow the same methodology as in the Twitter dataset.

5 Experiments and Results

Synthetic Data. To test the change detection method, we generate two-dimensional data located in a unit box in a chessboard pattern with $n_c \times n_c$ squares, where n_c is a tunable parameter, as shown in the left panel of Fig. 1. These data are uniformly distributed in red squares when $t \leq t_0$, and green squares for $t > t_0$, with $0 \leq t \leq 1$. Our motivation for this synthetic example is that as n_c increases, it becomes harder to distinguish the change in data. Therefore, the data is not meant to be realistic but simple to construct with a change detection task difficulty controlled by n_c. For first part of this experiment, we set $t_0 = 0.5$, the size of the data $N = 8K$, and n_c equal to either 2, 6, or 10. For second part of this experiment, we fix $n_c = 6$ and we vary t_0 between 0.2 and 0.8. If t_0 differs from 0.5, the population before and after the change will be unbalanced, which makes the task of inferring t_0 more challenging [14].

Fig. 1. The figure on the left illustrations of synthetic data, where observations have two features x_1 and x_2. Data lies in red squares before changes at t_0 and moves to green squares after t_0. In the illustration the data is an $n_c \times n_c$ grid, with $n_c = 6$; The 3 figures on the right shows word cloud for Covid-19 tweets in period 01/21–01/30, 01/30–02/04 and 02/04–02/11. (Color figure online)

We ran six trials for our method and competing algorithms. For MtChD and vanilla confusion [22], 50% of the data is used as training, 30% for validation, and 20% for testing, where training and validation data is randomly sampled during each trial. Validation data is used to tune hyperparameters of the classifiers used. In competing methods, we randomly sample 70% of data in each trial, except for Bayesian change detection, where 18.8% of data (around 1.5K) is sampled due to computational limitations (this method takes much longer than other competing methods).

The methods tested are as follows: the vanilla confusion method of Nieuwenburg et al. [22], GLR, dynamic programming segmentation (DP) with different loss kernels [19], and Bayesian detection with different priors and likelihood functions [1]. While alternative segmentation methods can be used besides DP, such as binary segmentation [8], bottom up methods [13] and window based methods [21], DP is found to outperform these alternatives (not shown). For Bayesian change point detection, a conditional prior on change point and a likelihood function needs to be defined. We used uniform and geometric distributions as priors, and applied the Gaussian, individual feature model [23], and full covariance model [23] as likelihood functions.

The results are shown in Table 1. We see that for $n_c = 2$, optimal segmentation methods ($DP+RBF$) perform as well as ours. Vanilla confusion performs well when change point is in the center of data ($t_0 = 0.5$). Otherwise our method outperforms competing methods, especially when the change point is different from 0.5. Of the two classifiers used by MtChD, random forest performs best.

Online Discussions About Covid-19. Now that we show our method works well for single-change detection, we move on to more complicated multi-change detection in empirical data, with results shown in Table 3. We start by identifying shifts in tweets about Covid-19 (embedded into tf-idf vectors), where the word cloud of hashtags for the first three periods are shown in the right panels of Fig. 1. We ran binary segmentation using MtChD with a Random Forest classifier, maximum segmentation depth of three and minimum time length between changes set to four days. The dates of the change points identified by the method are listed in Table 3 . The time intervals between changes match with the period of a typical news cycle [15], which is between 5 to 9 days. Results are robust in the way that when the minimum length is increased to of 5 days, a subset of changes (01/30, 02/11, 02/16, 02/21 and 02/28) are found. Next, we analyze the discovered change points and interpret the findings by highlighting topics that shift the collective attention. To validate the results, we compare the change points found with the news events, as shown in Table 2.

Table 1. Comparison of performance of MtChD and competing methods. The entries show the mean change point $\mu(t_0)$ in comparison to the actual change point t_0. *(RF)* and *(MLP)* correspond to using a random forest and multilayer perceptron classifiers for confusion. Row 3 and row 4 show performance of naive confusion and GLR test. Row 5 to row 7 shows performance of DP, a non-approximate segmentation method, while other segmentation methods that perform worse are not shown (see main text). The cost functions used are *RBF* (RBF kernel), *L1* (L_1 loss function), and *L2* (L_2 loss function). The last four rows are for Bayesian change detection with a *uniform* prior or *Geo* (geometric) prior. *Gaussusian* stands for Gaussian likelihood function, *IFM* is the individual feature model and *FullCov* is the full covariance model. Bold values indicate change points that are closest to the correct value.

		2×2 $t_0 = 0.5$	10×10 $t_0 = 0.5$	6×6 $t_0 = 0.2$	6×6 $t_0 = 0.8$
MtChD (RF)	$\mu(t_0)$	**0.500 ± 0.003**	**0.496 ± 0.005**	**0.195 ± 0.005**	**0.802 ± 0.002**
	$\mu(\alpha)$	**0.949 ± 0.008**	**0.66 ± 0.02**	**0.65 ± 0.03**	**0.66 ± 0.03**
MtChD (MLP)	$\mu(t_0)$	**0.503 ± 0.003**	0.58 ± 0.06	0.56 ± 0.05	0.5 ± 0.1
	$\mu(\alpha)$	**0.96 ± 0.01**	0.009 ± 0.008	0.005 ± 0.004	0.02 ± 0.02
Confusion (RF)	$\mu(t_0)$	**0.497 ± 0.002**	**0.4973 ± 1E−4**	0.23 ± 0.04	0.54 ± 0.09
GLR	$\mu(t_0)$	**0.5003 ± 4E−4**	0.6 ± 0.3	0.24 ± 0.04	0.81 ± 0.03
DP+RBF	$\mu(t_0)$	**0.5002 ± 4E−4**	0.3 ± 0.2	0.4 ± 0.3	0.84 ± 0.07
DP+L2	$\mu(t_0)$	0.95 ± 0.01	0.5 ± 0.4	0.4 ± 0.3	0.3 ± 0.4
DP+L1	$\mu(t_0)$	0.957 ± 0.007	0.4 ± 0.3	0.6 ± 0.4	0.2 ± 0.3
Uniform+Gaussian	$\mu(t_0)$	0.5 ± 0.2	0.5 ± 0.2	0.6 ± 0.3	0.5 ± 0.3
Uniform+IFM	$\mu(t_0)$	0.997 ± 0.003	0.998 ± 0.003	0.999 ± 0.002	0.999 ± 0.001
Uniform+FullCov	$\mu(t_0)$	**0.4985 ± 2E−4**	0.9989 ± 9E−4	0.99 ± 0.01	0.997 ± 0.004
Geo+Gaussian	$\mu(t_0)$	0.028 ± 0.004	0.028 ± 0.004	0.033 ± 0.006	0.025 ± 0.004

Reddit Stories. We also applied our method to horror stories posted on reddit.com, the subreddit *r/nosleep*, with stories embedded using tf-idf. We find variations in the topics of stories, such as Jul 17 to September 25, 2019 ("camping" and "summer") appear, reflecting recreation activities in the US. The next change on September 25 to November 17 ("halloween") signals the topic of Halloween and November 17 to January 2nd ("santa" and "christmas"), corresponds to the holidays. Potentially inspired by Covid-19 restrictions, there were stories about "quarantine" from March 29 to May 4, 2020. Finally, quarantining became old news again, and discussions shifted in the final months until June 2020 back to stories on "rules". As a baseline, we used GLR and DP with an RBF kernel. Due to the limitations of memory, we first perform truncated SVD [9] to transform the tf-idf vector into a 64-dimensional vector. Then we down-sampled to 8K observations from the full dataset since dynamic programming runs in $O(T^2)$. We find that not only is our method able to process the full dataset, it can find more physically meaningful change points.

Table 2. Change points automatically identified in Covid-19 tweets and important events occurring on those dates.

Change point		Events
Date	α	
01–30	0.355	First confirmed case of person-to-person transmission of the "Wuhan Virus" in the US
02–04	0.341	Diamond Princess cruise ship quarantined. Ten people on cruise ship near Tokyo have virus
02–11	0.327	WHO announced official name for "COVID-19"
02–16	0.243	More than 300 passengers from the Diamond Princess are traveling in the US chartered planes
02–21	0.441	1st Covid-19 death in Italy (02–22)
02–28	0.366	First Covid-19 death in US (02–29)
03–04	0.447	California declares state of emergency. South Korea confirms 3 new deaths and 438 additional cases of novel coronavirus
03–09	0.269	Italy lockdown; Grand Princess cruise ship docks in Oakland
03–15	0.303	First lockdown orders in parts of California; national emergency declared (3/13)
03–24	0.146	US sees deadliest day with 160 deaths

Table 3. Comparison with baseline change detection. (Left) Tweets and (right) r/nosleep.

Covid-19 tweets			Reddit stories		
Our result	GLR	DP+RBF	Our result	GLR	DP+RBF
01-30-20	02-07-20	01-27-20	03-10-19	03-26-19	04-10-19
02-04-20	02-08-20	01-28-20	06-05-19	06-03-19	04-12-19
02-11-20	02-08-20	01-31-20	07-17-19	08-11-19	11-06-19
02-16-20	02-08-20	02-13-20	09-25-19	11-05-19	01-13-20
02-21-20	02-09-20	02-15-20	11-17-19	12-20-19	01-29-20
02-28-20	02-09-20	02-26-20	01-02-20	01-30-20	02-19-20
03-04-20	02-17-20	02-29-20	02-21-20	03-02-20	03-10-20
03-09-20	02-17-20	03-02-20	03-29-20	04-03-20	03-31-20
03-15-20	02-17-20	03-07-20	05-24-20	04-09-20	04-07-20
03-24-20	02-27-20	03-13-20			

6   Conclusions

In this paper, we aim to identify and understand the shifts of conversation on social media. In contrast to emergent topic detection, which detects new topics of interest, our method identifies when the distribution of features in high-dimensional streams of text changes. We create a method to robustly detect multiple changes within these conversations which appear to represent intuitive and realistic changes in conversations. Moreover, quantitative and qualitative comparisons to baseline methods show improved detection of changes. Our method has a unique feature – it allows us to quantify the fraction (parameter α) of data which shows observable changes. This parameter can be interpret as the significance of a certain change.

There are, however, important limitations of our approach. First, multiple changes are found with a simple binary segmentation, which is only meant to find approximate change points [21]. While this allows us to dramatically speed up computation, it may compromise on accuracy. Next, the social media data we explore has no ground truth about changes except for daily news. So we cannot assess whether our method, or competing methods, correctly found all change points. This may affect conclusions about what are the most important changes within social media early in the Covid-19 pandemic. A more detailed analysis involving tweets from different languages and from across the globe would be a promising candidate for future research.

These limitations, however, point to promising future work. For example, it will be important to explore advancing on the binary segmentation approach in order to sacrifice some potential speed for greater accuracy or precision. Next, we should compare against realistic data with a fixed number of known change points to determine the overall accuracy of this method. Finally, these results should be extended to other high-dimensional datasets, including video.

Acknowledgments. This work was funded in part by DARPA (W911NF-17-C-0094 and HR00111990114) and AFOSR (FA9550-20-1-0224).

References

1. Adams, R.P., MacKay, D.J.: Bayesian online changepoint detection. arXiv preprint arXiv:0710.3742 (2007)
2. Alkhodair, S.A., Ding, S.H., Fung, B.C., Liu, J.: Detecting breaking news rumors of emerging topics in social media. Inf. Process. Manag. **57**(2), 102018 (2020)
3. Arlot, S., Celisse, A., Harchaoui, Z.: A kernel multiple change-point algorithm via model selection. JMRL **20**(162), 1–56 (2019)
4. Barber, J.: A generalized likelihood ratio test for coherent change detection in polarimetric SAR. IEEE GRSL **12**(9), 1873–1877 (2015)
5. Bardet, J.M., Kengne, W.C., Wintenberger, O.: Detecting multiple change-points in general causal time series using penalized quasi-likelihood. arXiv preprint: arXiv:1008.0054 (2010)
6. Blei, D.M., Lafferty, J.D.: Dynamic topic models. In: ICML, pp. 113–120 (2006)
7. Chen, E., Lerman, K., Ferrara, E.: Tracking social media discourse about the COVID-19 pandemic: development of a public coronavirus Twitter data set. JPHS **6**(2), e19273 (2020)
8. Fryzlewicz, P., et al.: Wild binary segmentation for multiple change-point detection. Ann. Stat. **42**(6), 2243–2281 (2014)

9. Halko, N.: Finding structure with randomness: stochastic algorithms for construct-
    ing approximate matrix decompositions. arXiv:0909.4061 (2009)
10. Hido, S., Idé, T., Kashima, H., Kubo, H., Matsuzawa, H.: Unsupervised change
    analysis using supervised learning. In: Washio, T., Suzuki, E., Ting, K.M.,
    Inokuchi, A. (eds.) PAKDD 2008. LNCS (LNAI), vol. 5012, pp. 148–159. Springer,
    Heidelberg (2008). https://doi.org/10.1007/978-3-540-68125-0_15
11. Hodas, N.O., Lerman, K.: How limited visibility and divided attention constrain
    social contagion. In: SocialCom 2012 (2012)
12. Jiang, J., Chen, E., Yan, S., Lerman, K., Ferrara, E.: Political polarization drives
    online conversations about COVID-19 in the United States. Hum. Behav. Emerg.
    Technol. **2**, 200–211 (2020)
13. Keogh, E., Chu, S., Hart, D., Pazzani, M.: An online algorithm for segmenting
    time series. In: ICDM, pp. 289–296. IEEE (2001)
14. Leichtle, T., Geith, C., Lakes, T., Taubenböck, H.: Class imbalance in unsupervised
    change detection: a diagnostic analysis from urban remote sensing. Int. J. Appl.
    Earth Obs. Geoinf. **60**, 83–98 (2017)
15. Leskovec, J., Backstrom, L., Kleinberg, J.: Meme-tracking and the dynamics of the
    news cycle. In: KDD, pp. 497–506 (2009)
16. Page, E.S.: Continuous inspection schemes. Biometrika **41**(1–2), 100–115 (1954)
17. Pedregosa, F., et al.: Scikit-learn: machine learning in Python. J. Mach. Learn.
    Res. **12**, 2825–2830 (2011)
18. Raghavan, V., Galstyan, A., Tartakovsky, A.G.: Hidden Markov models for the
    activity profile of terrorist groups. Ann. Appl. Stat. **7**(4), 2402–2430 (2013)
19. Rigaill, G.: A pruned dynamic programming algorithm to recover the best segmen-
    tations with 1 to K_max change-points. J. de la Société Française de Stat. **156**(4),
    180–205 (2015)
20. Siegmund, D., Venkatraman, E.: Using the generalized likelihood ratio statistic for
    sequential detection of a change-point. Ann. Stat. **23**(1), 255–271 (1995)
21. Truong, C., Oudre, L., Vayatis, N.: Selective review of offline change point detection
    methods. Sig. Process. **167**, 107299 (2020)
22. Van Nieuwenburg, E.P., Liu, Y.H., Huber, S.D.: Learning phase transitions by
    confusion. Nat. Phys. **13**(5), 435–439 (2017)
23. Xuan, X., Murphy, K.: Modeling changing dependency structure in multivariate
    time series. In: IMLS, pp. 1055–1062 (2007)

wapr.tugon.ph: A Secure Helpline for Detecting Psychosocial Aid from Reports of Unlawful Killings in the Philippines

Maria Regina Justina E. Estuar[1]([✉]) [iD], John Noel C. Victorino[2],
Christian E. Pulmano[1], Zachary Pangan[1], Meredith Jaslyn B. Alanano[1],
Jerome Victor C. Celeres[1], John Loyd B. de Troz[1], Marlene M. De Leon[1],
Yvonne McDermott Rees[3], Riza Batista-Navarro[4], and Lucita Lazo[5]

[1] Ateneo de Manila University, Quezon City 1108, Philippines
{restuar,cpulmano}@ateneo.edu, {zachary.pangan,meredith.alanano,
jerome.celeres,john.detroz}@obf.ateneo.edu
[2] XForm Inc., Quezon City 1108, Philippines
xform@tugon.ph
[3] Swansea University, Swansea, UK
[4] University of Manchester, Manchester, UK
[5] World Association for Psychosocial Rehabilitation - Philippines,
Manila, Philippines
https://wapr.tugon.ph/

Abstract. Availability and affordance of information communications technology has provided additional medium to monitor human rights violation. Reporting, extraction, collection and verification of reports through natural language processing and machine learning techniques can now be integrated into one system. The processed narratives becomes readily available for response, monitoring, interventions, policy-making and even as evidence in court. This paper discusses the design and development of wapr.tugon.ph, a block-chain enabled NLP-based platform that provides a simple yet effective way of reporting, validating and securing human rights violation reports from victims or witnesses. wapr.tugon.ph allows for SMS-based and web-based reporting of human rights violation. Reports are processed for detection of emotions using NRCLex, and behaviors using Stanford Parser and modified Multi-Liason algorithm from narratives which serve as input to assess wellness. A total of 5,418 records were obtained from Reddit's subreddits and HappyDB corpus to serve as baseline corpora for our model. Our best psychosocial wellness detection model produced an accuracy and F1 score of 84% on validation set ($n = 1,426$) and 87% on test set ($n = 666$). An ethereum private blockchain is implemented to record all transactions made in the system for authenticity tracking. Findings underscore the importance of providing a system that assists in determining the appropriate psychosocial intervention to victims, families and witnesses of human rights violation. Specifically, the study contributes a framework in embedding a combined sentiment and behavior model that outputs: sentiments that are used to assess mental wellness, behaviors that are used to assess

© Springer Nature Switzerland AG 2021
R. Thomson et al. (Eds.): SBP-BRiMS 2021, LNCS 12720, pp. 235–244, 2021.
https://doi.org/10.1007/978-3-030-80387-2_23

physical needs, and detection of wellness that serves as input to refer victims, families of victims and witnesses to appropriate agencies.

Keywords: Human rights · Natural language processing · Psychosocial intervention

1 Introduction

The 1987 Constitution of the Republic of the Philippines incorporates, in Article III, a Bill of Rights. The Bill of Rights includes many established fundamental rights, such as the rights to life, liberty, and equality before the law (Sec. 1); the right to be protected against unlawful searches and seizures (Sec. 2), and due process rights (Secs. 12 and 14). These rights are also protected by international human rights treaties, principally the International Covenant on Civil and Political Rights and the International Covenant on Economic, Social and Cultural Rights, both of which have been ratified by the Philippines.

Since the Government launched its campaign against illegal drugs in 2016, the prevalence of extra-judicial killings and other serious human rights violations, including breaches of due process rights and attacks against journalists and human rights defenders, have been well-documented. The United Nations Office of the High Commissioner noted in 2020, that official government figures indicate that there were at least 8,663 killings related to the anti-illegal drugs campaign since 2016, with the likely real death toll to be much higher [14]. In 2018, the Supreme Court of the Philippines noted that there had been a 'total of 20,322 deaths during the Duterte Administration anti-drug war from July 1, 2016 to November 27, 2017, or an average of 39.46 deaths every day' and that this unusually high number of deaths required a deeper understanding of the impact of the anti-drug war legislation and its impact on fundamental constitutional rights [26]. In October 2020, the United Nations Human Rights Council passed a resolution urging the Philippines to 'ensure accountability for human rights violations and abuses, and in this regard to conduct independent, full, and transparent investigations and to prosecute all those who have perpetrated serious crimes, including violations and abuses of human rights' [6].

Traditional psychosocial intervention poses complications in distance, timing, scarcity of specialists, cost and stigma [4]. There is a need to study alternative methods that will provide a secure and safe environment for reporting and managing psychosocial interventions. As such, accessible technology that facilitates the submission of reports, process to determine sentiments and behaviors that aid in detecting the wellness as well as appropriate psychosocial intervention is deemed relevant.

The World Association for Psychosocial Rehabilitation (WAPR), Philippines is an affiliate organization of the global organization of the same name. WAPR is one such non-government organization that advocates for the rights of those with mental health conditions and seeks to be a technical resource with the capacity to provide evidence-based information on strategies for promoting mental health and providing psychiatric and psychosocial interventions. Its programs consist of providing technical assistance to selected communities in order to develop capacity for delivering mental health services at the community and providing direct assistance to individuals and communities in the form of livelihood and skills training projects, and care farming projects. Psychosocial intervention requires identification of individuals requiring mental health and psychosocial support, referral of psychosocial and mental health clients for psychiatric treatment, and direct assistance to individuals and groups requiring psychosocial assistance.

There is much evidence in the development of models that detect mental health concerns formed from narratives social media and other data collection platforms. However, there are a few that discuss its integration in an actual platform. The study situates itself in addressing the need of WAPR for a secured and NLP-based system to monitor and manage reports on human rights violation.

2   Review of Related Literature

2.1   Origins of wapr.tugon.ph

wapr.tugon.ph is the third generation of a crowdsourcing platform developed for participatory governance. In 2015, the eBayanihan platform was launched to address the need for an automated crowdsourcing platform to assist responders in the collection and verification of reports extracted in twitter platform [9]. The system included building blocks for bilingual sentiment analysis including understanding behavior to twitter users to differentiate victims from observers [19], topic modeling [31], data visualization [15], validation of crowdsourced reports [32], and agent-based modeling [10]. In 2017, the platform was used as a crowdsourced validation platform for agos.ph [28]. Using elements of the eBayanihan framework, a version of tugon.ph, wapr.tugon.ph was designed to address collection of crowdsourced and volunteer information for human rights violation reporting and monitoring.

2.2   Providing Psychosocial Intervention During Crisis

The concept of psychosocial intervention materialized in the early 1990s with the intention of shifting the focus from the vulnerability of the concerned party to the emphasis of their resiliency [5]. Assessment is used to survey the effects of the crisis to the affected individual where questions are framed to determine the best form of intervention. Selected interventions are then mapped and training is provided to communities who will assist in psychosocial support. Monitoring and evaluation are regularly performed to measure the effectiveness of the intervention to the psychosocial well-being of the affected person [13]. Psychological first aids are designed to accommodate the variety of trauma survivors' reactions, whether physical, psychological, behavioral, cognitive, spiritual, and other responses [23]. Specifically, behavior is observed to determine if help is needed. At the same time, emotions are also monitored to measure signs of emotional distress [23]. Psychosocial interventions to victims and witnesses of human rights violation require a multidimensional approach to ensure that all aspects of the person are diagnosed properly and appropriate interventions are applied. Assessment should also be performed in a timely manner and preferably at carefully determined periods in time for continuity of care. However, physical meetings are hampered by fear, anxiety and inability to travel to areas where psychosocial interventions are available. Remotely delivered interventions can then be an alternative solution to geographical barriers [4].

2.3 Digital Psychosocial Intervention

Digital psychosocial intervention for victims of crime, family members and friends of victims of crime and witnesses of crime have been implemented in various mediums [3]. Psychological and emotional assistance, legal assistance, financial assistance, counseling, and referral to other appropriate services are some of the interventions offered. In the Philippines, the Commission on Filipinos Overseas (CFO) and Philippine Overseas Employment Administration (POEA) offer free online legal counseling issues concerning human trafficking and OFWs, respectively [1,24]. A mobile application for crime reporting has also been launched by the Philippine National Police [27]. As for victims and/or families of victims of EJKs specifically, there are psychosocial support services provided by NGOs, however, they are not delivered online. As such, there is opportunity to develop a human rights violation reporting platform that also provides initial assessment to provide possible psychosocial interventions. In addition, support services given online have been increasing with the improvement of technology. These tools bridge gaps in several barriers that consumers encounter, such as problems in distance, timing, scarcity of specialists, cost and stigma [22]. Nonetheless, the study on the efficacy of psychosocial interventions delivered digitally is still limited as various resources focus more on the effects of psychological support more than societal support [22].

2.4 Natural Language Processing Text Classification for Detecting Elements of Psychosocial Wellbeing

Several studies present applications for text classification modeling using NLP and machine learning methods including detecting sensitive comments in YouTube videos [8], detect behavioral information from Facebook text data [18] and predicting digital health interventions [11] indicating that NLP frameworks have potential in therapeutic decision-making. Methods include development of lexicons derived from answered questionnaires [17] through preprocessing, boundary detection, phrase matching, detection of pronouns and negation, and score computation. Another approach explores the features of the text data such as unigrams, pronoun counts, punctuation counts, average sentence length, and word sentiments [30] before feeding into classification algorithms. In an information extraction architecture, part-of-speech (POS) tagging allows for the syntactic categorization of words. This method allows for the identification and

extraction of relationships between words in a narrative. The Stanford Parser
is one example that determines typed dependencies between words using POS
tagging. The Stanford Parser uses an enhanced English Universal Dependencies representation of words to find implicit relations between words. On the
other hand, decision tree methods are one of the machine learning techniques
that establish classification systems and prediction algorithms. Decision tree
algorithms classifies data as nodes or branches stemming from one root node.
Decision tree methodologies are advantageous to use alongside ensemble learning
methods because, given enough data, it is able to explore and represent a larger
representation of functions [7].

2.5 Securing Human Rights Violation Reports and Data Admissibility Through Blockchain

Information on unlawful killings and related human rights violation are deemed
to contain personal, privilege and sensitive information. Therefore, trust and
confidence to these reporting tools should be established aside from addressing
the need for remote submission, screening and processing of reports. Blockchain
is a distributed network that aims to provide data verifiability and traceability
by sharing copies of a ledger of transactions to multiple participants. Records are
immutable which means that data are considered final and cannot be tampered
with once it is already committed to the blockchain [12]. Distributing multiple
copies of the same ledger and having multiple participants constantly verify it
makes it free from tampering of records. This feature of blockchain makes it an
appropriate option for storing data that needs to be secured, transparent and
verifiable. Blockchain can be used for evidence admissibility by being able to
provide computer-generated record receipts from the data that are stored in the
blockchain [12].

3   Framework and Methodology

The wapr.tugon.ph system has three main components: (1) the SMS and web-
based platform that is used for submission, verification and validation of human
rights violation reports, (2) the models that are used to classify and extract
the sentiments, behavior and wellness from the reports and (3) the blockchain
component that handles the securing of transactions.

3.1   warp.tugon.ph Architecture

wapr.tugon.ph is built using a combination of web technologies. It uses Ruby on
Rails (Rails) as the programming framework. HTML, Javascript, CSS were used
for the front-end technologies. PostgreSQL was used for the database. Python
and other Python libraries were used to develop the models for sentiment analysis, behavioral analysis, and wellness analysis. These are accessible in a separate
server via an HTTP REST service. Transactions logs are being submitted to a

private Ethereum blockchain network for the decentralization and verification of records. A smart contract, written in Solidity, is deployed in the blockchain to handle the incoming transaction logs. This smart contract is being invoked every time new records come in. All servers run on Amazon Web Services's (AWS) Elastic Compute Cloud (EC2).

3.2   Psychosocial Wellness Detection Model

Dataset Description. The psychosocial wellness detection model used datasets from Reddit, and HappyDB corpus [2]. Subreddits *r/depression, r/Anxiety, r/survivorofabuse, r/abuse r/CPTSD* and *r/rape* contributed the posts from February 14, 2020 until February 14, 2021. Three annotators labeled these posts using the list of posttraumatic stress reactions from the Psychological First Aid Field Operations Guide [23]. A total of 2,376 posts were annotated as "unwell" posts while invalid posts were discarded. On the other hand, 2,376 records were randomly picked from the HappyDB corpus. It is a corpus of 100,000 happy moments crowdsourced from workers in Amazon's Mechanical Turk [2]. A train-validate-test split divided the 4,752 combined datasets. The training and validation of the model were composed of 3,326 and 1,426 records, respectively. The test set had a total of 666 records, with 317 posts from a different subreddit (*r/abuse*), and 349 records from the HappyDB corpus.

Model Architecture. An ensemble binary classification approach was implemented to predict if a given narrative was "well" or "unwell". The ensemble consisted of doc2vec, emotion, and behavior sub-models. Text data were cleaned and vectorized using the doc2vec method within the GenSim library to convert texts to vectors [29]. All three submodels used doc2vec to transform the text data. Then, a linear SVM model with default hyperparameters from the Scikit-learn library [25] was used to train a binary wellness classifier. This base configuration was considered as the doc2vec sub-model.

The emotion sub-model used the transformed data from doc2vec to produce an emotion vector. The NRCLex Python package approximates fear, anger, anticipation, trust, surprise, sadness, disgust, joy, positive and negative from the text [21]. Instead of the whole doc2vec output, the emotion sub-model used the emotion vector to train the linear SVM model.

On the other hand, the behavior sub-model extracted behaviors from the vectorized text data. The Stanford CoreNLP toolkit [20] extracted the sentence dependency and the part-of-speech tags. Then, a modified Multi-Liason algorithm [16] obtained the behaviors from the verbs and adjectives with their respective dependencies. Similar to the rest of the sub-models, a linear SVM model applied the extracted behaviors.

Finally, a majority rule among the three sub-models was applied to decide the final label. Figure 1 summarizes the ensemble binary classification approach.

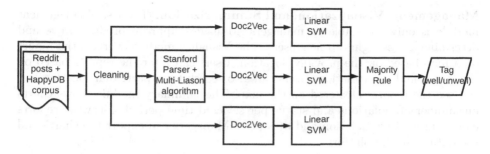

Fig. 1. Ensemble binary classification

4 Results and Discussion

4.1 Features and Functionalities of wapr.tugon.ph

The system offers various features to manage reports. The platform allows for different types of users (victims, witnesses, volunteer reporters, verifiers and managers) and corresponding access level. The platform consists of Reports, User Management, Visualize, and Summaries modules.

SMS and Web-Based Reporting Module. The Reports module is the main module of the platform. A registered user can create a new report via the web interface or via SMS. Reporters are limited to submitting and updating reports assigned to them. A higher level access, verifiers use the system to update the contents of the report after conducting offline verification. Verifiers can only update the status of the received report from open to verified or invalidated. Managers view verified reports and reclassifies the report to: in-progress or resolved after processing reports to produce sentiments, behavior and wellness. Reports that have been used for referral are re-classified as resolved.

Fig. 2. Sample report

The analyze button provides managers access to the prediction model server. This function sends the report to the prediction model server. Afterward, the analysis outputs the emotions and behaviors with corresponding scores and wellness classification. Figure 2 shows an example of an analyzed report.

Management, Visualization, and Summarization. The User Management module is only accessible to managers. Managers approve or disapprove and determine access rights to new users. The Visualize module overlays the report on a map based on the provided location. Users can filter the reports based on the status and the violation. Users can also search for a particular location and can see the reports clustered on the map. The Summary module shows charts on number of violations and status per selected time period. Likewise, reports can be filtered by violation and by status. Managers can export the charts and the data from the platform.

Securing the Information Through Blockchain. The creation and update of a report trigger the blockchain component of the platform. The central platform submits new and updated reports to the blockchain network. Figure 2 also shows the blockchain transaction ids. Aside from the blockchain network, the platform provides an audit trail for a report. Users can view previous versions of the report and compare them with the latest version.

4.2   Psychosocial Wellness Detection Model Performance

The ensemble binary classification model considered the whole narrative, the emotions, and behaviors found in the narrative. The doc2vec, behavior, and emotion sub-models produced accuracy scores of 92%, 75%, and 72% respectively, on the validation set. Finally, the ensemble approach produced an accuracy and f1 score of 84% on the validation set and 87% on the test set. Results show acceptable performance of the sub-models as well as the ensemble model.

5   Conclusion

Most studies in making meaning out of qualitative data focus and stop at finding the best NLP and machine learning techniques that improves model accuracy. This study continues where most people stop by implementing an ensemble model into a full platform. The design and implementation of human rights violation reporting tool also takes into consideration the authenticity of reports through technologies that enable secured transactions and crowdsourced validation methods. wapr.tugon.ph allows registered communities to verify and validate submitted reports by adding additional sources of similar information. Once a report has been submitted, a block for that particular report is created. All transactions that are related to that report is recorded in the block as additional transactions thereby creating an audit trail. wapr.tugon.ph platform provides analysis of narratives through identification of sentiments and behaviors and detection of wellness that serve as input to determination of appropriate psychosocial intervention. The automation and implementation of these models in a secured human rights violation reporting platform offer hope in assisting humanitarians who work on normalizing lives of victims of unlawful killings, facilitating in psychosocial aid and intervention, especially in areas where challenge in performing traditional aid remain at large.

References

1. Philippine legal employment administration. http://legalassistance.poea.gov.ph/. Accessed 10 Aug 2020
2. Asai, A., et al.: HappyDB: a corpus of 100,000 crowdsourced happy moments. arXiv preprint arXiv:1801.07746 (2018)
3. Action for the Rights of Children (ARC): Victims counselling and support service. https://www.raq.org.au/services/victims-counselling-and-support-service. Accessed 10 Aug 2020
4. Bragesjö, M., Larsson, K., Nordlund, L., Anderbro, T., Andersson, E., Möller, A.: Early psychological intervention after rape: a feasibility study. Front. Psychol. **11**, 1595 (2020)
5. Action for the Rights of Children (ARC): ARC - Foundation Module 7: Psychosocial support. SIDA, Swedish Agency for International Development Cooperation European Commission Humanitarian Aid Department (2009)
6. United Nations General Assembly Council: Report of the human rights council, forty-fifth session, 14 September–7 October 2020, un doc. a/75/53/add. United Nations General Assembly (2020)
7. Dietterich, T.: Ensemble methods in machine learning, pp. 1–15, January 2000
8. Dinakar, K., Reichart, R., Lieberman, H.: Modeling the detection of textual cyberbullying. In: Fifth International AAAI Conference on Weblogs and Social Media (2011)
9. Estuar, M.R.E.: Ebayanihan platform. https://ebayanihan.ateneo.edu. Accessed 16 Apr 2021
10. Estuar, M.R.J.E., Rodrigueza, R.C., Victorino, J.N.C., Sevilla, M.C.V., De Leon, M.M., Rosales, J.C.S.: Agent-based modeling approach in understanding behavior during disasters: measuring response and rescue in ebayanihan disaster management platform. In: International Conference on Social Computing, Behavioral-Cultural Modeling and Prediction and Behavior Representation in Modeling and Simulation, pp. 46–52. Springer (2017)
11. Funk, B., et al.: A framework for applying natural language processing in digital health interventions. J. Med. Internet Res. **22**(2), e13855 (2020)
12. Guo, A.: Blockchain receipts: patentability and admissibility in court. Chi.-Kent J. Intell. Prop. **16**, 440 (2016)
13. Hansen, P.: Psychosocial Interventions: A Handbook. International Federation Reference Centre, International Federation of Red Cross and Red Crescent Societies (2014)
14. United Nations High Commissioner for Human Rights: Report of the united nations high commissioner for human rights, situation of human rights in the Philippines, un doc. a/hrc/44/22. https://www.ohchr.org/Documents/Countries/PH/Philippines-HRC44-AEV.pdf. Accessed 29 June 2020
15. Isla, J.T., Estuar, M.R.E.: Real-time visualization of disaster behavior. In: 2014 International Conference on IT Convergence and Security (ICITCS), pp. 1–4 (2014). https://doi.org/10.1109/ICITCS.2014.7021804
16. Jivani, A., Jivani, G.: The multi-liaison algorithm. IJACSA Int. J. Adv. Comput. Sci. Appl. **2**, 130–134 (2011)
17. Karmen, C., Hsiung, R.C., Wetter, T.: Screening internet forum participants for depression symptoms by assembling and enhancing multiple NLP methods. Comput. Methods Programs Biomed. **120**(1), 27–36 (2015)

18. Katchapakirin, K., Wongpatikaseree, K., Yomaboot, P., Kaewpitakkun, Y.: Facebook social media for depression detection in the Thai community. In: 2018 15th International Joint Conference on Computer Science and Software Engineering (JCSSE), pp. 1–6. IEEE (2018)
19. Lee, J.B., Ybañez, M., De Leon, M.M., Estuar, M.R.E.: Understanding the behavior of Filipino Twitter users during disaster. GSTF J. Comput. (JoC) **3**(2) (2014)
20. Manning, C.D., Surdeanu, M., Bauer, J., Finkel, J., Bethard, S.J., McClosky, D.: The Stanford CoreNLP natural language processing toolkit. In: Association for Computational Linguistics (ACL) System Demonstrations, pp. 55–60 (2014). http://www.aclweb.org/anthology/P/P14/P14-5010
21. Mohammad, S.M., Turney, P.D.: Crowdsourcing a word-emotion association lexicon. Comput. Intell. **29**(3), 436–465 (2013)
22. Morland, L.A., Greene, C.J., Rosen, C.S., Kuhn, E., Hoffman, J., Sloan, D.M.: Telehealth and ehealth interventions for posttraumatic stress disorder. Curr. Opin. Psychol. **14**, 102–108 (2017)
23. National Child Traumatic Stress Network and National Center for PTSD: Psychological first aid field operations guide. https://www.cidrap.umn.edu/sites/default/files/public/php/146/146_guide.pdf. Accessed 15 Dec 2020
24. ABS-CBN News: 'itanong mo kay ato' now online, August 2015. https://news.abs-cbn.com/global-filipino/08/04/15/itanong-mo-kay-ato-now-online. Accessed 10 Aug 2020
25. Pedregosa, F., et al.: Scikit-learn: machine learning in Python. J. Mach. Learn. Res. **12**, 2825–2830 (2011)
26. Supreme Court of the Philippines: Supreme court of the Philippines (2018). petitioners V. Dela Rosa et al. https://lawphil.net/sc_res/2018/pdf/gr_234359_2018.pdf. Accessed 03 Apr 2018
27. Philippine National Police: Crime reporting app. https://pnp.gov.ph/news/463-pnp-launches-crime-reporting-mobile-app. Accessed 10 Aug 2020
28. Rappler: Agos platform. https://agos.rappler.com/respond. Accessed 16 Apr 2021
29. Rehurek, R., Sojka, P.: Gensim-python framework for vector space modelling. NLP Centre, Faculty of Informatics, Masaryk University, Brno, Czech Republic 3(2) (2011)
30. Saleem, S., et al.: Automatic detection of psychological distress indicators and severity assessment from online forum posts. In: Proceedings of COLING 2012, pp. 2375–2388 (2012)
31. Santos, J.S.: Summarization algorithms performance for topic clustered twitter microblogs. Ateneo de Manila University
32. Victorino, J.N.C., Estuar, M.R.J.E., Lagmay, A.M.F.A.: Validating the voice of the crowd during disasters. In: Xu, K.S., Reitter, D., Lee, D., Osgood, N. (eds.) Social, Cultural, and Behavioral Modeling, vol. 9708, pp. 301–310. Springer, Cham (2016). https://doi.org/10.1007/978-3-319-39931-7_29

Human and Agent Modeling

Social-Judgment-Based Modeling of Opinion Polarization in Chinese Live Streaming Platforms

Yutao Chen[✉]

The University of Chicago, Chicago, IL 60637, USA
cytwill@uchicago.edu

Abstract. Live streaming platforms have become popular media for young people in China to interact with others. While the synchronized interaction with live streamers through comments is said to make users feel more engaged than simply watching a video, it also brings the risk of online violence when the words become highly negative. By integrating the social-judgment-based opinion dynamic model (SJBO) with other theories like information overload and entertainment motivation, I modeled the opinion evolution during live streaming to analyze the opinion polarization. When no extremists exist initially, the simulation results reveal the possibility of opinion polarization when the audiences have low thresholds for repulsion. When positive and negative extremists are involved initially, the results suggest that their proportions among the audiences are responsible for the ratios of the positive or negative polarization of the ordinary audiences, which verifies the rationale of controlling the number of extremists to reduce opinion polarization. The impacts from other model parameters are also discussed in the research.

Keywords: Live streaming · Opinion polarization · Social judgment theory · Agent-based modeling

1 Introduction

Since the middle of the 2010s, live streaming has become increasingly popular in China's social media. Users can interact with the streamers by sending comments during the live. Many Chinese live-streaming platforms also allow the viewers to send virtual gifts to the hosts when watching or gain membership in the hosts' channel. These behaviors require audiences to pay for virtual gifts and fandom membership, bringing substantial income to the hosts and the platforms. Therefore, live streaming also grows to be a commercial mode, attracting more people, especially youngsters, to become part of the industry due to its relatively low threshold to enter and promising profitability.

A significant difference between the live streaming platforms and traditional social media is the synchronicity of viewers' interactions with the streamer.

© Springer Nature Switzerland AG 2021
R. Thomson et al. (Eds.): SBP-BRiMS 2021, LNCS 12720, pp. 247–256, 2021.
https://doi.org/10.1007/978-3-030-80387-2_24

All the audiences watching the same live streaming could send out their comments concurrently during the process, and they can view each other's comments scrolling on the screen. While some research suggests the synchronized interactive mode improves audiences' engagement [7], it might trigger problems like online violence towards the hosts during the streaming if extremists or anti-fans guide the opinions [9]. Besides, the insulting or discriminatory language used in extreme comments would impact the mental health of the hosts in the long term and give harmful instructions to the teenagers, the primary audiences of live streaming, on how they should behave in online communities.

Based on these concerns, I proposed a model to explore the conditions and possible reasons that cause opinion polarization in live streaming. The model is developed by integrating cognitive theories, and the properties of synchronized interactions into the social judgment theory-based (SJBO) model [1], an agent-based model describing continuous opinion dynamics. Section 2 illustrates the mathematical formulation and corresponding theories. The simulation setup and the measurements of opinion polarization are proposed in Sect. 3, and Sect. 4 discusses the results.

2   Model Formulation

Before introducing the model, I would notify the assumptions of the live streaming setting: the structure of the network between the audiences is assumed to be fully connected and static during the live streaming, which means each audience can see all others' comments; the content of live streaming is quantified by a triangular distribution and is assumed not to be impacted by the audience's opinions as most hosts would prepare their contents before and make sure the live streaming progresses well. The basic model to simulate the updates of the audiences' opinions is the SJBO model proposed in [1]. It is developed on the framework of the Bounded Confidence (BC) model [3] by adding the scenario of opinion repulsion, or Boomerang effect [6] from the social judgment theory [4]. The model specifies the opinion of each agent i at any time t (x_i^t) as continuous values. The update rules indicate that if the opinions of two agents are close enough, they tend to assimilate with each other; if the distance is too large, they tend to repulse each other; otherwise, the opinions would remain unchanged. The audiences are impacted by the live streaming content and other people's opinions simultaneously. Therefore, we use X_i^t to represent the overall external opinion that agent i perceives at time t. The mathematical form of the model is shown below:

$$x_i^{t+1} = \begin{cases} x_i^t + f\left(x_i^t, X_i^t\right)\left(X_i^t - x_i^t\right) & i \neq host \\ Triang\left(l, m, 1\right) & i = host \end{cases}, \tag{1}$$

$$f\left(x_i^t, X_i^t\right) = \begin{cases} \frac{\alpha\left(1-\left|x_i^t\right|\right)}{2}, & \left|x_i^t - X_i^t\right| < \varepsilon \\ 0, & \varepsilon \leq \left|x_i^t - X_i^t\right| \leq r \\ \frac{-\beta\left(1-\left|x_i^t\right|\right)}{2}, & \left|x_i^t - X_i^t\right| > r \end{cases}, \tag{2}$$

where $f\left(x_i^t, X_i^t\right)$ refers to the kernel function defined in the family of the continuous opinion dynamic models [2]. The formula of $\left(1 - |x_i^t|\right)/2$ is used in the original paper to constrain the opinion within the range $[-1, 1]$ after repulsion. α, β are the coefficients of assimilation and repulsion between $[0, 1]$, and ε, r are the thresholds for assimilation and repulsion, defined between $[0, 1]$ and $[1, 2]$, respectively. l and m are the lower limit and the mode of the live streaming content values.

As mentioned above, audiences' opinions are influenced by both the host and other audiences' opinions, so we need to specify how these two information sources contribute to the updates of audiences' opinions differently. The content of live streaming fully impacts the audiences' opinions when no comments are sent. As the comments increase, audiences' attention would be distracted by them, and the streaming content would have less impact as a result of information overload [8,12]. Besides, when too many comments scroll on the screen, it is unrealistic for audiences to capture all of them. Thus, I define that for each time t, agents can only interact with $\min(k, |S^t| - 1)$ randomly picked comments, where $|S^t|$ denotes the total number of comments at t including the live streaming content. We can formulate the process as below.

$$X_i^t = \sum_{j \in S_i^t} w_{ij}^t x_j^t, \tag{3}$$

$$w_{ij}^t = \begin{cases} b(|S^t|-1)^\lambda & j = host \\ \frac{1 - b(|S^t|-1)^\lambda}{\min(|S^t|-1, k)} & j \neq host \end{cases}, \tag{4}$$

where S_i^t is the subset of comments (including the streaming content) picked by agent i at time t. x_j^t is the opinion from agent j that agent i interacts with at t. b is the attention base for the live streaming, and λ is the distraction coefficient defined in $[0, 1]$, where a larger value means being more distracted. w_{ij}^t denotes the attention weight that agent i assigns to the perceived opinion from agent j at t.

Audiences usually do not always send out their comments, so we need to design a mechanism to make the agents express their opinions sometimes rather than constantly. The theory of entertainment motivation suggests that online behaviors like leaving comments or posting tweets can bring them relaxation and pleasure, which drive them to engage in online communities actively [10]. I applied this theory to determine whether the agents would send a comment at time t by defining a Bernoulli variable τ as following.

$$\tau_i^t = Bernoulli\left(\delta\right), \tag{5}$$

$$S^t = \left\{i | \tau_i^t = 1\right\}, \tag{6}$$

where δ is the parameter of entertainment motivation for audiences, which also denotes their probability of sending comments. The streamer is always active, so $\delta_{host} = 1$. S^t denotes the set of agents who send comments (including the streaming content) at time t.

3   Simulation Setup and Measurements

In all simulation tests, the population N is set to be 500, and the total number of time steps is 600^1. The streaming content generated by the hosts is positive by law requirement and usually attractive, so I set $m = 0.75, l = 0.5$ for all scenarios. Besides, previous studies [5] suggest that parameters α, β are insensitive to the polarization tendencies, so I set them to be 0.5 for all scenarios.

The influence from extremists is found important in previous studies of opinion polarization [2,3,5], thus Scenario 2 is designed to understand their impacts. According to [11], extremists can be characterized and quantified with the following rules: *having more extreme opinions than ordinary people:* 0.8 is set to be the threshold of the absolute values of extreme opinions according to [3], and the absolute values of extremists are in $[0.8, 1]$; *being more difficult to change their opinions to a milder direction:* extremists have lower assimilation threshold ($\epsilon_e < \epsilon$) and lower repulsion threshold ($r_e < r$); *being easier to take actions to advocate their opinions:* extremists have higher entertainment motivation ($\delta_e > \delta$).

Two simulation scenarios are designed to separately test the impact of agents' inner state and the introduction of extremists. In Scenario 1, all agents have the same parametric settings and can be described as pseudo-extremists (having ϵ, r values like extremists but other conditions unlike), conservatives, versatiles, or moderatists based on their values of ϵ, r (see Fig. 1). The agents have initial opinions uniformly drawn from $[-0.8, 0.8]$. In Scenario 2, the agents are divided into ordinary audiences and extremists. The initial opinions of positive extremists are uniformly drawn from $[0.8, 1]$, and negative extremists' opinions are drawn from $[-1, -0.8]$. $\epsilon_e = 0.1$ and $r = 1.1$ is set for all extremists. Ordinary audiences have the same initial opinion distribution as in Scenario 1 and have $\epsilon = 0.5$, $r = 1.5$, and $\delta = 0.02$ to mimic the overall state of a non-extreme audience in reality.

To measure the polarization of the opinions at the end of a simulation, the Polarization Rate (PR) is proposed, denoting the ratio of non-extreme audiences that have extreme opinions ($|x_i^T| > 0.8$). PR could be decomposed to positive polarization rate (PPR) and negative polarization rate (NPR). Besides, I also calculated the average final opinion value of the audiences in some cases to help explain the results. All statistics are averaged through 30 times of simulations.

4   Results and Discussion

4.1   Scenario 1

According to the polarization tendency shown in Fig. 2, the pseudo-extremists reveal a strong tendency of bi-polarization. Based on the simulation process, we can foresee that more agents would have extreme opinions in the following time

---

[1] Qualitative polarization tendencies are shown after 600 steps in most cases, and a fixed period also ensures the validity of the comparison between different parametric settings.

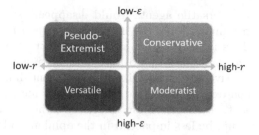

Fig. 1. Agent categorization based on thresholds of assimilation and repulsion

units. This phenomenon verifies the quantification of the extremists in Scenario 2 and suggests that ordinary audiences with an inner condition of a pseudo-extremist are likely to become true extremists later. The conservatives would not go to the extreme since the agents are the most difficult to change their opinions either by assimilation or repulsion. Especially for those whose initial opinions are negative than -0.5, their opinions remain unchanged. The versatile state also makes the agents' opinions polarized. However, the ratio of positive polarization and negative polarization looks imbalanced. Most of the agents have positive opinions, which is likely caused by the impact of the streaming content. The moderatists show a strong tendency of opinion convergence and also the absence of extreme opinions. In particular, their final state opinion is close to 0.75, which is similar to the content value, which indicates that the streamers would have more impact on audiences' opinions when the audiences are more moderate.

The extent of polarization might change according to the variation of other parameters. For example, when $k = 2$, the versatiles would not have positive polarization (see Fig. 3), but instead, have a positive convergence like in the moderatist state. But in general, as ϵ increases (moderatist and versatile), the impact of the streaming content become more significant as the majority of the agents would be assimilated to a level close to the content values; as r increases (conservative and moderatist), agents are less likely to go extremes. These trends remain stable in different parametric settings. Also, the result here indicates that it is still possible to develop audiences of polarized opinions without extremists at the beginning.

We also see that agents are less likely to grow extreme opinions as parameter k increases from the experiment results. Agents need to have inner conditions closer to pseudo-extremists to finally have extreme opinions as k increases (see Fig. 3). Moreover, the polarization ratio also decreased as k increases for agents of the same ϵ, r. This is in line with our intuition that when people interact with more opinions, they tend to develop balanced opinions rather than go extreme. Parameters b, λ, δ determine how much attention that agents assigned to the content. In Fig. 3, as the attention for the streamer increases (larger b, smaller λ), the agents need lower repulsion thresholds to develop extreme opinions in the end. Especially, I found that the positive polarization that happened in pseudo-extremist or versatile agents would disappear (see Fig. 4) while the negative polarization remains when the attention assigned to the hosts is high ($b = 0.7, \lambda = 0.3$). This indicates that having more attention to the live contents might rationalize some initially positive audiences but not help to reduce the generation of negative extremists like haters or anti-fans, who can easily go extreme from a non-extreme negative opinion. As δ increases, more concurrent comments are sent, so the attention assigned to the live content would be decreased. Thus it should be less impactful in the opinion updates. In the figure of the final average opinion (see Fig. 5), we can see that as the state of the agents become moderate, the value of the average opinion becomes higher and closer to the live content value (brighter color), indicating an increasing convergence to the streamer. As δ increases, the overall brightness in the heatmap decreases, suggesting a decline in the impact of the streamer.

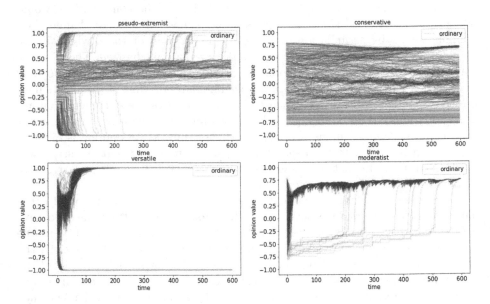

Fig. 2. Polarization tendency of different agent states ($k = 1, \delta = 0.02, b = 0.6, \lambda = 0.4$): pseudo-extremist: $\epsilon = 0.1, r = 1.1$, conservative: $\epsilon = 0.1, r = 1.9$, versatile: $\epsilon = 0.9, r = 1.1$, moderatist: $\epsilon = 0.9, r = 1.9$

4.2 Scenario 2

For this Scenario, the analysis focuses on the polarization under different proportions of positive and negative extremists. The result shows the overall trend is consistent with our intuition (see Fig. 6[2]). When the percentage of positive extremists increases, the PPR value becomes larger (brighter color), suggesting more ordinary agents have extreme positive opinions and vice versa.

---

[2] The heat-maps for NPR show the opposite trend, omitted due to the page limit.

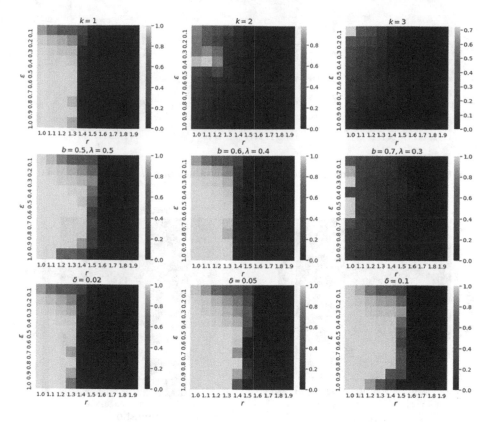

Fig. 3. Top: Polarization rate with different values of k ($\delta = 0.02, b = 0.6, \lambda = 0.4$); Middle: Polarization rate with different values of b, λ ($k = 1, \delta = 0.02$); Down: Polarization rate with different values of δ ($k = 1, b = 0.6, \lambda = 0.4$)

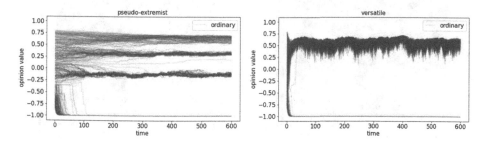

Fig. 4. Polarization tendency of pseudo-extremist and versatile agents with high attention to the live content ($k = 1, \delta = 0.02, b = 0.7, \lambda = 0.3$)

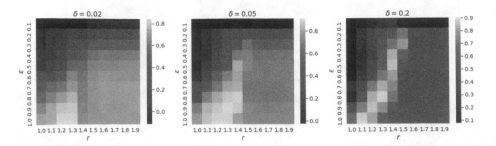

Fig. 5. Average final opinion value with different values of δ ($k = 1, b = 0.6, \lambda = 0.4$)

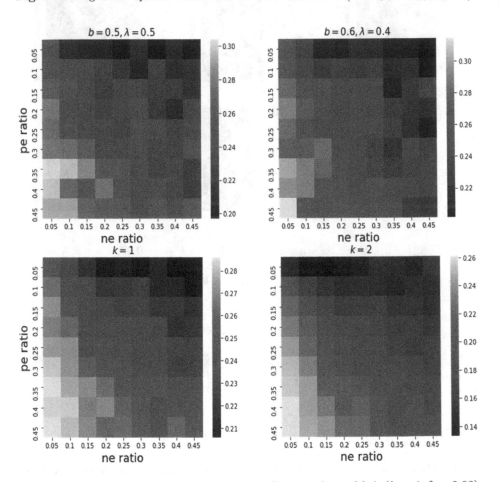

Fig. 6. Top: Positive polarization rate with different values of b, λ ($k = 1, \delta = 0.02$); Down: Positive polarization rate with different values of k ($b = 0.6, \lambda = 0.4, \delta = 0.02$), pe and ne in the axis labels mean positive and negative extremists

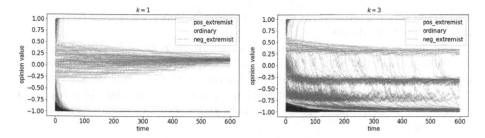

Fig. 7. Polarization tendency comparison between $k = 1$ and $k = 3$ with a negative extremist dominance ($b = 0.6, \lambda = 0.4, \delta = 0.02, \delta_e = 0.8$)

This phenomenon suggests that the involvement of extremists contributes to the opinion polarization in live streaming, just like in other social media. It also explains why some streamers would buy pseudo-fans to make extremely positive comments when streaming because it might develop more positively polarized audiences, who have more potential to become fans and profit the hosts later. Besides, controlling the number of antagonistic extremists like anti-fans or internet trolls would be helpful to prevent ordinary audiences from going to the negative extreme.

While in Scenario 1 parameter k shows the effect of reducing opinion polarization, it is a different story when there is a dominance of one type of extremists. In this situation, the ordinary audiences would have more interactions with that type of extreme opinions and would be more likely to develop more extreme opinions. An example comparing the case of $k = 1$ and $k = 3$ for a situation dominated by negative extremists is shown in Fig. 7. As the number of the two types of extremists becomes balanced, the effect of k would become more similar to that in Scenario 1.

5   Conclusion

The study of extremism in opinion dynamic models has developed for decades. However, previous work seldom extended these models to interactions in real social media. In this research, I modeled the opinion dynamics in Chinese live-streaming platforms using the framework of the SJBO model and considered the effect of information overload, which is caused by the concurrent comments during live streaming, and the entertainment motivation which drives the audiences to send comments.

The simulation results show that opinion polarization could happen without the introduction of extremists if the audiences have low repulsion thresholds, which are described as pseudo-extremists and versatiles in this study. When the attention assigned to the host increases, the majority of the audiences would converge to the opinion value representing the live streaming content. However, this would not prevent the generation of extreme negative opinions. When extremists exist at the start of live streaming, results show that their proportions would bring them relative advantages in turning more ordinary audiences to their side than the counterparts do.

The number of comments that audiences can interact with at each moment also plays an essential role in the process. When there no dominance of positive or negative extremists during the live streaming, the increase of this number could help to reduce opinion polarization. However, when there is a bias of a specific type of extremists, interacting with more opinions would increase that type of polarization.

References

1. Chau, H., Wong, C., Chow, F., Fung, C.H.F.: Social judgment theory based model on opinion formation, polarization and evolution. Physica A **415**, 133–140 (2014)
2. Deffuant, G.: Comparing extremism propagation patterns in continuous opinion models. J. Artif. Soc. Soc. Simul. **9**(3), 8 (2006). http://jasss.soc.surrey.ac.uk/9/3/8.html
3. Deffuant, G., Amblard, F., Weisbuch, G., Faure, T.: How can extremism prevail? A study based on the relative agreement interaction model. J. Artif. Soc. Soc. Simul. **5**(4), 1 (2002)
4. Doherty, M.E., Kurz, E.M.: Social judgement theory. Think. Reason. **2**(2–3), 109–140 (1996)
5. Fan, K., Pedrycz, W.: Evolution of public opinions in closed societies influenced by broadcast media. Physica A **472**, 53–66 (2017)
6. Hart, P.S., Nisbet, E.C.: Boomerang effects in science communication: how motivated reasoning and identity cues amplify opinion polarization about climate mitigation policies. Commun. Res. **39**(6), 701–723 (2012)
7. Hilvert-Bruce, Z., Neill, J.T., Sjöblom, M., Hamari, J.: Social motivations of live-streaming viewer engagement on Twitch. Comput. Hum. Behav. **84**, 58–67 (2018). https://doi.org/10.1016/j.chb.2018.02.013, https://www.sciencedirect.com/science/article/pii/S0747563218300712
8. Jacoby, J.: Perspectives on information overload. J. Consum. Res. **10**(4), 432–435 (1984)
9. Jane, E.A.: 2. Hating 3.0, pp. 42–61. New York University Press (2019). https://doi.org/10.18574/9781479866625-003
10. Järvinen, A., Heliö, S., Mäyrä, F.: Communication and community in digital entertainment services. Univ. Tampere Hypermedia Lab. Net Ser. **2**, 9–43 (2002)
11. Martins, A.C.: Extremism definitions in opinion dynamics models. arXiv preprint arXiv:2004.14548 (2020)
12. Rodriguez, M.G., Gummadi, K., Schoelkopf, B.: Quantifying information overload in social media and its impact on social contagions. In: Proceedings of the International AAAI Conference on Web and Social Media, vol. 8 (2014)

Having a Bad Day? Detecting the Impact of Atypical Events Using Wearable Sensors

Keith Burghardt[✉], Nazgol Tavabi, Emilio Ferrara, Shrikanth Narayanan, and Kristina Lerman

USC Information Sciences Institute, Marina del Rey, CA 90292, USA
keithab@isi.edu

Abstract. Life events can dramatically affect our psychological state and work performance. Stress, for example, has been linked to professional dissatisfaction, increased anxiety, and workplace burnout. We therefore explore the impact of atypical positive and negative events on a number of psychological constructs through a longitudinal study of hospital and aerospace workers. We use causal analysis to demonstrate that positive life events increase positive affect, while negative events increase stress, anxiety and negative affect. While most events have a transient effect on psychological states, major negative events, like illness or attending a funeral, can reduce positive affect for multiple days. These findings provided motivation for us to train machine learning models that detect whether someone has a positive, negative, or generally atypical event. We show that wearable sensors paired with embedding-based learning models can be used "in the wild" to help detect atypical life events of workers across both datasets. Extensions of our results will offer opportunities to regulate the negative effects of life events through automated interventions based on physiological sensing.

Keywords: Psychological constructs · Atypical events · Causal modeling · Wearable sensors · Machine learning

1 Introduction

As organizations prepare their workforce for changing job demands, worker wellness has emerged as an important focus, especially since the COVID-19 pandemic. Worker wellness is central to organization missions to maintain optimal job performance by developing a healthy and productive workforce. This goal is especially important in high-stakes jobs, such as healthcare providers and other frontline workers, where job-related stress often leads to burnout and poor performance [11,15,22], and is one of the most costly modifiable health issues at the workplace [10]. An additional challenge faced by workers is balancing job demands with equally stressful events in their personal life. Adverse events—such as illness or death of a family member and the death of a pet—may amplify worker stress and

© Springer Nature Switzerland AG 2021
R. Thomson et al. (Eds.): SBP-BRiMS 2021, LNCS 12720, pp. 257–267, 2021.
https://doi.org/10.1007/978-3-030-80387-2_25

potentially harm job performance. On the other hand, positive life events—such as getting a raise, getting engaged, or taking a vacation—may decrease stress and improve well-being. The ability to detect such atypical life events in a workforce can help organizations better balance tasks to reduce stress, burnout, and absenteeism and improve job performance. Until recently, detecting such life events automatically, in real time and at scale, would have been unthinkable. Recent advances in accurate and relatively inexpensive wearable sensors, however, offer opportunities for unobtrusive and continuous acquisition of diverse physiological data. Sensor data allows for real-time, quantitative assessment of individual's health and psychological well-being [12, 18] . It could also provide insights into atypical life events that individual workers experience and could affect their well-being and job performance, while keeping information about these personal events private from the organization. However, the connection between sensor data, atypical life events, and individual well-being, has not been demonstrated for such dynamic environments, especially in real-world scenarios.

In this paper, we ask what effects do atypical events have on workers, and can we detect these events with non-invasive wearable sensors? We study two large longitudinal studies of hospital and aerospace workers who wore sensors and reported ecological momentary assessments (EMAs) over the course of several months. Workers also reported whether they had experienced an atypical event. We apply difference-in-difference analysis, a causal inference method [20], to measure the effect of atypical events—either positive or negative events—on individual psychological states and well-being. We show atypical events have dramatic effects on psychological states, which motivates our event detection method. Namely, negative events increase self-reported stress, anxiety, and negative affect by 10–20% or more, while decreasing positive affect over multiple days. Positive life events, meanwhile, have little effect on stress, anxiety, and negative affect, but boost positive affect on the day of the event. Overall, negative atypical events have a greater impact on worker's psychological states than positive events, which is in line with previous findings [2]. Next, we show that it is possible to detect these events from a non-invasive wristband sensor. We propose a method that learns a representation of multi-modal physiological signals from sensors by embedding them in a lower-dimensional space. The embedding provides features for classifying when atypical events occur. Detection results are improved over baseline F1 scores by up to a factor of nine, and achieve ROC-AUC of between 0.60 and 0.66.

Physiological data from wearable sensors allows for studying individual response to atypical life events in the wild, creating opportunities for testing psychological theory about affect and experience. In addition, sensors data creates the possibility of passive monitoring to detect when individuals have stressful or negative experiences, while preserving the privacy of these events from the organization. Informed consent of workers, and strong data protections will still be necessary, but organizations can improve the health and well-being of their workforce and reduce their detrimental effects on vulnerable populations through detecting such experiences.

2   Related Work

In this paper, we explore the effect of positive and negative events on human behavior, and how to detect these events with wearable sensors. There exists extensive research on how sensors can be used to detect patterns and changes in human behavior , including psychological constructs such as stress, anxiety, and affect (cf. literature review of wearable sensors [1]). These papers most often explore detection of stress either induced exogenously (cf. [12]), or (as in our work) endogenously (cf. [3]). There is also literature on detecting bio-markers associated with other psychological constructs, such as anxiety [13], positive and negative affect [21], and depression [4].

Past research has often suffered from two limitations. First, research has focused on either short time intervals (up to two weeks) and very small sample sizes (on the order of tens of subjects) [3,12,18], or collected data sporadically (once every several months) [6]. Second, previous literature has typically detected very short-term stresses (e.g., stresses that affect people on minute level [3,12,18]) rather than individual stressful events that impact someone over the longer term, such as funerals. Our work differs from these previous studies through continuous evaluation over several weeks of hundreds of subjects, allowing us to robustly uncover effects in diverse populations. Moreover, we uncover patterns associated with unusually good or bad events that can affect multiple psychological constructs over multiple days.

3   Data

The data used in this paper comes from two studies aimed at understanding the relationship between individual variables, job performance, and wellness [17]. The studies took place in high-stress environments, but with diverse workforces. Conclusions and methods that generalize across these studies offer hope these results can generalize to a variety of workplaces. Both studies had similar longitudinal design and collected similar data, despite being conducted in different locations and recruited different populations. The *hospital* workforce data was collected during a ten-week long study that recruited 212 hospital workers. The *aerospace* workforce data was collected from 264 subjects, and most details of their data collection, including survey questions asked and the use of wearable devices, match [17]. The studies administered daily surveys to collect self-assessments of participant stress, sleep, job performance, organizational behavior, and other personality constructs. We focus on positive affect, negative affect, anxiety and stress, which we discuss in greater detail in the psychological construct section.

In this paper, we use data collected from *Fitbit* wristbands. We focus on this modality since it was common to both studies. The Fitbit wristband captures dynamic heart rate and step count, and offers daily summary data based on heartrate and step count that includes: daily minutes in bed, daily minutes asleep, daily sleep efficiency, sleep start and end time, and time spent in "fat burn," "cardio," or "out of range" heartrate zones.

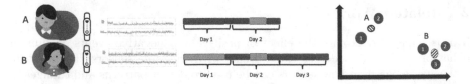

Fig. 1. Overview of the modeling framework. Sensor data collected from participants A and B (left two panels) is fed into a non-parametric HMM model that outputs state sequences (middle panel). Output from the HMM model is used to learn embeddings for each day of each participants (right panel). The daily embedding (lighter and darker-colored circles) and the average embedding for each participant (hashed circles) are used as features. These features, and daily atypical event labels, are then fed into an SVM classifier to predict whether any given day is atypical.

Psychological Constructs. The data used for this study includes daily self-assessments of psychological states provided by subjects over the course of the study. These constructs include self-assessments of stress, anxiety, positive affect, and negative affect, which were found to significantly change during an atypical event. In contrast, job performance, personality, alcohol and tobacco use, and sleep quality did not significantly change. Stress and anxiety were measured by responses to questions that read, "Overall, how would you rate your current level of stress?" and "Please select the response that shows how anxious you feel at the moment" respectively and have a range of 1–5. Positive and negative affect were measured based on 10 questions from [16] (five questions for measuring positive affect and five for measuring negative affect) and have a range of 5–25. **Atypical Event** In addition to these constructs, subjects were also asked, "Have any atypical events happened today or are expected to happen?". If subjects replied yes, they had the option of add free-form text describing the atypical event. In the hospital dataset, there are 8,155 days of recorded data, of which 958 days have atypical events (11.7%). The aerospace dataset has 10,057 days of data, of which 1,503 are considered atypical (14.9%). We have access to the free-form text in the hospital data, which was filled out by participants in 87% of all atypical events, and is categorized as positive, negative, or very negative events [17]. Surprisingly, the severity of the event could not be easily gleaned from sentiment analysis, such as VADER [14], as these tools gave neutral sentiment to text samples that were clearly negative (e.g., "at a funeral"). Of all categorized atypical events, 210 (24%) were positive, 626 (71%) were minor negative events, and 39 (4.5%) were major negative events.

4   Methods

4.1   Causal Inference Method

Atypical events are often described in free-form text as exogenous, e.g., an injured family member or unusually heavy traffic. We can therefore conjecture that atypical events create an as-if random assignment of any given subject over time (these events typically happen at random). This is not always true, as in the case of subjects who report being on vacation multiple days, or are at different stages of burying a loved one. These are, however, relatively rare instances, with sequential events occurring in less than 15% of atypical events in either dataset and exclusion of this data does not significantly affect results. To determine the effect of atypical events on subjects, we use a difference-in-difference approach to causal inference. Specifically, we look at all subjects who report an atypical event and then look at a subset who report stress, anxiety, negative affect, or positive affect the prior day (83% of all events). We finally take the difference in their self-reported constructs from the day before the event. If subjects report construct values after the event we report the difference between these values and the day prior to an atypical event. We contrast these measurements with a *null model*, in which we find subjects who did not report an atypical event on the same days that other subjects reported an atypical event, and find the change in their construct values from the prior day. The difference between construct values associated with the event and the null model is the *average treatment effect* (ATE).

4.2   Representation Learning

We detect atypical events by embedding individuals' heartrate and step count time series data into a vector space, using the framework proposed in [19]. We then train models to identify where in this space atypical events occur. Based on [19], each subject's physiological data is interpreted as a multivariate time series, as described in Fig. 1, left panels. The time series are transformed into sequences of hidden Markov states using a Beta Process Auto Regressive HMM (BP-AR-HMM) [7] (Fig. 1, center panel). Unlike classical hidden Markov models, BP-AR-HMM is flexible by allowing the number of hidden states to be inferred from the data. Based on these datasets the model found 73 states in the hospital data, and 130 states in the aerospace dataset, i.e., we find 73–130 general states/activities that subjects perform, although a subset of these activities are observed in a day. These states are shared among all subjects, rather than being subject specific. This makes it feasible to embed data across different subjects and across different days. After the states are learned, we calculate the stationary distribution of time spent in each state to embed the daily data into the activity space (Fig. 1, right panel). This can be calculated from the HMM transition matrix by finding the eigenvector corresponding to the largest eigenvalue of the matrix.

5   Results

Causal Effect of Atypical Events. How do atypical events affect individual's psychological states? We apply a difference-in-difference approach to measure the impact of atypical events on self-reported psychological constructs. We first look at the effect of atypical events across all our datasets, as shown in Fig. 2.

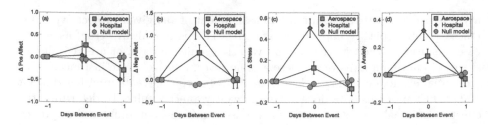

Fig. 2. Effect of atypical events among the datasets studied. (a) Positive affect, (b) negative affect, (c) stress, and (d) anxiety. Green squares and red diamonds show aerospace and hospital datasets, respectively, and gray circles are null models. (Color figure online)

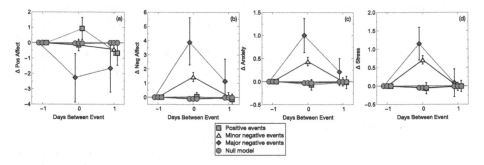

Fig. 3. Effect of different types of atypical events. (a) Positive affect, (b) negative affect, (c) stress, (d) anxiety. Green squares are positive events, white triangles are minor negative events, red diamonds are major negative events, and gray circles are null models. (Color figure online)

Atypical events, on average, have a relatively small effect on positive affect the day of the event (difference from null = 0.09, 0.33; p-value = 0.6, 0.009, for hospital and aerospace data, respectively). We notice a decrease in positive affect from the day of the event to the day after the event (difference = 0.54, 0.55; p-values = 0.0015, 0.017 for aerospace and hospital data, respectively). On the other hand, there is a substantial increase in negative affect, stress, and anxiety (p-values < 0.001), although changes are smaller in the aerospace dataset.

For the hospital data, we categorized atypical events as positive, minor negative, or major negative events using free text descriptions. Namely, negative life events (e.g., funerals) were classified as major negative events; sickness, daily hassles (e.g., traffic), and unpleasant work events were classified as minor negative events, and positive life and work events were classified as positive events. Each category (e.g., positive or negative life events) is defined in [17]. Most atypical events are negative, such as a fight with the spouse, traffic, or deaths. Examples of reported positive events include passing a test or a promotion. The effect of different types of atypical events is shown in Fig. 3. We find that positive events increase positive affect (p-value = 0.009), but, surprisingly, have no statistically significant effect on negative affect, stress, or anxiety (p-value ≥ 0.3). Minor negative events do not substantially change positive affect on the day of the event (difference from null $= -0.15$, p-value $= 0.57$), and have a small effect on positive affect the day after the event (difference from null $= -0.42$, p-value $= 0.04$). On the other hand, they significantly increase negative affect, anxiety, and stress (p-value < 0.001). Finally, major negative events both decrease positive affect the day of the event and the day after the event (p-value $= 0.005$, 0.03 respectively). These results point to a strong diversity in atypical events, and support the idea that "bad is stronger than good" [2]: adverse, or negative, events have a stronger effect on people than positive events, and are reported as atypical events more often.

Detecting Atypical Events. We evaluate performance of three classification tasks using sensor data: (1) detecting whether an atypical event occurred on that day; (2) detecting whether subjects experienced a good day; or (3) detecting whether subjects experienced a bad day. For (2) and (3) the classification task was "1" if subjects experienced a good or bad day, respectively, and "0" otherwise. Hence we simplify all tasks into a binary detection task. We emphasize that these last two tasks are only available for the hospital data.

We use ten-fold cross-validation. Performance metrics are averaged across all held-out folds. Two type of experiments are presented in the paper. In the first set (Table 1) datapoints are split at random, In the second set (Table 2) subjects are split into training and testing sets to approximate a cold-start scenario, where model is trained on one cohort of subjects and tested on another cohort. The challenge of the latter detection task is that we need to classify if a subject has a good or bad day despite not being trained on any previous data from that subject. We use three performance metrics for evaluation. Area under the receiver operating characteristic (ROC-AUC), F1 score, and precision. These metrics are chosen because the data is highly imbalanced.

Table 1. Performance of atypical event detection from sensors in both datasets with randomly sampled cross-validation.

Dataset	Construct	Model	ROC-AUC	F1	Precision
Hospital workforce	Atypical event	Random	0.50	0.12	0.12
		Aggregated	0.57	0.24	0.15
		Embedding	**0.66**	**0.37**	**0.32**
	Positive event	Random	0.50	0.03	0.03
		Aggregated	**0.63**	0.08	0.04
		Embedding	0.62	**0.27**	**0.30**
	Negative event	Random	0.50	0.08	0.08
		Aggregated	0.57	0.17	0.10
		Embedding	**0.61**	**0.27**	**0.24**
Aerospace workforce	Atypical event	Random	0.50	0.15	0.15
		Aggregated	**0.59**	0.31	0.21
		Embedding	**0.60**	**0.32**	**0.36**

We compare detection quality for two types of models: models using features from statistics of aggregated data, and models using features based on time series embeddings. For the aggregated model, we create several features based on aggregated statistics of signals and static modalities. These statistics included the sum, mean, median, variance, kurtosis, and skewness of signals the day before, the day of, and the day after each day, as well as all FitBit daily summary data. Missing data is substituted with mean statistic value within the training or testing set. Statistics before and after each day were created because some physiological features, such as mean heart rate, might change before an atypical event, and some may change after, such as sleep duration. We use Minimum Redundancy Maximum Relevance [5] to select the best features (23 and 26 for the aerospace and hospital data respectively). Important features in the hospital data relate to sleep (the top feature was tomorrow's minutes in bed). Important features in the aerospace dataset tend to relate to heart rate (the top feature was the number of minutes in the "fat burn" heart rate zone in the past day).

Representations from HMM embeddings were learned for the day of, and the day after each day (no summary features are used). We also include the centroid of embeddings for each person in the training data as features, to control for subject-specific differences in behavior. We did not use any additional feature selection because embedding naturally reduces the feature dimensions. Imputation is also not needed because the HMM learns states based on the amount of data available for that day.

We train several candidate classification methods using aggregate features: logistic regression, random forest, support vector machines, extra trees [9], AdaBoost [8], and multi-layered perceptrons. The majority class (no atypical event) is downsampled such that the number of datapoints in each class are equal, which improves the models over using the raw data or upsampling the minority class Based on ten-fold cross-validated F1, ROC-AUC, and precision, we find atypical events in the hospital dataset are best modeled with random forests, while the aerospace workforce dataset is best modeled with logistic regression. In comparison, positive events are best modeled with random forests but negative events are best modeled with extra trees. Finally for embedding features, we used the SVM classifier with no down-sampling, which follows the original work [19].

We demonstrate the first set of results in Table 1. We find that the HMM embedding-based model outperforms other models. The ROC-AUC for the HMM-based model is 0.60 for the aerospace workforce and 0.66 for the hospital workforce. Positive and negative events similarly have an ROC-AUC of 0.61–0.63. F1 and precision exceed random baselines by factors of two to nine. The seemingly low F1 and precision are due to the rarity of atypical events, especially for positive events, which only happen on 3% of days, and negative events which only happen in 8% of all days. A detection therefore represents a "warning sign" that a worker may have had an negative event that day. Overall, detecting atypical events shows promise. Alternatively, a model may be trained on one dataset and tested on another (cold-start scenario). These results are presented in Table 2. Atypical events can be detected 91–220% above baselines based on

F1 score, but results are more modest than in the Table 1, with a reduction in ROC-AUC from 0.66 to 0.58 for hospital atypical events. These results are alike to other recent papers, which split subjects into training and testing and found relatively poor model performance (cf. [18]). These results suggest that models will perform best when personalized to subjects or transfer learning methods are developed for these data.

Table 2. Performance of atypical event detection from sensors in both datasets with subject held-out detection.

Dataset	Construct	Model	ROC-AUC	F1	Precision
Hospital workforce	Atypical event	Random	0.50	0.12	0.12
		Aggregated	0.55	0.22	0.14
		Embedding	**0.56**	**0.23**	**0.16**
	Positive event	Random	0.50	0.03	0.03
		Aggregated	0.57	0.065	0.035
		Embedding	**0.58**	**0.08**	**0.05**
	Negative event	Random	0.50	0.08	0.08
		Aggregated	**0.57**	0.15	0.09
		Embedding	0.56	**0.16**	**0.10**
Aerospace workforce	Atypical event	Random	0.50	0.15	0.15
		Aggregated	**0.58**	**0.30**	**0.20**
		Embedding	0.54	0.25	0.17

6 Conclusion

We discover that atypical events and negative events substantially increase stress, anxiety, and negative affect. Major negative events are found to reduce positive affect over multiple days, while positive events improve positive affect that day. We also demonstrate that wearable sensors can provide important clues about whether someone is experiencing a positive or negative event. We find atypical events can be predicted with ROC-AUC of up to 0.66 with relatively little hyperparameter tuning. More improvements are therefore possible to predict atypical events. These results point to the importance and detectability of atypical events, which offer hope for remote sensing and automated interventions in the future.

Acknowledgements. The research was supported by the Office of the Director of National Intelligence (ODNI), Intelligence Advanced Research Projects Activity (IARPA), via IARPA Contract No. 2017-17042800005.

References

1. Banaee, H., Ahmed, M.U., Loutfi, A.: Data mining for wearable sensors in health monitoring systems: a review of recent trends and challenges. Sensors **13**(12), 17472–17500 (2013)
2. Baumeister, R., Bratslavsky, E., Finkenauer, C., Vohs, K.: Bad is stronger than good. Rev. Gene. Psychol. **5**, 323–370 (2001)
3. Can, Y.S., Chalabianloo, N., Ekiz, D., Ersoy, C.: Continuous stress detection using wearable sensors in real life: algorithmic programming contest case study. Sensors **19**(8), 1849 (2019)
4. Canzian, L., Musolesi, M.: Trajectories of depression: Unobtrusive monitoring of depressive states by means of smartphone mobility traces analysis. In: UbiComp (2015), pp. 1293–1304. ACM, New York, USA (2015)
5. Ding, C., Peng, H.: Minimum redundancy feature selection from microarray gene expression data. J. Bioinform. Comput. Biol. **3**(02), 185–205 (2005)
6. Edwards, D., Burnard, P., Bennett, K., Hebden, U.: A longitudinal study of stress and self-esteem in student nurses. Nurse Educ. Today **30**(1), 78–84 (2010)
7. Fox, E.B., et al.: Joint modeling of multiple time series via the beta process with application to motion capture segmentation. Ann. Appl. Stat. **8**(3), 1281–1313 (2014)
8. Freund, Y., Schapire, R.E.: A decision-theoretic generalization of on-line learning and an application to boosting. J. Comput. Syst. Sci. **55**(1), 119–139 (1997)
9. Geurts, P., Ernst, D., Wehenkel, L.: Extremely randomized trees. Mach. Learn. **63**(1), 3–42 (2006)
10. Goetzel, R.Z., et al.: Ten modifiable health risk factors are linked to more than one-fifth of employer-employee health care spending. Health Aff. **31**(11), 2474–2484 (2012)
11. Gray-Toft, P., Anderson, J.G.: Stress among hospital nursing staff: its causes and effects. Soc. Sci. Med. A **15**(5), 639–647 (1981)
12. Healey, J.A., Picard, R.W.: Detecting stress during real-world driving tasks using physiological sensors. IEEE Trans. Intell. Transp. Syst. **6**(2), 156–166 (June 2005)
13. Huang, Y., et al..: Discovery of behavioral markers of social anxiety from smartphone sensor data. In: DigitalBiomarkers (2017), pp. 9–14 (2017)
14. Hutto, C.J., Gilbert, E.: Vader: A parsimonious rule-based model for sentiment analysis of social media text. In: Eighth International AAAI Conference on Weblogs and Social Media (2014)
15. Kyriakou, K.: Detecting moments of stress from measurements of wearable physiological sensors. Sensors **19**(17),(2019)
16. Mackinnon, A., Jorm, A.F., Christensen, H., Korten, A.E., Jacomb, P.A., Rodgers, B.: A short form of the positive and negative affect schedule: evaluation of factorial validity and invariance across demographic variables in a community sample. Person. Individ. Diff. **27**(3), 405–416 (1999)
17. Mundnich, K.: Tiles-2018: a longitudinal physiologic and behavioral data set of hospital workers. Sci. Data **7**(354) (2020)
18. Smets, E.: Large-scale wearable data reveal digital phenotypes for daily-life stress detection. npj Digit. Med. **1**(67) (2018)
19. Tavabi, N.: Learning behavioral representations from wearable sensors. arXiv preprint arXiv:1911.06959 (2019)
20. Varian, H.R.: Causal inference in economics and marketing. PNAS **113**(27), 7310–7315 (2016). https://doi.org/10.1073/pnas.1510479113

21. Yan, S., Hosseinmardi, H., Kao, H., Narayanan, S., Lerman, K., Ferrara, E.: Estimating individualized daily self-reported affect with wearable sensors. ICHI **2019**, 1–9 (2019). https://doi.org/10.1109/ICHI.2019.8904691

22. Zamkah, A., Hui, T., Andrews, S., Dey, N., Shi, F., Sherratt, R.S.: Identification of suitable biomarkers for stress and emotion detection for future personal affective wearable sensors. Biosensors **10**(4), 40 (2020)

Trusty Ally or Faithless Snake: Modeling the Role of Human Memory and Expectations in Social Exchange

Jonathan H. Morgan[1]([✉]) [iD], Christian Lebiere[2] [iD], James Moody[1], and Mark G. Orr[3] [iD]

[1] Department of Sociology, Duke University, Durham, USA
`jhm18@duke.edu`
[2] Department of Psychology, Carnegie Mellon University, Pittsburgh, USA
[3] Biocomplexity Institute, University of Virginia, Charlottesville, USA

Abstract. Exchange is a foundational form of human interaction underlying more complex forms of cooperation and collaboration. Exchange scholars have demonstrated that both the structure of exchange relationships, and the cultural logics that govern them influence the benefits that exchange partners contribute and receive. These factors influence behavior by shaping expectations; but, the cognitive process involved in forming expectations over the course of repeated exchanges is less well understood. We introduce a cognitive model of social exchange implemented in ACT-UP, employing instance-based learning (Gonzalez et al. 2003). In this paper, we focus on how the logical structure of exchange relationships influences exchange behaviors by comparing simulation outcomes to experimental data collected by (Molm et al. 2013). We find that the cognitive model is able to replicate well the outcomes of negotiated exchanges but that reciprocal exchanges pose a greater challenge.

Keywords: Cognitive modeling · Social simulation · Sociology · Psychology · Economics

1 Introduction

Exchange is an important form of interaction that influences the formation of micro-social orders both in markets and in other institutional contexts, frequently resulting in systematic and varied payoffs between group/team members. Past work has identified how the structure of exchange relationships and the norms that govern them influence the formation of expectations about and commitment to exchange partners (Lawler 2001; Molm 2003). For example, the structure of exchange relationships influences the relational dependence between actors, and thus how power is exercised in the relationship (Molm 1997). Whereas the form of the exchange (norms associated with making, and receiving offers) influences the tendency toward strategies that prioritize reward maximization on the part of individuals rather than group-level outcomes (Molm 2003; Lawler et al. 2008).

© Springer Nature Switzerland AG 2021
R. Thomson et al. (Eds.): SBP-BRiMS 2021, LNCS 12720, pp. 268–278, 2021.
https://doi.org/10.1007/978-3-030-80387-2_26

We seek to develop these intuitions via a formal and computationally explicit understanding of such hypothesized mental constructs, and their relation to exchange behaviors in the context of an empirical social exchange experiment. Our approach leverages a well-established cognitive architecture, ACT-R, which can model behavior based on expectations of outcomes. Using expectations generated from experience is a key component of Instance-Based Learning (Gonzalez et al. 2003), an approach to decision-making that has been used for domains ranging to game playing (West and Lebiere, 2001; Lebiere et al. 2003), game theory (Lebiere et al. 2000), social dilemmas (Martin et al. 2014; Juvina et al. 2015), and probabilistic choice (Erev et al. 2010).

1.1 Forms of Exchange and Their Consequences

Ties of mutual dependence underlie all social structures (Homans 1961; Emerson 1962). One important mechanism for managing these dependencies is social exchange: patronage for votes, esteem for advice, loyalty for friendship to name a few examples. Mutual dependencies can be a source of cohesion as each actor can potentially benefit from the combined efforts of the group. One actor's dependence; however, is also another's power. Mutual dependence ensures that both actors have some power over one another. Asymmetries of dependence result in asymmetries in power in terms of the benefits that each actor contributes and receives.

Managing relational dependencies takes on many forms. These forms are cultural frameworks that frame cultural expectations about how power in exchange relationships can be legitimately exercised; these forms can be classified into two broad categories direct and indirect (Molm 2003). Direct exchange refers to forms of exchange where each actor's outcomes directly depend on the actions of their exchange partners. Indirect exchange, such as the practice of potlatch among the first peoples of the Pacific Northwest, refers to forms of exchange between three or more actors. In these types of exchange, benefits received by one individual, B, from another, A, do not depend on B giving to A, but rather on B giving to another actor in the network. Direct exchanges can be further divided into two types: negotiated and reciprocal. Negotiated exchanges entails explicit bargaining about the terms and outcomes of the transaction. Reciprocal exchanges refer to non-negotiated interactions where one individual performs a beneficial act for another such as giving assistance with no guarantee that this gesture will be repaid in kind. This form of exchange is referred to as reciprocated in the literature because it evolves from cultural norms of reciprocity—people tend to do nice things for people who are nice to them.

In this paper, we focus on negotiated and reciprocal exchange because these are the most widely studied types, making the natural starting point for developing a cognitive model of social exchange. Negotiated and reciprocal exchange have some commonly observed properties. Negotiated exchanges in small group settings tend to promote reward maximization strategies, resulting in greater differences between more advantaged and less advantaged parties (Molm 2003; Molm et al. 2013). Differences in outcomes occur in reciprocated exchanges as well; but emergent norms of reciprocal giving tend to reduce these differences.

2 Methods

We conduct a set of simulation experiments using the ACT-R cognitive architecture that were designed for direct comparison to prior human experimental data (Molm et al. 2013). We describe the empirical data and the simulation in separate sections below. ACT-R (Anderson and Lebiere 1998; Anderson et al. 2004) is a cognitive architecture that implements a unified theory of cognition. ACT-R is composed of a set of functional modules, including procedural control, declarative memory, working memory, perception and action, interacting asynchronously in a distributed architecture. Declarative representations consist of symbolic chunks encoding simple combinations of values (or other chunks). Access to those chunks is controlled by a subsymbolic activation calculus that reflects learned statistical regularities such as the power law of practice, the power law of decay, and similarity-based generalization. Chunks are learned automatically from experience. Information can be retrieved from memory by selecting the most active chunk given the current context, or through a blended retrieval process that returns an aggregate consensus of relevant memories, weighted by their probability of retrieval: that information can then be used by simple production rules to make decisions and control behavior. No claims of optimality or rationality are made. Decisions are instead prone to suboptimality and cognitive biases (Lebiere et al. 2013).

2.1 Human Experimental Data

To evaluate the proposed model, we compare model results to experimental data collected from 2009–2010 by Linda Molm and colleagues (Molm et al. 2013). We would like to thank David Melamed for these data. In this paper, we focus on replicating the distributions of points earned by three and four-person networks participating in either negotiated exchange or reciprocal exchange. We limit our focus in this paper to these two types in order to more fully evaluate and compare multiple decision logics and their consequences. When comparing models to experimental data, we analyze ratios between people and agents occupying more and less advantaged structural positions in each configuration.

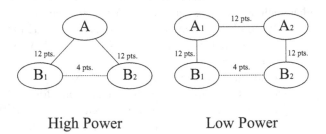

High Power   Low Power

Fig. 1. Exchange networks: high (left) and low (right) relational power.

To provide a better sense of the experimental data, we briefly discuss Molm et al.'s experimental design. Participants attempted to accrue points by making bids to exchange partners. Molm and colleagues varied how the exchanges were organized (who could exchange with whom) and the rules that governed the exchanges. The experimental design consists of two networks and four exchange types, resulting in eight configurations. We focus on negotiated and reciprocal exchange in three and four-person networks. We discuss the network manipulation first followed by the exchange types.

Each network links each participant to two interaction partners or alters. Each network also consists of two positions: A and B (illustrated in Fig. 1). The positions differ from one another in terms of their topological advantage. Participants in the A position are able to make high-value exchanges with both of their alters. High value exchanges can produce a total of 12 points for both actors. Participants in the B position, however, have access to only one high-value relation (their relation with A). Their other relation, their tie to the other B, is a low-value relation capable of only producing 4 points for both exchange partners. Position A, thus, has a topological advantage because all actors have an incentive to exchange with participants occupying A positions, whereas there is less incentive for people occupying B positions to exchange with other Bs.

The networks differ in terms of structural power imbalance between participants occupying A and B positions. In the three-person, high-power, network, only one potential agreement (between any two of the three participants) can be made. In contrast, in the four-person, low-power, network, two agreements (between two pairs of the four actors) are possible. Consequently, participants in low-power networks have more exchange options at each turn. In addition, people occupying the A position in the high-power network receive bids from both B positions whose only opportunity for a high-value payoff depends on A's cooperation. Under these conditions, people occupying the B position can inadvertently engage in a race to the bottom with the other B as both Bs offer increasingly favorable bids to A that net them fewer points in hopes of getting more than four points. In contrast, in the low power networks, people in the A positions do not benefit from the Bs competing with each other for their cooperation.

Participants were randomly placed in their positions. A participant occupied only one position for the duration of the experiment, and they did not know whether they occupied a high or low-power position. They only knew that they were in a three or four-person group, that they could exchange with two alters, and what the payoffs were associated with each alter.

Participants in high and low-power networks engaged in either negotiated or reciprocal exchanges. Molm et al. (2013) ran 12 groups for each network/task combination (e.g., twelve 3-person networks engaging in negotiated exchanges) in two-session blocks, resulting in 48 group outcomes for each configuration. Table 1 lists the number of exchanges or trials individuals in each configuration participated in, the mean points of each configuration, and the SD of each.

Table 1. Descriptive statistics for each configuration (12 groups per configuration).

Configuration	Num. of exchanges	Mean points	SD
High power (Neg.)	50	170.7	108.8
High power (Recp.)	150	756.2	331.06
Low power (Neg.)	30	148.2	51.37
Low power (Recp.)	150	827.02	212.5

Negotiated exchanges involved participants negotiating the division of a fixed number of points, 12 or 4 points for the high or low-value exchange respectively. Each negotiation consisted of up to five rounds. On each round, all actors in the network simultaneously made offers to both of their alters. After the first round, actors could accept an offer, repeat their last offer, or make a counteroffer. Negotiations continued until all potential agreements were made or the five rounds were up. As soon as an agreement was reached, both actors received the amounts agreed upon, making all agreements binding. Participants knew the range of points they could request, and that the more points they received the less their exchange partner received. They, however, neither knew that a fixed amount was being divided nor the exact amount received by their exchange partner.

Reciprocal exchange, in contrast, consisted of each actor giving a fixed number of points to one of his or her partners on each exchange opportunity. Giving points to a partner added to the partner's total points without subtracting from the actor's own points. As in the negotiated exchanges, participants knew only the number of points they could receive from others, not the number of points they could give to others. All participants chose simultaneously and independently a partner to give points to, without knowing whether or when their partner might reciprocate. Participants were then informed that each of their partners either added to their earnings or not, and the participant's cumulative points were updated. To hold constant the potential joint benefit of negotiated versus reciprocal forms of exchange, the number of points that each actor could give to an alter per opportunity was fixed (2 points in the low-value relations, and 6 points in the high-value relations), or one-half the total points (4 and 12) that two actors could divide in a negotiated exchange.

Because negotiated exchange potentially involved five rounds of negotiation compared to one round per reciprocal exchange, Molm et al. (2013) weighted the conditions to balance the amount of time and effort required for each condition. The ratio for 3-person networks was 3 reciprocal exchanges to one negotiated exchange, while for 4-person networks the ratio was 5 reciprocal exchanges to one negotiated exchange. These ratios are reflected in the number of trials for high-power negotiated (50) and reciprocal (150) conditions, and the low-power negotiated (30) and reciprocated (150) conditions respectively.

2.2 Cognitive Model

Each player is represented by a single model instantiation implemented in ACT-UP, a scalable, rapid-prototyping implementation of ACT-R (Reitter and Lebiere 2010). We will first describe the model for negotiated exchanges. The model's representation is as simple as possible. Past experiences are encoded in memory as chunks associating the name of the partner, the number of points obtained from that partner, and the round at which that offer was accepted. At each round, the player generates an expectation of points from each partner. The highest offer made in the previous round is accepted if it meets the highest expectation across all partners. Otherwise, a request is made to each partner that reflects the expectation of points from that partner at that round. The expectation reflects the history of past exchanges with that partner, with the more recent ones having higher activation and thus a more prominent role in determining the consensus value. Each model is initialized by adding two chunks for each partner at the start: one contains the maximum amount that could be obtained from that relationship (12 in high-power relationships; 4 in low-power relationships) associated to the first round and one containing an amount of 0 associated to the fifth (last) round. This will result in an approximate strategy of asking high amounts in the initial round, then gradually lowering demands and typically accepting any offer in the last round. However, actual outcomes are added after each trial, gradually shaping the expectation function for each partner, and in turn the amounts requested from that partner.

The model's representation for reciprocal exchanges is the same, with the simplification that the round value is always the same. As for negotiated exchanges, an expectation is generated for each partner. However, since points offered were fixed rather than determined by each partner, the expectation generated is an estimate of the probability of receiving an offer from that partner rather than its amount. The model then offers its points to the partner with the highest expectation. As is common in IBL models (e.g., Cranford et al. 2020), the expectation generated in the process of making the decision enters memory on the same basis as the actual outcome, leading to a form of confirmation bias where the outcome of the initial experiences get repeatedly reinforced in future trials. All ACT-R parameters are left at the architecture's default values.

3   Results

We first summarize the human experimental results from Molm et al.'s (2013) study, followed by the ACT-R simulation results as a point of comparison. The patterns we find in the data are quite similar to prior studies (Molm 1999; Molm 2003). As the outcomes of individuals occupying A and B positions is of central interest to exchange scholars, we compare outcomes of individuals in each position for each configuration. Figure 2 compares the ratio of points earned by each position out of the total points earned by the group across experimental and simulated groups for each configuration: 3-person and 4-person negotiated and reciprocal exchanges. Experimental and simulation conditions are represented by grouped bars. Each position is indicated by a bar, with warm colors indicating A and cold colors B positions. The x-axis of each plot indicates each condition, the y-axis the proportion of points earned. As the point totals are, in part, a consequence of the weighting, we focus on the proportion of points earned by participants in each position rather than their point totals.

Consistent with past studies, participants occupying the A position in high-power (3-person) negotiated exchange networks earn the most points relative to their B position

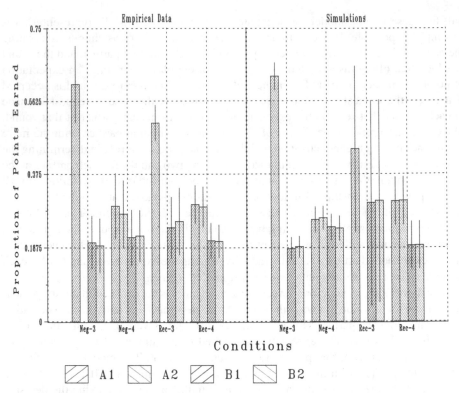

Fig. 2. Position ratios by condition for simulated and experimental groups

exchange partners, 3 times as many points on average. In contrast, participants occupying A positions in high-power reciprocal networks earn on average 2 times as many points. Also consistent with past studies, the addition of a second A position significantly weakens the relative advantage of A over B. Individuals occupying B positions in lower-power (4-person) negotiated exchange networks earned about 75% of the points of individuals occupying A positions. In the low-power reciprocal networks, individuals in the B positions earned about 70% of the points of individuals in the A positions. Although players in reciprocal exchanges earned more points, the presence of an additional exchange partner made negotiated exchange a more attractive option in terms of relative return for each exchange partner in the low-power networks.

When comparing the simulation results to the experimental findings, we ensured for each configuration that agents had the same number of exchange opportunities as the participants, and aggregated the results over a sample of forty-eight simulated groups (again matching the experimental design). We find that the model nicely captures the exchange dynamics of negotiations in 3-person exchange networks. The model also captured the general tendencies of the 4-person negotiations. In both experimental and simulated groups, A positions received more points than B positions, but the relative differences between the A and B positions were somewhat less in some conditions. Another difference between the model and experimental results is variation in the outcomes.

We can better understand both the experimental and simulation outcomes if we consider the task. In negotiations with the 3-person network, the player in position A has two opportunities to achieve a high-power deal, while the players in position B only have one opportunity. To achieve a low-power deal between the two players in position B, the outcome would have to be better than the expectation for each one of their potential deals with the player in position A. Unless that expectation drops below 3 points, that will preclude a low-power deal since the players in position B get an average of 2 points each, roughly corresponding to a 9-3 split of the 12 points at stake in a high-power deal, which is a ratio of about 3, the value we observe. The same negotiation dynamic in a 4-person network yields very different outcomes. In that case, the alternative for the players in position A to make a deal with their respective player in position B is to make a deal with the other player in position A. But that is a relation of equals, resulting in splitting the 12 points at stake evenly. Thus, the players in position A settle for only a slightly better outcome (close to 7-5) with their respective players in position B despite their relative advantage. In negotiated exchange networks, the vast majority of deals concluded occur in high-power relations. Seldom do the two players in position B conclude a low-power deal.

Reciprocal exchanges involve different dynamics. Comparing experimental and simulated reciprocal exchanges, we find in both cases that the relative proportion of points earned by actors in the A position is far less. In both cases, A retains an advantage, with that advantage being reliably greater in the experiments than in our simulations. To understand these results, it is important to remember that in reciprocal exchange participants and agents must establish how reliable their exchange partners are, as there is no guarantee of a positive outcome. Consequently, the model's focus is on establishing reliable reciprocity rather than maximizing points. Reliable reciprocal relations will, thus, be established largely independently of the degree of power in the relation, partly by leveraging the reinforcing of expectations in an established relationship. The resulting ratios, therefore, reflect the distribution of power at each node, yielding for both 3-person and 4-person networks a roughly 3-2 ratio.

Variations in the model results for the negotiated exchanges are consistently smaller than corresponding variations in human data. That is often the case as models only capture a subset of human complexity. Moreover, we assume here that all models are identical before accumulating their individual experiences that modulate their behavior. An additional feature of negotiated exchanges that limits variation in the model behavior is the near-continuous nature of the outcome in terms of splitting a total amount of points available for each relation. The (biased) averaging nature of the blending retrieval process that generates expectations from experience that are the basis for offers between partners typically results in fairly smooth, stable behavior. By contrast, reciprocal exchanges result in discrete, winner-take-all outcomes that tend to emphasize variations: a player can only offer a set amount or nothing at all to each partner and has to choose between partners at each trial. While the model matches the human variations fairly well for the 4-person reciprocal exchanges, it generates much larger variations than the humans for the 3-person reciprocal exchanges. The tendency of the model to form mutually stable relations results in large variations between runs when a given partner manages to achieve consistent exchanges and those in which they are shut out by the other two partners. The lower variation in the human data suggests that relationships are not as stable but can instead shift, perhaps as a result of a deliberate strategic attempt by the player being shut out to convince one of the other partners to defect. These kinds of sustained calculated metacognitive strategies can also be modeled using instance-based learning (Reitter et al. 2010; Romero and Lebiere 2014).

4 Discussion

The primary objective of this article is to provide insight into the potential cognitive mechanisms that underlie the human experimental results. We see a strong correspondence between the experimental and simulation data when the task primarily involves forming expectations about concrete outcomes. In negotiations, there is far greater certainty that there will be a reward, and the amount of any particular round's reward can be logically deduced within one or two turns. The model struggles, as do participants, when the task involves first ascertaining how reliable one's exchange partners are. Because there is no certainty that a player's offer will be reciprocated, signaling takes on greater importance, with actions occurring in the first few rounds often determining the pattern of subsequent exchanges. Because players occupying B positions are highly dependent on successfully signaling to players in the A positions that they are a reliable partner to earn points, their outcomes vary more as both participants and agents are more or less successful at this task. Our agents struggle more with this task than human players. One reason is that we do not incorporate any affective information that would either make agents more likely to reciprocate in the first few turns of an exchange, or to develop positive sentiments towards their exchange partners. Incorporating such information in a theoretically plausible way is possible (Jung et al. 2016).

Further, and more generally, research in the domain of exchange emphasizes two important aspects not addressed in our work. First, social exchange theorists think deeply about the time constraints when constructing experiments; but this aspect of exchange rarely is incorporated into theorizing about the exchange process. Participants must quickly decide what to do; and they frequently make these decisions based on the first few turns. Those constraints are quite compatible with an Instance Based Learning (IBL) approach. Second, exchange theorists such as Edward Lawler and colleagues note that these tasks also take on emotional connotations. Negotiated exchange tends to dampen participants' attachment to others in the network. Lawler proposes that group members' orientations are heavily influenced by the nature of the task, namely whether the outcome depends on a diffuse collective effort or on discrete comparable tasks for which participants can take credit. Although again plausible, applying Lawler's concept of jointness to interpret the differences in emotional attachment resulting from reciprocal versus negotiated exchange is dissatisfying because these tasks lack an interdependent collective achievement that all group members can relate to. Nevertheless, emotion does seem to play a role in the exchange process.

This method of building simulations of human social phenomena to look at etiological cognitive and psychological processes is not new, but it is rare. It flows from the prior work in our group and has potential for not only elicitation of cognitive underpinnings of social phenomena but also for increasing our understanding of cognitive processes. See Orr et al. (2019) for a general treatment of this idea.

References

Anderson, J.R., Lebiere, C.: The Atomic Components of Thought. Lawrence Erlbaum Associates, Mahwah (1998). ISBN 9780805828177

Anderson, J.R., Bothell, D., Byrne, M.D., Douglass, S., Lebiere, C., Qin, Y.: An integrated theory of the mind. Psychol. Rev. **111**(4), 1036–1060 (2004). https://doi.org/10.1037/0033-295X.111.4.1036

Cranford, E.A., Aggarwal, P., Gonzalez, C., Cooney, S., Tambe, M., Lebiere, C.: Adaptive cyber deception: cognitively-informed signaling for cyber defense. Presented at the Cyber Deception for Defense Track at Hawaii International Conference on System Sciences (HICSS 2020). Maui, Hawaii, 7–10 January 2020. https://doi.org/10.24251/HICSS.2020.232

Emerson, R.: Power-Dependence Relations. Am. Sociol. Rev. **27**(1), 31–41 (1962). https://doi.org/10.2307/2089716

Erev, I., et al.: A choice prediction competition: Choices from experience and from description. J. Behav. Decis. Making **23**(1), 15–47 (2010). https://doi.org/10.1002/bdm.683

Gonzalez, C., Lerch, F.J., Lebiere, C.: Instance-based learning in dynamic decision making. Cogn. Sci. **27**(4), 591–635 (2003)

Homans, G. C.: Social Behavior: Its Elementary Forms. New York: Harcourt, Brace (1961)

Jung, J.D., Hoey, J., Morgan, J.H., Schröder, T., Wolf, I.: Grounding social interaction with affective intelligence. In: Khoury, R., Drummond, C. (eds.) AI 2016. LNCS (LNAI), vol. 9673, pp. 52–57. Springer, Cham (2016). https://doi.org/10.1007/978-3-319-34111-8_7

Jung et al. 2016 and Schroeder, Hoey, and Rogers 2018: but our aim here was to see to what extent memory combined with the task's structure influences exchange outcomes, as these factors are less fully developed in the exchange literature

Juvina, I., Lebiere, C., Gonzalez, C.: Modeling trust dynamics in strategic interaction. J. Appl. Res. Mem. Cogn. **4**(3), 197–211 (2015). https://doi.org/10.1016/j.jarmac.2014.09.004

Lawler, E.J.: An affect theory of social exchange. Am. J. Sociol. **107**(2), 321–352 (2001). https://doi.org/10.1086/324071

Lawler, E.J., Thye, S.R., Yoon, J.: Social exchange and micro social order. Am. Sociol. Rev. **73**(4), 519–542 (2008). https://doi.org/10.1177/000312240807300401

Lebiere, C., Wallach, D., West, R.L.: A memory-based account of the prisoner's dilemma and other 2x2 games. In: Proceedings of International Conference on Cognitive Modeling 2000, pp. 185–193. Universal Press (2000). http://act-r.psy.cmu.edu/?post_type=publications&p=13786

Lebiere, C., Gray, R., Salvucci, D., West, R.: Choice and learning under uncertainty: a case study in baseball batting. In: Proceedings of the 25th Annual Meeting of the Cognitive Science Society, pp. 704–709. Erlbaum, Mahwah (2003). https://escholarship.org/uc/item/5p25k0wg

Lebiere, C., et al.: A functional model of sensemaking in a neurocognitive architecture. Comput. Intell. Neurosci. (2013). https://doi.org/10.1155/2013/921695

Martin, J., Gonzalez, C., Juvina, I., Lebiere, C.: A description-experience gap in social interactions: information about interdependence and its effects on cooperation. J. Behav. Decis. Mak. **27**, 349–362 (2014). https://doi.org/10.1002/bdm.1810

Molm, L.D.: Coercive Power in Social Exchange. Cambridge University Press, New York (1997). https://doi.org/10.1017/CBO9780511570919

Molm, L.D.: Theoretical comparisons of forms of exchange. Sociol Theory **21**(1), 1–17 (2003). https://doi.org/10.1111/1467-9558.00171

Molm, L.D., Melamed, D., Whitham, M.M.: Behavioral consequences of embeddedness: effects of the underlying forms of exchange. Soc. Psychol. Q. **76**(1), 73–97 (2013). https://doi.org/10.1177/0190272512468284

Molm, L. D., Peterson, G., Takahashi, N.: Power in negotiated and reciprocal exchange. Am. Sociol. Rev. 876–890 (1999). https://doi.org/10.2307/2657408

Orr, M.G., Lebiere, C., Stocco, A., Pirolli, P., Pires, B., Kennedy, W.G.: Multi-scale resolution of neural, cognitive and social systems. Comput. Math. Organ. Theory **25**(1), 4–23 (2019). https://doi.org/10.1007/s10588-018-09291-0

Reitter, D., Lebiere, C.: Accountable modeling in ACT-UP, a scalable, rapid-prototyping ACT-R implementation. In: Proceedings of the 2010 International Conference on Cognitive Modeling, Philadelphia, PA (2010). http://act-r.psy.cmu.edu/wordpress/wp-content/uploads/2012/12/935 Reitter.pdf

Reitter, D., Juvina, I., Stocco, A., Lebiere, C.: Resistance is futile: winning lemonade market share through metacognitive reasoning in a three-agent cooperative game. In: Proceedings of the Behavior Representation in Modeling and Simulations (BRIMS 2010) Conference, Charleston, SC (2010)

Romero, O., Lebiere, C.: Simulating network behavioral dynamics by using a Multi-agent approach driven by ACT-R cognitive architecture. In: Proceedings of the Behavior Representation in Modeling and Simulation Conference (BRIMS 2014), Washington, DC, April 2014 (2014). https://www.cs.cmu.edu/~oscarr/pdf/publications/2014_simulating_network.pdf

Schröder, T., Hoey, J., Rogers, K.B.: Modeling dynamic identities and uncertainty in social interactions: Bayesian affect control theory. Am. Soc. Rev. **81**(4), 828–855 (2016)

West, R.L., Lebiere, C.: Simple games as dynamic, coupled systems: randomness and other emergent properties. J. Cogn. Syst. Res. **1**(4), 221–239 (2001). https://doi.org/10.1016/S1389-0417(00)00014-0

Examining the Effects of Race on Human-AI Cooperation

Akil A. Atkins, Matthew S. Brown, and Christopher L. Dancy^(⊠)

Bucknell University, Lewisburg, PA 17837, USA
{aaa014,msb027,christopher.dancy}@bucknell.edu

Abstract. Recent literature has shown that racism and implicit racial biases can affect one's actions in major ways, from the time it takes police to decide whether they shoot an armed suspect, to a decision on whether to trust a stranger. Given that race is a social/power construct, artifacts can also be racialized, and these racialized agents have also been found to be treated differently based on their perceived race. We explored whether people's decision to cooperate with an AI agent during a task (a modified version of the Stag hunt task) is affected by the knowledge that the AI agent was trained on a population of a particular race (Black, White, or a non-racialized control condition). These data show that White participants performed the best when the agent was *racialized* as White and not racialized at all, while Black participants achieved the highest score when the agent was racialized as Black. Qualitative data indicated that White participants were less likely to report that they believed that the AI agent was attempting to cooperate during the task and were more likely to report that they doubted the intelligence of the AI agent. This work suggests that racialization of AI agents, even if superficial and not explicitly related to the behavior of that agent, may result in different cooperation behavior with that agent, showing potentially insidious and pervasive effects of racism on the way people interact with AI agents.

Keywords: Human-AI · Race · Stag hunt · Cooperation

1 Introduction

There has been a variety of discussions related to the ways in which implicit racial biases affect interactions between people (e.g., [8, 11]) and interactions between people and artifacts (e.g., [1, 3, 12]). Correll et al. [8] analyzed a specific situation where these biases could play a critical role: study participants were tasked with "shooting" armed targets and to "not shoot" those who were unarmed. White participants were quicker to shoot armed targets if they were Black but were quicker to make the decision to not shoot an unarmed target if they were White [8].

Stanley et al. [11] gave participants $10 and asked them to give whatever amount of money they wished to a partner; the partner would receive quadruple the amount and had already decided whether they would split their earnings evenly or keep it to themselves. During the experiment, participants were shown an image of either a Black man or White

© Springer Nature Switzerland AG 2021
R. Thomson et al. (Eds.): SBP-BRiMS 2021, LNCS 12720, pp. 279–288, 2021.
https://doi.org/10.1007/978-3-030-80387-2_27

man to represent their partner [11]. Before completing the tasks, participants were also given implicit bias tests to evaluate whether they were "pro-black", "pro-white", or neither [11]. The researchers found that those who were identified as pro-white, which was about eighty percent of the participants, were more likely to trust their partner and therefore give them more money if they were white and vice versa [11]. These findings demonstrated that people were more likely to trust and work with those that align with their own racial identity.

As our technology continues to become more ubiquitous in society, there will be different ways in which racism can continue to foundationally affect behavior. Focusing on human interaction with robots, Strait et al. [12], conducted a study and found that robots racialized as Black and Asian were subject to almost double the dehumanizing comments that the robot racialized as White received [7]. This example shows that the social construct of race plays a critical role in the ways in which not only people are treated but other agents as well. In another study that used the shooter bias incident model, but with robots as well as humans, Bartneck et al. [3] showed that even though those potentially being shot in the study were robots, the results were almost identical to the human-based study described in Correll et al. [8]. Participants were still quicker to shoot Black agents that were armed, whether they were a human or a robot [3].

The present study investigates how participants will interact with an AI agent that has been explicitly racialized, or not racialized at all. To measure their level of cooperation with the AI agent, participants were tasked to play a game called *catch the pig* (a modified version of the Stag Hunt task [15]), where the goal was to score as many points as possible. While completing the task, participants' in-game movements and scores were tracked to measure level of cooperation and after completing the trials participants were asked 3 qualitative questions regarding any strategies they had and their perceived level of intelligence that the AI agent pertained.

2   Methods

2.1   Participants

The participant sample included 186 participants, all recruited through Prolific.co. Participants were told they would be paid $6.50 per hour for their participation in the study and they could be eligible for up to $7.50 per hour based on their performance in the study (in reality all participants were given $7.50 regardless of performance). Prolific's automatic participant filters were used to specify participants' demographics (their reported race as well as being from and located within the United States) to guarantee a balanced sample of people who identified as either "Black/African American" or "White/Caucasian".

2.2   Design

Each participant was placed into one of three groups (all containing the same total amount of participants): (1) A *Black* treatment condition where participants were told that the AI agent, they would be cooperating with was trained on a predominantly Black population; (2) A *White* treatment condition where participants were told that the AI agent they would be cooperating with was trained on a predominantly White population; (3) A *control (no race)* condition where the race of behavioral data was not mentioned. Despite what the participants were led to believe, the AI had not been trained on any human behaviors and used an A* algorithm to complete the task. During the task, individual task related behavior was collected during each round. Task-related behavior included score during each trial, the keys pressed for each trial, as well reaction times.

2.3 Procedure

Once participants started the experiment through Prolific, they were directed to a Qualtrics survey. The Qualtrics survey contained the condition-specific text for their randomly assigned condition and instructions on navigating the game. Participants were instructed that the first three trials in the study were for practice to ensure they understood the game, and they were also told to immediately exit through the rightmost gray square on the eighth trial to ensure that they were paying attention.

After participants read through the instructions, they began the experiment. Participants were tasked with controlling a blue, triangular game piece and to work with an "AI controlled" orange, triangular game piece to catch a pink, circular game piece that represented a pig. All game pieces were placed in a 9 × 9 grid in which they could only move within a 5 × 5 square within the grid. Inside the 5 × 5 grid, participants could move on any of the green squares, with each movement subtracting 1 point from the score. There were two methods for participants to score points which were to either work with the AI agent to catch the pig or exit through one of the gray squares on either side of the board. For each trial, the human controlled piece was placed in the upper right corner, the "AI" was placed in the lower left corner, and the pig was placed in the middle of the game board, as seen in Fig. 1.

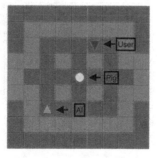

Fig. 1. The starting positions for the players (human and AI) and the pig did not change. (Color figure online)

Once in the game, the participant could choose to work with the AI agent to attempt to catch the pig, in which case if they were successful, they would be awarded 25 points. Successfully catching the pig is only possible while cooperating with the AI agent as the Human and AI agent must surround the Pig on a block where the Pig has no valid moves. However, participants could also choose to not work with the AI agent by electing to exit through the grey blocks on either side of the board, in which case they would only be awarded 5 points.

After participants completed all 15 trials within Pavlovia, they were instructed to return to Qualtrics to answer three qualitative questions about their actions in the experiment. Participants were asked "Did you think the AI agent was using a certain strategy to play the game? If so, could you explain it?", "Generally, how did you choose your own behavior during the trials?" and lastly, they were asked to "Rate the level of intelligence the AI exhibited during the experiment:" using a slider with the leftmost value representing "no intelligence" and the rightmost value representing "very intelligent".

3 Results

To understand how the experiment participants performed in cooperation with the AI agent given the treatment condition, we analyzed recorded task performance across trials. We also collected answers to open-ended survey questions to understand why they may have exhibited that behavior.

3.1 Task Performance

A 2 × 3 ANOVA (participant demographic, treatment group) of participants' final scores showed a significant effect of participant demographic ($F = 5.81$, $p < .05$), but not treatment ($F = .45$, $p = .64$) or demographic × treatment ($F = 1.81$, $p = .17$). Figure 2 shows all of the average running total scores for each demographic-by-treatment interaction and a discernable difference between Black participants' and White participants' average running total scores across trials.

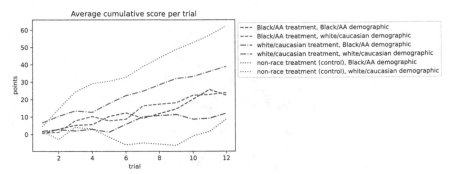

Fig. 2. Average of running total score for each treatment group × demographic interaction.

Looking at the ordering of the average running total scores within each participant demographic (Fig. 2), those data show a more consistent ordering across trials between treatment groups for the White participants (with the control condition showing the highest score and the Black condition showing the lowest score), though Black participants did show a less consistent opposite ordering (with the Black condition showing the highest score and the control condition showing the lowest score).

Though we did not find significant results with the treatment or demographic x treatment in the ANOVA, a more direct comparison between participants from different demographics may prove more apt given the trends shown in Fig. 2. There was little difference between the average running total score across blocks for self-identified Black participants and self-identified White participants who were in the Black/Africana American treatment group (Fig. 3). Black participants and White participants in this treatment group showed an average final score of 22.6 and 24.1 (respectively).

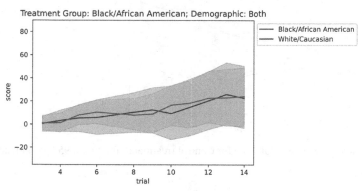

Fig. 3. Average of running total score for Black treatment group with 95% confidence interval bands.

When looking at Black and White participants in the White treatment group, the running total score begins to diverge as the task progresses. Despite Black and White participants showing a greater distance between average final scores (12.3 and 39.1, respectively), those data still showed an overlap in confidence intervals (albeit lesser when compared to the Black treatment) (Fig. 4).

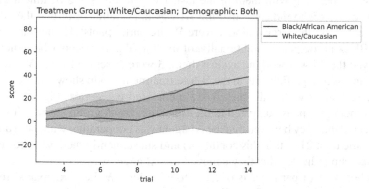

Fig. 4. Average of running total score for White treatment group with 95% confidence interval bands.

When comparing the average running score of Black participants and White participants in the control (non-race) condition, these data show a greater divergence in performance than the other two conditions. Indeed, by the end of the final trial, average Black and White participant score was 8.9 and 62.6 (respectively) (Fig. 5).

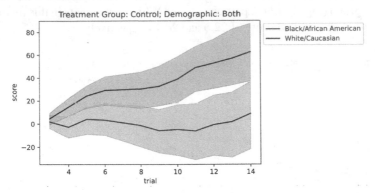

Fig. 5. Average of running total score for Control treatment group with 95% confidence interval bands.

3.2 Qualitative Data

The qualitative survey dataset provides more insight into how the treatment group might have affected the participants' sentiments towards the AI agent. Similar to the task-related performance data in the previous section, Black participants appeared to have more consistency across some key categories of responses. Figure 6 shows that of the 19 participants in the Black treatment group whose qualitative responses indicated that they believed that the AI was cooperating with the participant, a large proportion were Black (15 total as opposed to the 4 White participants who indicated that they believed the AI was cooperating with them in the task.) Within this same treatment group, of the 15 participants who indicated that the AI was not intelligent or that the AI had no pattern of behavior, the large majority were White participants (12 total) as compared to the (3) Black participants. Additionally, of the 4 total participants who indicated that they believed the AI was working against them, 3 were Black and 1 was White.

Black and White participants in the White treatment group show a more equal proportion of responses within the previously mentioned categories (Fig. 7). Within the White treatment group, as compared to the Black treatment group, more White participants indicated that they believed that the AI agent was cooperating with them during the task (11 of the total 21 within this condition) and subsequently there were fewer White participants who indicated that they believed the AI agent was not intelligent or had no pattern of behavior (4 participants of the 8 total). Thus, White participants' responses aligned more with responses from Black participants. Interestingly, while none of the Black participants indicated that they believed the AI was working against them in this condition, 3 White participants indicated as such in their survey response.

Figure 8 shows that while there were a higher number of responses in the control group that indicated a belief that the AI was cooperating with the participant as compared

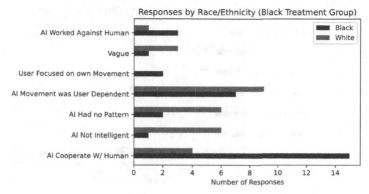

Fig. 6. Counts of categorized written responses from participants in the Black treatment group.

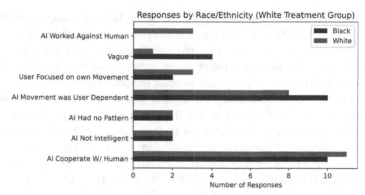

Fig. 7. Counts of categorized written responses from participants in the White treatment group.

to the other two treatment groups (25 total), Black participants showed a higher number of responses in this category (15) than White participants (10). Black participants in the control group had a number of responses that indicated they did not believe the AI agent was intelligent or that it did not have a discernable pattern (2 for both) that was similar to the number of responses by Black participants in the Black and White treatment groups.

3.3 Discussion

The task performance results showed an interesting difference between Black and White participants: though White participants (on average) performed better on this task, this appeared to be somewhat mediated by the treatment. Though Black and White participants showed nearly identical average running total scores within the Black/African American treatment group, the treatments show opposite ordering for the White and control conditions. Thus, while Black participants in the Black treatment group performed the best among the Black participants, White participants in the Black treatment group performed the *worst* among the White participants. Given that a participant achieves a higher score when they work with the AI agent to "capture the pig" (as opposed to

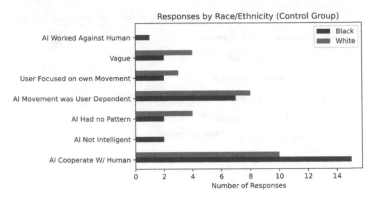

Fig. 8. Counts of categorized written responses from participants in the control treatment group.

just exiting the screen), the lowest average score by the White participants in the Black treatment group likely indicates a decreased willingness to cooperate with the AI agent (as compared to other treatment groups), and conversely the highest average score by Black participants in the Black treatment group indicates an increased willingness to cooperate with the AI agent (as compared to other treatment groups).

Interestingly, the qualitative data indicate that White participants were much less likely to judge the AI agent as cooperative than both the White participants in other treatment groups and the Black participants in their same treatment group and much more likely to judge the AI agent as not intelligent or as not showing a discernable pattern (with external judgement of discernable patterns being another way one might judge *intelligence*, Haugeland [10], p. 5). While the Black participants showed a fairly consistent (and low) proportion of responses that fell in the "cooperate" and "not intelligent" or "no pattern" categories, the White participants proportions of responses that consider the AI agent not intelligent appear to be more similar between the White and control treatment conditions, with the Black treatment condition standing out as high among all the demographic, treatment interactions.

Cave [6] describes intelligence as a value-laden term and argues that throughout history it has been used to preserve the power of the "white, male elite". The implementation of IQ tests and SAT provide context for these statements as these two forms of testing were used to justify the oppression of Black people, as well as other people from historically marginalized communities [6]. These methods of measuring intelligence have become dominant in our society; however, they represent a very narrow view of what intelligence actually can be. Given the historical use of intelligence tests to justify oppression through the painting of certain groups of people as less intelligent, some participants (particularly the White participants given the performance and qualitative data) may have assumed the Black AI would be less intelligent and would not allow for participants to score the maximum number of points within the study.

Stanley et al. [11] found that people were more willing to work with and trust those they relate to more (including by race), which is another way in which the results can be explained. Another explanation may be tied to the "dominant mode" of *being human* (that is, what it means to be human) being ascribed to White Western male, a "genre of the human" that excludes Black people [14]. The historical, sociocultural knowledge that defines *the human* in terms of whiteness perhaps led to participants in the study to relate the *whiteness* of the AI agents to the *humanness* and the *intelligence* of that agent [14]. Thus, White participants may have trusted the AI agent racialized as White more than the AI agent racialized as Black. This also could explain why White participants were more likely to work with the AI agent in the control condition, than in the Black condition. Though participants weren't told that the AI agent was trained on a "specific demographic" in the control condition, AI agents have been popularized as *White*, perhaps increasing the White participants likelihood to see the AI agent as trustworthy and cooperative during the task, and consequently making the Black participants less likely to cooperate with the agent during the task (even if those participants weren't more likely to report that the AI agent was *not intelligent* or *without a pattern*) [7].

4 Conclusion

Future work should expand upon these findings to dig deeper into these race/ethnicity and treatment effects. Other traditional categories of race within the existing (US) system of racism should be included to understand how *proximity to* whiteness (e.g., see Bonilla-Silva [5]) may affect interaction with knowledge of how the agent was trained (i.e., the *Whiteness* of the AI agent). More work could also be done to understand whether the effect of the treatment may be modulated by the presentation of the knowledge of the demographics of the people that had their behavior recorded to train the agent. In this study, the treatment was a statement on how the agent was trained that may not have proved particularly salient (or may not have had as large of an effect as possible on the experiment). Given the likely decay of the *subsymbolic* value of memory associated with the treatment (e.g., Anderson [2]), increasing the effects of the treatment on memory may be possible by repeating how the agent was trained during the task.

The current study makes use of a collaborative task, to examine whether participants would be more likely to trust and cooperate with an AI racialized as White and an AI that isn't explicitly racialized rather than one that is racialized as Black. Moreover, instead of selecting a phenotypical representation for the racialization (e.g., changing the color of the agent), we sought to see how *knowledge of* racialization may affect cooperation behavior. We found that the demographics of the participant and the racialization of the agent affected not only task performance, but also how the participants *perceived* AI agent behavior (particularly for white participants in the latter case). This work suggests that racialization of AI agents, even if superficial and not explicitly related to the behavior of that agent, may result in different cooperation behavior with that agent, showing potentially insidious and pervasive effects of racism on the way people interact with AI agents.

Acknowledgments. This work was supported by the National Science Foundation under grant No. 1849869.

References

1. Addison, A., Bartneck, C., Yogeeswaran, K.: Robots can be more than black and white: examining racial bias towards robots. In: Proceedings of the AAAI/ACM Conference on Artificial Intelligence, Ethics, and Society, Honolulu, HI, pp. 493–498 (2019)
2. Anderson, J.R.: How Can the Human Mind Occur in the Physical Universe? OUP, New York (2007)
3. Bartneck, C., et al.: Robots and racism. In: Proceedings of the 2018 ACM/IEEE International Conference on Human-Robot Interaction (HRI 2018), New York, NY (2018)
4. Benjamin, R.: Race After Technology: Abolitionist Tools for the New Jim Code. Polity Press, Medford (2019)
5. Bonilla-Silva, E.: The structure of racism in color-blind, "post-racial" America. Am. Behav. Sci. **59**(11), 1358–1376 (2015)
6. Cave, S.: The problem with intelligence: its value-laden history and the future of AI. In: Proceedings of the AAAI/ACM Conference on AI, Ethics, and Society, New York, NY, USA, pp. 29–35 (2020)
7. Cave, S., Dihal, K.: The whiteness of AI. Philos. Technol. **33**(4), 685–703 (2020). https://doi.org/10.1007/s13347-020-00415-6
8. Correll, J., Park, B., Judd, C.M., Wittenbrink, B.: The police officer's dilemma: using ethnicity to disambiguate potentially threatening individuals. J. Pers. Soc. Psychol. **83**(6), 1314–1329 (2002)
9. Costanza-Chock, S.: Design Justice: Community-Led Practices to Build the Worlds We Need. MIT Press, Cambridge (2020)
10. Haugeland, J.: Mind Design II: Philosophy, Psychology, Artificial Intelligence. MIT Press, Cambridge (1997)
11. Stanley, D.A., Sokol-Hessner, P., Banaji, M.R., Phelps, E.A.: Implicit race attitudes predict trustworthiness judgments and economic trust decisions. Proc. Natl. Acad. Sci. **108**(19), 7710 (2011)
12. Strait, M., Ramos, A.S., Contreras, V., Garcia, N.: Robots racialized in the likeness of marginalized social identities are subject to greater dehumanization than those racialized as white. In: Proceedings of the 2018 27th IEEE International Symposium on Robot and Human Interactive Communication (RO-MAN), 27–31 August 2018, pp. 452–457 (2018)
13. Beasley, M.: There is a supply of diverse workers in tech, so why is silicon valley so lacking in diversity? Center for American Progress (2017). https://www.americanprogress.org/issues/race/reports/2017/03/29/429424/supply-diverse-workers-tech-silicon-valley-lacking-diversity/
14. Wynter, S.: Unsettling the coloniality of being/power/truth/freedom: towards the human, after man, its overrepresentation - an argument. CR New Centennial Rev. **3**(3), 257–337 (2003)
15. Yoshida, W., Dolan, R.J., Friston, K.J.: Game theory of mind. PLoS Comput. Biol. **4**(12), e1000254 (2008)

Tweeting While Pregnant: Is Online Health Communication About Physical Symptoms and Emotions About the Tweeter?

Julia M. Hormes[1], Laurie Beth Feldman[1], Eliza Barach[1],
Vidhushini Srinivasan[2], and Samira Shaikh[3(✉)]

[1] State University of New York - University at Albany, Albany, USA
[2] Oracle Cloud Infrastructure, Seattle, WA, USA
[3] University of North Carolina at Charlotte, Charlotte, USA
samirashaikh@uncc.edu

Abstract. We analyze word usage patterns in posts on Twitter to learn
about information and support seeking related to the common experience
of food cravings in pregnancy. We apply information-theoretic measures
of lexical diversity across tweets to understand information sharing about
the body and its conditions (measured by use of embodiment verbs), and
about emotions (measured by degree of positive or negative valence) conditioned on degree of self-disclosure (choice of pronoun). Descriptions of
food cravings were consistent with face-to-face work and included frequent
references to sweets and fruits and sometimes included face emojis signaling humor and self-deprecation. Tweet content varied with degree of self-
disclosure such that messages formed around variants of the pronoun I
included a wider variety of emotion-related words but a more limited variety of words referencing the body than did those without that pronoun.
Patterns of reduced word redundancy suggest greater elaboration with
respect to emotional affect in the presence of I pronouns while enhanced
redundancy indicates shared endorsements and concerns pertaining to the
body in the presence of I pronouns. Findings are discussed in the context
of prior work on communication in online and face-to-face communities.

Keywords: Pregnancy · Twitter · Food cravings · Embodiment ·
Pronoun use · Egocentric perspective · Self-disclosure

1 Introduction

Women often manage the uncertainty and anxiety associated with the transition
to motherhood by "performing" pregnancy in accordance with prevailing societal
norms and practices [1]. Much of the research on how women seek out information
and approval for their "performance" of pregnancy has been qualitative, surveying
small samples of participants in great detail. While this approach has substantial
merits, the representativeness of these samples and the generalizability of findings

© Springer Nature Switzerland AG 2021
R. Thomson et al. (Eds.): SBP-BRiMS 2021, LNCS 12720, pp. 289–298, 2021.
https://doi.org/10.1007/978-3-030-80387-2_28

are limited. We adopt an alternative methodology that uses information pregnant women share on an online social media website to gain a better understanding of how they navigate changes characteristic of pregnancy, specifically those involving food, emotions, and their bodies. Social media posts reveal what people are interested in and concerned about in domains as varied as politics [2], consumer behavior [3], and health care [4,5]. For example, posts to Twitter by users who self-report a diagnosis of ADHD exhibit distinct language differences relative to controls, providing insights into their thoughts and emotional experiences [6]. An analysis of Twitter posts about antibiotics revealed shared patterns of misinformation and misuse of medications [7]. A study of an asynchronous Rheumatoid Arthritis online forum showed how users engage in task and rapport oriented functions, thereby contributing to the construction of a collective identity [8].

We demonstrate here how post produced on the Twitter online social media platform provide a unique opportunity for researchers to gain novel insights into concerns and questions pertaining to pregnancy that have been documented from traditional responses to structured survey questions by more limited samples. Our approach is a template for the study of similar health phenomena using data generated from social media.

Food cravings and embodiment in pregnancy: We define the corpus of messages that forms the basis for the analyses presented here around the topic of food cravings, strong and subjectively irresistible urges to consume certain foods [9]. A significant increase in the frequency and intensity of food cravings has been linked to certain female reproductive states, including pregnancy, making it an important part of the "performance" of pregnancy and a common shared experience that pregnant women communicate about, including in an online setting [10,11]. We identify the types of foods craved in pregnancy, as indicated by the frequency with which relevant words are mentioned in an ongoing blog and link the pattern to survey responses by pregnant women. We also examine affective reactions to cravings via analyses of emotionally valenced words and use of emojis, which were previously shown to be commonly used to express affect in food-related tweets [12].

Pregnant women face an increasingly complex relationship not only with their changing bodies, but also with the real and perceived reactions of society to those changes, a process often referred to as pregnant embodiment [1,13]. More generally, embodiment refers to a sense of the physical self, based on what we perceive and perhaps attend to, as well as to how we relate to others [14]. In the domain of language study, word meaning varies systematically along many dimensions, including embodiment (e.g., SLEEP, SEE, STARVE and SWEAT have high ratings of embodiment) [15]. We hypothesize that embodiment taps into a core shared experience among women who are pregnant, in particular as it relates to craving, and therefore identify and include embodiment terms in the analyses presented here.

Pronoun use, tweet valence, and lexical diversity: Lexical diversity is a relatively novel measure that captures the degree to which tweets differ from one another with respect to redundancy of word choice [16]. It is based on information theory and widely used in computational linguistics [17–19]. Lexical diversity derives from both the number of word types and the number of different instances (tokens) within a type. Low diversity is associated with a small set of equiprobable words. High diversity reflects a larger set of equiprobable words.

Prior work by our group has detected differences between the pattern of word use in tweets constructed with the first-person singular pronouns (I, ME, MY, MINE) and those constructed with other personal pronouns [20–22]. Tweets with the pronoun I contained more redundant emotion vocabulary reflective of decreased lexical diversity, compared to tweets with WE or no pronoun [20]. Here, we ask whether in the context of pregnancy-related social media posts, communication with emotionally valenced or high embodiment words similarly varies in a manner that can be associated with an aspect of tweeting style relevant to self disclosure, specifically the use of a particular pronoun.

Goals of this study: This study sought to replicate and expand upon previous findings regarding patterns of word usage in online communication by examining a corpus of tweets produced around the shared experience of food cravings in pregnancy. We expand upon simple descriptions of general word and hashtag use by examining emoji use, which was previously shown to reflect and foster group cohesion [23–25]. We expand upon responses to traditional structured survey questions about cravings by tracking the content of tweets.

Our predominant analytic method in the present study is to examine redundancy of word usage with lexical diversity analyses derived from information theory [26]. We test the previously articulated Pronoun Disclosure Hypothesis and extend the scope of prior work on lexical diversity beyond positive and negative emotional affect to embodiment terms, that is, words that refer to actions or states of the body or a part thereof, in part because they may be especially relevant to the experience of pregnancy [20–22].

2   Method

This study did not involve direct contact with participants and no institutional ethics approval was required. Given its focus, those who contributed tweets of interest to the current investigation are likely to be women of child-bearing age who are currently pregnant. We collected data via the TWITTER GNIP service between 06-15-2016 and 06-30-2016, a period chosen because it included no major food-related events (e.g., Thanksgiving). A list of targeted hashtags was developed collaboratively by members of the research team, based on pilot work on the nature, types, and frequency of food cravings commonly reported by women during pregnancy in online forums [11]. The ten most common were #pregnancyproblems, #pregnancy, #diet, #postpartum, #newborn, #fitmoms, #pregnancycravings, #pregnancy, #pregnancylife, #cravings. We extracted the following metadata from the 1712 tweet downloaded using these hashtags: location, text, timestamp, and a list of hashtags. We first determined which words occurred most frequently in the full corpus of tweets defined by the set of relevant hashtags. Forms of the same word were combined (e.g., the frequencies of BABY and BABIES were combined in the word form BABI).

We classified foods mentioned in tweets examined here based on the Food Craving Inventory (FCI), which defines four categories of cravings: (1) high fats, (2) sweets, (3) carbohydrates/starches, and (4) fast-food fats [27]. Any uncategorized food words were assigned either to a category of (5) fruits and vegetables or to a category of (6) other, as appropriate.

Tweets were then categorized according to whether they included variants of the first-person singular pronoun. We tracked the number of occurrences of each word with strongly positive and strongly negative emotional affect based on the valence ratings of Warriner et al. [33] and of embodiment terms [15], and then determined whether diversity measures for word valence and embodiment interacted with pronoun choice. Similar to our recent work on emotion words in an emergent online community [20], we focused on those words with the most extreme ratings (i.e., top 25%) of valence and embodiment (top 20%) and performed two types of analyses. First, we compared weighted means for the most extremely valenced and embodied words in tweets with versus without first-person singular pronouns. We then compared lexical diversity in tweets with and without first-person singular pronoun when embodiment and valence (positive words such as FAMILY, HOPE, and negative words such as SWOLLEN, CRIME) were extreme.

3   Results

Lexical measures: The typical tweet had between zero and two hashtags in addition to the target hashtag. The median message was comprised of 15 words. "BABY" "TIPS," "WEIGHT" and "LOSING" were the most frequently used words in the corpus of tweets examined here (Fig. 1). All but one of the most common emojis to appear in the tweets collected here were faces, with the face emoji with a stuck-out tongue and winking eye being used most frequently (12%; Table 1). Emojis depicting food items were more rarely used and included a chocolate bar (3%), strawberry (2%), and donut (1.5%).

Food cravings by category. The most commonly mentioned foods in tweets about cravings in pregnancy were those categorized as "other" (34%) and included items such as "ICE," "SALT," AND "MILK" (Table 2). Next most frequently mentioned were "sweets" (22%), "fast foods fats" (14%), and "high fat foods" (13%). "Fruits/vegetables" were mentioned in only 9% of tweets analyzed and "carbohydrates/starches" were least likely to be cited (7%). In terms of number of different foods, tweets about sweet cravings ranked highest in variability, followed by tweets about fast foods and high fat foods. Results are generally consistent with those from survey data of pregnant women [11].

3.1   Pronoun Use and Lexical Diversity

Valence. A total of 695 tweets contained a variant of I (I, ME, MY) and 1017 did not; 3022 unique words occurred in the tweets that did not contain I pronouns, compared to 2329 words in tweets without an I pronoun. Valence ratings from

Table 1. Frequency of use of face and food emojis.

Emoji	😂	😆	😫	😍	😋	🍫	😊	😎	😳
%	12%	8%	6%	4%	4%	3%	2%	2%	2%

Table 2. Percent mentions and food types by food category in tweets about pregnancy cravings

	High fat	Sweets	Carbs/Starches	Fast Food/Fats	Fruits/Veggies	Others
Examples	Ribs, cheese, chicken	Ice cream, chocolate, cake	Bread, pasta, cereal	Pizza, fries, burger	Fruit, orange, salad	Ice, salt, milk
Mentions (%)	13%	22%	7%	14%	9%	34%
Num. unique words	35	57	25	40	36	35

Fig. 1. Most common words in all tweets

Fig. 2. Relative frequency of high embodiment terms for tweets with (a) and without (b) variants of the pronoun I

Warriner et al. [33] were slightly less extreme for strongly (i.e., top 25%) negative and positive words in tweets that co-occur with as compared to without a variant of I. Lexical diversity was higher in the presence of a variant of the I pronoun (ME, MY, MINE, I), relative to non I tweets both when polarity was positive and when it was negative (see Table 3).

Embodiment. Words with the highest embodiment values in our corpus included EAT (6.23), KICK (6.32), SLEEP (5.97), WEAR (5.68), GO (5.13), and CRAVE (4.42) [14]. Lexical diversity among terms high on embodiment was lower when they included a variant of I than when they did not (Table 3; Fig. 2). At the same time, the mean of I embodiment terms tended to be more strongly embodied. Of note, the word CRAVE frequently co-occurred with I (Fig. 2a), as contrasted with FEEL, EAT, LOVE and GO in its absence (Fig. 2b).

Table 3. Number of tweets, weighted valence and lexical diversity of the 25% most strongly valenced and 20% most strongly embodied words in tweets co-occurring with and without variants of I

	Pronoun	Num. of tweets	High valence tokens	Average valence	Lexical diversity	Confidence interval
Negative valence	I tweets	368	49%	2.64	6.11	5.70–6.51
	Non I tweets	439	51%	2.59	4.45	4.34–4.61
Positive valence	I tweets	327	45%	7.01	6.68	6.52–6.85
	Non I tweets	578	55%	7.04	5.86	5.79–5.97
Embodiment	I tweets	695	40%	3.98	5.24	5.08–5.41
	Non I tweets	1017	60%	2.24	5.42	5.29–5.59

4   Discussion

The goal of this study was to analyze word usage patterns in tweets about food cravings in pregnancy. We specifically sought to replicate prior work regarding the relationship between pronoun use, word valence, and lexical diversity in an online health context. We expanded on previous research by examining emoji use in addition to word frequency. We also examined the relationships between pronoun type, lexical diversity, and use of embodiment terms about the body and its states, a topic especially relevant to information seeking and sharing about the experience of pregnancy.

Analyses of word frequency when tweeting are largely consistent with the survey literature on food cravings during pregnancy. Both suggest frequent urges for sweets as well as fruits during gestation [11] and support our assumption that most tweeters we sampled were women. Findings related to emojis in tweets about food are consistent with general usage patterns and document the generally higher prevalence of emojis depicting faces as compared to food [34]. The most frequently used face emoji was the one with a stuck-out tongue and winking eye, indicating that women who posted on the topic of pregnancy cravings seem to approach those cravings with a certain sense of humor and self-deprecation [28–30]. In summary, emoji use in tweets about food cravings are consistent with those in consumer research about food, in both faces with tears of joy and savoring delicious food are very common [31].

A primary goal of the present study was to test the predictions of the recently formulated Pronoun Disclosure Hypothesis in a health-related context. We used indices of lexical diversity to determine the extent to which tweets in the current corpus recycled the same words. We detected systematic variation between patterns of word use and pronoun choice here as in other contexts. Interestingly, in contrast to our prior work on tweets shared in the aftermath of a terrorist attack [19], in social media posts about food and pregnancy, tweets in I showed *greater* lexical variation of strongly valenced words than comparison tweets. Differences in lexical diversity between I and non-I tweets was more pronounced in negatively than positively valenced messages, but a similar pattern was evident for both. In other words, we observed more original word choice and less coordination among participants whose tweet were strongly affect laden with forms of I. Evidently in a health related context, contributors tended not to borrow emotion words from others.

Findings suggest that the relation between lexical diversity in emotion words and patterns of pronoun use varies across corpora. We interpreted our previous finding as evidence of pressures against self-disclosure of personal emotions in the context of a global crisis, especially when expressing negative affect [19]. In online communication about food cravings and pregnancy, by contrast, lexical diversity pertaining to affect patterns according to pronoun use in a consistent way, one that is indicative of greater elaboration when including the first-person pronoun, regardless of affective valence. It appears that any pressure not to self-disclose what one is feeling in online communication does not generalize to the food and pregnancy context. Instead, elaborated sharing of emotion seems consistent with the "performance" of pregnancy in accordance with perceived social norms and encouraged in the context of an online community. Whether this reflects relative homogeneity of concerns and needs and rituals among the members of a community, the longevity of the community and the trust that ensues, or some combination of (possibly other) factors cannot be determined from the present study and warrants further investigation.

To answer a question of particular relevance to pregnancy, we also examined the interaction of lexical diversity by pronoun use with words whose meaning pertains to embodiment. We found that diversity among high embodiment words was lower in tweets that included variations of the I pronoun than in ones that did not. Until now, we have interpreted reduced lexical variability as an indication that tweeters were recycling words and repeating the statements of others and have assumed that this arises because of pressures against self-disclosure. In the present context we introduce another possible factor that can reduce lexical diversity, namely, pregnant women are seeking information or are recounting similar experiences about their own bodies. In the context of online heath communication, we propose that reduced lexical diversity involving the body may reflect seeking or sharing similar experiences and information rather than any pressures against self-disclosure.

Limitations. Some significant limitations to this research must be noted. First, despite many promising parallels to real-life interactions, online communication can change the nature of some social relationships in fundamental ways. The extent to which our findings enhance an understanding of face-to-face communication remains open. Second, analysis of large data sets generated by social media does not consider individual differences that could influence the frequency and content of tweets. Some contributors may primarily seek information while others are predominantly offering support. Further, some are more direct and personally revealing in their communication than are others. At present, our findings should be interpreted as primarily reflecting collective interests and concerns. Another limitation of this method of analysis is that it does not consider if tweets were posted in isolation or as part of an ongoing exchange among two or more communication partners who may or may not have a history of interactions with one another. Verbal communication is a highly social process, and research suggests that much like the actual physical presence of a communication partner, online communications can similarly elicit coordination especially when participants feel close [32]. Future research should take these factors into account and assess the extent to which tweet patterns change as a function of the life span and dynamics of the online community in which they are elicited.

Finally, results with embodiment terms seem to pose a challenge to the claim that the reduction always reflects a failure to disclose. An alternative explanation is that reduced lexical diversity may reflect stereotypy of concerns conveyed by consistent use of the same terms as for information seeking. Future work should explore the relation between lexical diversity and communicative context.

References

1. Neiterman, E.: Doing pregnancy: pregnant embodiment as performance. Women Stud. Int. Forum **35**, 372–83 (2012)
2. Himelboim, I., Hansen, D., Bowser, A.: Playing in the same Twitter network: political information seeking in the 2010 gubernatorial elections. Inf. Commun. Soc. **16**(9), 1373–1796 (2013)
3. Goel, S., Hofman, J.M., Lahaie, S., Pennock, D.M., Watts, D.J.: Predicting consumer behavior with Web search. Proc. Natl Acad. Sci. USA. **107**(41), 17486–17490 (2010)
4. Wang, J., Ashvetiya, T., Quaye, E., Parakh, K., Martin, S.S..: Online health searches and their perceived effects on patients and patient-clinician relationships: a systematic review. Am. J. Med. (2018); Epub ahead of print
5. Fielden, A.L., Sillence, E., Little, L., Harris, P.R.: Online self-affirmation increases fruit and vegetable consumption in groups at high risk of low intake. Appl. Psychol. Health Well Being **8**(1), 3–18 (2016)
6. Guntuku, S.C., Ramsay, J.R., Merchant, R.M., Ungar, L.H.: Language of ADHD in adults on social media. J. Attent. Disord. **23**(12), 1475–85 (2019)
7. Scanfeld, D., Scanfeld, V., Larson, E.L.: Dissemination of health information through social networks: twitter and antibiotics. Am. J. Infect. Control **38**(3), 182–188 (2010)
8. Angouri, J., Sanderson, T.: You'll find lots of help here' unpacking the function of an online Rheumatoid Arthritis (RA) forum. Lang. Commun. **46**, 1–13 (2016)
9. Hormes, J.M., Rozin, P.: Does, "craving" carve nature at the joints? Absence of a synonym for craving in most languages. Addict. Behav. **35**, 459–463 (2010)

10. Orloff, N.C., Flammer, A., Hartnett, J., Liquorman, S., Samelson, R., Hormes, J.M.: Food cravings in pregnancy: preliminary evidence for a role in excess gestational weight gain. Appetite **105**(1), 259–265 (2016)
11. Orloff, N.C., Hormes, J.M.: Pickles and ice cream! Food cravings in pregnancy: hypotheses, preliminary evidence, and directions for future research. Front. Psychol. **5**, 1076 (2014)
12. Vidal, L., Ares, G., Machin, L., Jaeger, S.R.: Using twitter data for food-related consumer research: a case study on "what people say when tweeting about different eating situations." Food Qual. Ref. **45**, 58–69 (2015)
13. Sweeney, S.T., Hodder, I.: The Body, Cambridge University Press; Cambridge (2002)
14. Dijkstra, K., Post, L.: Mechanisms of embodiment. Front. Psychol. **6**, 1525 (2015)
15. Sidhu, D.M., Kwan, R., Pexman, P.M., Siakaluk, P.D.: Effects of relative embodiment in lexical and semantic processing of verbs. Acta Psychol. **149**, 32–39 (2014)
16. Barach, E., Shaikh, S., Srinivasan, V., Feldman, L.B.: Hiding behind the words of others: does redundant word choice reflect suppressed individuality when tweeting in the first person singular? Adv. Hum. Fact. **783**, 603–614 (2019)
17. Lester, N.A., Du Bois, J.W., Gries, S.T.: Moscoso del Prado MF. Considering experimental and observational evidence of priming together, syntax doesn't look so autonomous. Behav. Brain Sci. **40**, e300 (2017)
18. Moscoso del Prado, M.F.: Vocabulary, grammar, sex, and aging. Cogn. Sci. **41**(4), 950–975 (2017)
19. Moscoso del Prado, M.F., Kostic, A., Baayen, R.H.: Putting the bits together: an information theoretical perspective on morphological processing. Cognition **94**(1), 1–18 (2004)
20. Barach, E., Shaikh, S., Srinivasan, V., Feldman, L.B.: Hiding behind the words of others: does redundant word choice reflect suppressed individuality when tweeting in the first person singular? In: Kantola, J.I., Nazir, S., Barath, T. (eds.) AHFE 2018. AISC, vol. 783, pp. 603–614. Springer, Cham (2019). https://doi.org/10. 1007/978-3-319-94709-9_59
21. Twenge, J.M., Campbell, W.K., Gentile, B.: Changes in pronoun use in American books and the rise of individualism, 1960–2008. J. Cross Cult. Psychol. **44**(3), 406–415 (2013)
22. Shaikh, S., Feldman, L.B., Barach, E., Marzouki, Y.: Tweet sentiment analysis with pronoun choice reveals online community dynamics in response to crisis events. Adv. Cross Cult. Decision Making **480**, 345–356 (2017)
23. Danesi, M.: Emoji in advertising. Int. J. Semi. Vis. Rhet. **1**(2), 1–12 (2017)
24. Vandergriff, I.: A pragmatic investigation of emoticon use in nonnative/native speaker text chat. Language@Internet **11** Article 4 (2014)
25. Graham, S.L.: A wink and a nod: the role of emojis in forming digital communities. Multilingua. (2019, online first)
26. Shannon, C.E.: A mathematical theory of communication. Bell Syst. Tech. J. **27**, 379–423 (1948)
27. White, M.A., Whisenhunt, B.L., Williamson, D.A., Greenway, F.L., Netemeyer, R.G.: Development and validation of the food-craving inventory. Obes. Res. **10**, 107–114 (2002)
28. Weissman, B., Tanner, D.: A strong wink between verbal and emoji-based irony: how the brain processes ironic emojis during language comprehension PLoS ONE **13**(8), e0201727 (2018)

29. Pohl, H., Domin, C., Rohs, M.: Beyond just text: semantic emoji similarity modeling to support expressive communication. ACM Trans. Comput. Hum. Inter. **24**(1), Article 6 (2017)
30. Sun, N., Lavoue, E., Aritajati, C., Tabard, A., Rosson, M.B.: Using and perceiving emoji in design peer feedback. In: Lund, K., Niccolai, G.P., Lavoue, E., Hmelo-Silver, C., Geweon, G., Baker, M. (eds.) 13th International Conference on Computer Supported Collaborative Learning (CSCL), pp. 296–303. International Society of the Learning Sciences, Lyon, France (2019)
31. Vidal, L., Ares, G., Jaeger, S.R.: Use of emoticon and emoji in tweets for food-related emotional expression. Food Qual. Prefer. **49**, 119–128 (2016)
32. Gonzales, A.L., Hancock, J.T., Pennebaker, J.W.: Language style matching as a predictor of social dynamics in small groups. Commun. Res. **37**(1), 3–19 (2009)
33. Amy Beth, W., Victor, K., Marc, B.: Norms of valence, arousal, and dominance for 13,915 English lemmas. Behav. Res. Methods **45**(4), 1191–1207 (2013). https://doi.org/10.3758/s13428-012-0314-x
34. Wijeratne, S., Saggion, H., Kiciman, E., & Sheth, A.P.: Emoji Understanding and applications in social media: lay of the land and special issue introduction. ACM Traans. Soc. Comput. **3**(2020)

On the Effect of Social Norms on Performance in Teams with Distributed Decision Makers

Ravshanbek Khodzhimatov[1]([✉]) [iD], Stephan Leitner[2][iD], and Friederike Wall[2][iD]

[1] Digital Age Research Center, University of Klagenfurt, 9020 Klagenfurt, Austria
ravshanbek.khodzhimatov@aau.at
[2] Department of Management Control and Strategic Management,
University of Klagenfurt, 9020 Klagenfurt, Austria
{stephan.leitner,friederike.wall}@aau.at

Abstract. Social norms are rules and standards of expected behavior that emerge in societies as a result of information exchange between agents. This paper studies the effects of emergent social norms on the performance of teams. We use the *NK*-framework to build an agent-based model, in which agents work on a set of interdependent tasks and exchange information regarding their past behavior with their peers. Social norms emerge from these interactions. We find that social norms come at a cost for the overall performance, unless tasks assigned to the team members are highly correlated, and the effect is stronger when agents share information regarding more tasks, but is unchanged when agents communicate with more peers. Finally, we find that the established finding that the team-based incentive schemes improve performance for highly complex tasks still holds in presence of social norms.

Keywords: Agent-based modeling and simulation · *NK*-framework · Emergence · Socially accepted behavior

1 Introduction

One of the main goals of team managers is to design framework conditions that allow teams to achieve a h igh performance. This, amongst others, includes the allocation of tasks and choice of the means of behavioral control, such as incentive schemes. Whenever people collaborate, however, there might be emergent social dynamics that probably interfere with an organization's or the managers' decisions on these framework conditions. In this paper, we address two questions related to emergent social norms and means of behavioral control: first, *how do emergent social norms affect a team's performance* and, second, *how is the efficiency of behavioral control mechanisms affected by the presence of social norms.* These questions are becoming even more apparent now, with continuous developments in digitalization and the shift of teams towards digital communication channels in time of a global pandemic.

© Springer Nature Switzerland AG 2021
R. Thomson et al. (Eds.): SBP-BRiMS 2021, LNCS 12720, pp. 299–309, 2021.
https://doi.org/10.1007/978-3-030-80387-2_29

In our research, we particularly address *descriptive* social norms, which Cialdini et al. [1] defined as the frames of reference that emerge when individuals observe and adopt the *actual behavior* of their peers, as opposed to normative moral claims that prescribe certain predefined actions to individuals. Priebe et al. [7] summarized the concept descriptive social norms as: *"When in Rome, do as the Romans do"*. An example of descriptive social norms is choosing LaTeX instead of PowerPoint to prepare presentations in the absence of any external requirement only because teammates do so.

The idea of descriptive social norms is not new but has been formalized, among others, by Cialdini et al. [1]. They argued that descriptive norms serve as "decisional shortcuts": *"If everyone is doing it, it must be a sensible thing to do"*. Pryor et al. [8] argued that another reason why people conform to descriptive norms is to diffuse responsibility for a risky action by imitating what they perceive is the status quo.

Since there are potentially many confounding variables, it is difficult to find an empirical evidence for whether and, if so, how individual's decisions are driven by social norms (in the above sense) [11]. Thus, in order to investigate this potential effect, we build a stylized agent-based model, in which descriptive norms are formed as individuals share and observe the behavior of their teammates.

In the context of agent-based simulations, the idea of *imitation* as a means to increase individual performance is not new. Rivkin [9] used the NK-framework [5,6] to model a firm that imitates an industry leader, and found that, in presence of complexity, the imitation does not result in an increase in performance. In this paper we follow a different approach: we do not model agents that directly imitate other agents but we focus on agents that conform with what appears to be socially acceptable behavior (which, of course, might be affected by imitation, amongst others). In particular, we model a team that works on a complex task and the team members are assigned parts of the task, which may or may not be interdependent. As the team members communicate, they share information about their past decisions, which forms the basis for the emergence of descriptive social norms. We investigate the effect of social norms on the functioning of the incentive schemes, which are adopted to control the team members' behavior, and observe the team's overall performance for environments with different task structures.

2   Model

In this section we introduce the agent-based model of a team of $P = 4$ individuals facing a complex task. The task environment is based on the NK-framework [5,6]. Agents make decisions to (a) increase their compensation and (b) comply with the descriptive social norms. Section 2.1 introduces the task environment, Sects. 2.2 and 2.3 characterize the agents and describe how social norms emerge, respectively. Section 2.4 describes the agents' search process, and Sect. 2.5 provides an overview of the sequence of events in the simulation.

2.1   Task Environment

We model an organization that faces a complex decision problem that is expressed as the vector of $M = 16$ binary choices. The decision problem is divided into sub-problems which are allocated to $P = 4$ agents, so that each agent faces an $N = 4$ dimensional sub-problem:

$$\mathbf{x} = (\underbrace{x_1, x_2, x_3, x_4}_{\mathbf{x}^1}, \underbrace{x_5, x_6, x_7, x_8}_{\mathbf{x}^2}, \underbrace{x_9, x_{10}, x_{11}, x_{12}}_{\mathbf{x}^3}, \underbrace{x_{13}, x_{14}, x_{15}, x_{16}}_{\mathbf{x}^4}), \qquad (1)$$

where bits $x_i \in \{0, 1\}$ represent single tasks. Every task x_i is associated with a uniformly distributed performance contribution as in Eq. 2. The decision problem is *complex* in that the performance contribution $\phi(x_i)$, might be affected not only by the decision x_i, but also by decisions x_j, where $j \neq i$. We differentiate between two types of such inter-dependencies: (a) *internal*, in which interdependence exists between the tasks assigned to agent p, and (b) *external*, in which interdependence exists between the tasks assigned to agents p and q for $p \neq q$. We control inter-dependencies by parameters K, C, S, so that every task interacts with exactly K other tasks internally and C tasks assigned to S other agents externally [4]:

$$\phi(x_i) = \phi(x_i, \underbrace{x_{i_1}, ..., x_{i_K}}_{\substack{K \text{ internal} \\ \text{interdependencies}}}, \underbrace{x_{i_{K+1}}, ..., x_{i_{K+C \cdot S}}}_{\substack{C \cdot S \text{ external} \\ \text{interdependencies}}}) \sim U(0, 1), \qquad (2)$$

where $i_1, \ldots, i_{K+C \cdot S}$ are distinct and not equal to i. We consider two cases: (i) *low complexity* (only internal interdependence: $K = 2, C = S = 0$) and (ii) *high complexity* (internal and external interdependence: $K = C = S = 2$).[1] Using Eq. 2, we generate *performance landscapes*[2] for all agents.

We are interested in the dynamics of the team's overall performance throughout $T = 500$ time periods. At each time period t, agent p's performance is a mean of performance contributions of tasks assigned to that agent:

$$\phi_{own}(\mathbf{x}_t^p) = \frac{1}{N} \sum_{x_i \in \mathbf{x}_t^p} \phi(x_i), \qquad (3)$$

and the team's overall performance is a mean of agents' performances:

$$\Phi(\mathbf{x}_t) = \frac{1}{P} \sum_{p=1}^{P} \phi_{own}(\mathbf{x}_t^p). \qquad (4)$$

---

[1] The exact choice of the coupled tasks is random with one condition: every task affects and is affected by exactly $K + C \cdot S$ other tasks.

[2] A performance landscape is a matrix of uniform random variables that correspond to every combination of $1 + K + C \cdot S$ decisions. We generate entire landscapes to find the overall global maximum and normalize our results accordingly, to ensure comparability among different scenarios.

The tasks allocated to agents can be similar or distinct. We model this using the pairwise correlations between performance landscapes[3]:

$$\text{corr}(\phi(\mathbf{x}_i^p), \phi(\mathbf{x}_i^q)) = \rho \in [0, 1], \tag{5}$$

for all $i \in \{1, 2, 3, 4\}$ and $p \neq q$. When $\rho = 0$ and $\rho = 1$, agents operate on perfectly distinct and perfectly identical performance landscapes, respectively.

2.2 Agents' Compensation

Agents p's compensation is based on p's own performance ϕ_{own}, and the residual performance ϕ_{res}, defined as the mean of performances of every agent other than p:

$$\phi_{res}^p = \frac{1}{P-1} \cdot \sum_{q \neq p} \phi_{own}(\mathbf{x}^q), \tag{6}$$

The compensation follows the linear incentive scheme[4]:

$$\phi_{inc}(\mathbf{x}_t^p) = \alpha \cdot \phi_{own}(\mathbf{x}_t^p) + \beta \cdot \phi_{res}^p, \tag{7}$$

where $\alpha + \beta = 1$.

2.3 Descriptive Norms

In this section we describe our model of descriptive norms[5]. First of all, we differentiate between two types of tasks, namely *private* and *social* tasks. Private tasks are specific to individual agents and are not subject to descriptive norms, while social tasks concern all agents. For example, in a team, the choice of a server operating system (Debian vs. RHEL) is a private task for the systems administrator, and the choice of a presentation software (LaTeX vs. PowerPoint) is a social task concerning all team members. In our formulation, agents want their decisions on social tasks to comply with those of their peers, but they do not have any preference for compliance in private tasks. Without loss of generality we assume that agents' task allocations are structured similarly and adhere to the following convention: the private tasks come first and social tasks come next. For example, if $N_S = 2$ is the number of social tasks allocated to each agent, then the first 2 tasks are private and the last 2 tasks are social:

$$\mathbf{x}^p = (\underbrace{x_1^p, x_2^p}_{\text{private}}, \underbrace{x_3^p, x_4^p}_{\text{social}}) \tag{8}$$

---

[3] See Verel et al. [10] for methodology.

[4] In our context linear incentives are as efficient as other contracts inducing non-boundary actions. See [2, p. 1461].

[5] We implement our version of the Social Cognitive Optimization algorithm. See Xie et al. [12] for the original version of the algorithm.

At every time step t, agents share their decisions on social tasks with $D = 2$ fellow agents, according to the network structure predefined by the modeler.[6] Every agent stores the shared information in the memory set L^p for up to $T_L = 50$ periods, after which the information is "forgotten" (removed from L^p). Descriptive norms are not formed until time period T_L.

The extent to which agent p's decisions $\mathbf{x}_t^p$ comply with the descriptive social norms is computed as the average of the matching social bits in the memory:

$$\phi_{soc}(\mathbf{x}_t^p) = \begin{cases} \dfrac{1}{|L_t^p|} \displaystyle\sum_{\mathbf{x}^L \in L_t^p} \dfrac{[x_3^p = x_3^L] + [x_4^p = x_4^L]}{2}, & t > T_L \\ 0, & t \leq T_L \end{cases} \qquad (9)$$

where $|L_t^p|$ is the number of entries in agent p's memory at time t, and the statement inside the square brackets is equal to 1 if true, and 0 if false [3].

2.4   Search Process

At time t, agents can observe their own performance in the last period, $\phi_{own}(\mathbf{x}_{t-1}^p)$, and the decisions of all team members in the last period *after* they are implemented, $\mathbf{x}_{t-1}$.

In order to come up with new solutions to their decision problems, agents perform a search in the neighbourhood of $\mathbf{x}_{t-1}$ as follows: agent p randomly switches one decision $x_i \in \mathbf{x}^p$ (from 0 to 1, or vice versa), and assumes that other agents will not switch their decisions[7]. We denote this vector with one switched element by $\hat{\mathbf{x}}_t^p$.

Next, the agent has to make a decision whether to stick with the status quo, $\mathbf{x}_t^p$, or to switch to the newly discovered $\hat{\mathbf{x}}_t^p$. The rule for this decision is to maximize the weighted sum of the performance-based compensation and the compliance with the descriptive social norms:

$$\mathbf{x}_t^p = \underset{\mathbf{x} \in \{\mathbf{x}_{t-1}^p, \hat{\mathbf{x}}_t^p\}}{\arg\max} \ w_1 \cdot \phi_{inc}(\mathbf{x}) + w_2 \cdot \phi_{soc}(\mathbf{x}), \qquad (10)$$

where $w_1 + w_2 = 1$.

2.5   Process Overview, Scheduling and Main Parameters

The simulation model has been implemented in *Python 3.8* and *Numba* just-in-time compiler. Every simulation round starts with the initialization of the

---

[6] We use the bidirectional *ring network*, in which each node is connected to exactly $D = 2$ other nodes with reciprocal unidirectional links, where nodes represent agents and the links represent sharing of information.

[7] Levinthal [6] describes situations in which agents switch more than one decision at a time as *long jumps* and states that such scenarios are less likely to occur, as it is hard or risky to change multiple processes simultaneously.

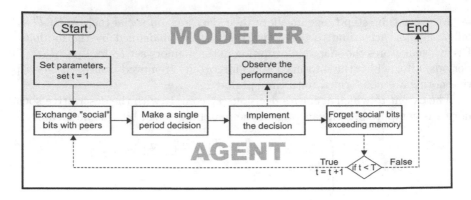

Fig. 1. Process overview

agents' performance landscapes, the assignment of tasks to $P = 4$ agents[8], and the initialization of an $M = 16$ dimensional bitstring as a starting point of the simulation run. After initialization, agents perform the *hill climbing* search procedure outlined above and share information regarding their social decisions in their social networks. The observation period T, the memory span of the employees T_L, and the number of repetitions in a simulation, R, are exogenous parameters, whereby the latter is fixed on the basis of the coefficient of variation. Figure 1 provides an overview of this process and Table 1 summarizes the main parameters used in this paper.

3   Results

We perform $R = 1000$ simulations for every combination of parameters presented in Table 1. Our main variable of interest is the dynamics of teams' normalized overall performance, Φ_t^r, for all $r \in \{1, 2, ..., R\}$ and $t \in \{1, 2, ..., T\}$[9]. We denote the normalized performance at time t, averaged over all simulation runs by:

$$\overline{\Phi_t} = \frac{1}{R} \sum_{r=1}^{R} \Phi_t^r \qquad (11)$$

The main results of the simulation for parameters defined in Table 1 are presented in Fig. 2. The figure contains 8 sub-figures for 2 levels of complexity

---

[8] For reliable results, we generate the entire landscapes before the simulation, which is feasible for $P = 4$ given modern computing limitations. Our sensitivity analyses with simpler models without entire landscapes suggest that the results also hold for $P = 5, 6, 7$.

[9] As the performance landscapes are randomly generated, we normalize the team performance by the maximum performance attainable in the current simulation run r to ensure comparability.

Table 1. Main parameters

Parameter	Description	Value
M	Total number of tasks	16
P	Number of agents	4
N	Number of tasks assigned to a single agent	4
$[K, C, S]$	Internal and external couplings	$[2, 0, 0]$, $[2, 2, 2]$
ρ	Pairwise correlation between landscapes	0.3
T_L	Memory span of agents	50
N_S	Number of social tasks	2
D	Number of connected peers (node degree)	1
T	Observation period	500
R	Number of simulation runs per scenario	1000
$[w_1, w_2]$	Weights for incentives ϕ_{inc} and compliance with the social norms ϕ_{soc}	$[1, 0]$, $[0.7, 0.3]$, $[0.5, 0.5]$
$[\alpha, \beta]$	Shares of own and residual performances	$[1, 0]$, $[0.75, 0.25]$, $[0.25, 0.75]$

described in Eqs. 2 and 3 incentive schemes with different weights of team performance described in Eq. 7. Each sub-figure includes 3 time series of normalized overall performance $\overline{\Phi_t}$ over $T = 500$ periods for different weights of social norms in agents' decision rules, defined in Eq. 10.

First of all, we observe that the performance is lower at all periods in environments with high complexity than in environments with low complexity (i.e. all values in the left column are greater than their counterparts in the right column), which is in line with the previous research [4, 6].

Second of all, we observe in all sub-figures that, as agents put more emphasis on social norms in their decision rules, the team's overall performance drops, which is more pronounced for environments with high complexity.

Third, we observe that in the absence of social norms (top curves in all sub-plots) the incentive schemes do not have a significant effect for tasks with low complexity, while incentive schemes that put more emphasis on team performance decrease the time it takes to converge for tasks with high complexity (i.e. the curves skew left as we move from upper to lower sub-plots in the right column, and do not change in the left column). This finding is in line with the existing literature [9].

Fourth, we find that in presence of social norms the incentive schemes do not have a significant effect on the long-run team performance most of the time, with two exceptions: (i) if the task has a low complexity and agents put a moderate emphasis on social norms, the long-run performance increases for incentive schemes that put less weight on team performance (i.e. curve denoted by ◆ in the left column shifts upward as we move from lower to upper sub-plots) and (ii) if the task has a high complexity and agents put a high emphasis on social norms, the long-run performance increases for incentive schemes that put more

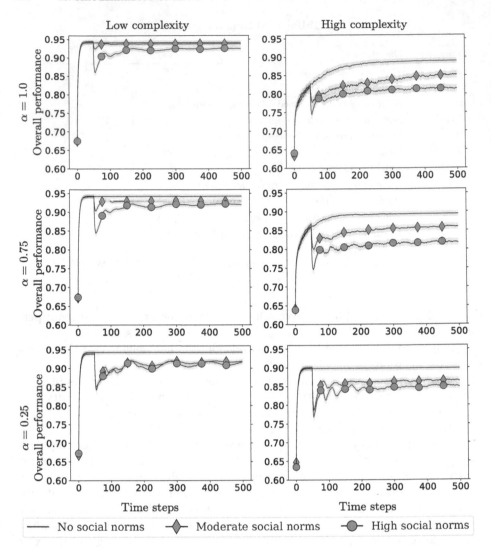

Fig. 2. Main results based on parameters defined in Table 1. The plots show normalized performance for $T = 500$ time steps, averaged over $R = 1000$ simulation runs with 99.9% confidence interval.

weight on team performance (i.e. curve denoted by ●; in the right column shifts upward as we move from upper to lower sub-plots).

Next, we perform sensitivity analyses on variables that are essential to our formulation of social norms: the fixed network degree D, the number of social tasks N_S, and the correlation between landscapes ρ. For the sensitivity analyses, we consider the baseline case with no team-based incentives ($\alpha = 1$), high social norms ($w_1 = w_2 = 0.5$). All other parameters are taken from Table 1. Moreover, we also include the case with no social norms ($w_1 = 1, w_2 = 0$) as a benchmark.

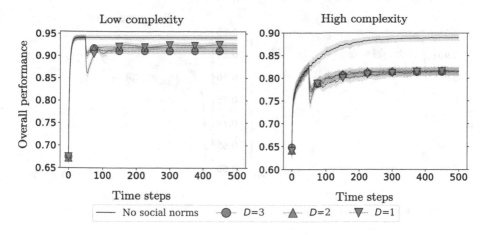

Fig. 3. Overall performance for different values of D (99.9% confidence interval)

Figure 3 illustrates that network degree D does not significantly change the effect of social norms on the overall performance. Figure 5 shows that as the number N_S of social bits increases, the effect of descriptive social norms gets more pronounced. Figure 4 shows that the correlation between landscapes does not significantly affect the overall performance in environments with high complexity. However, in environments with low complexity, not only does the positive correlation offset the performance drop caused by the descriptive social norms, but it also may improve the performance for $\rho \geq 0.9$.

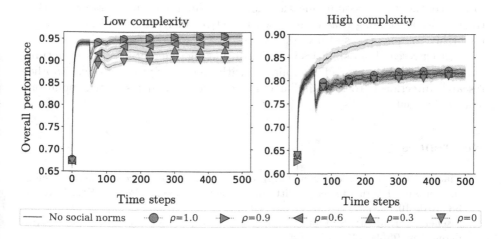

Fig. 4. Overall performance for different values of ρ (99.9% confidence interval)

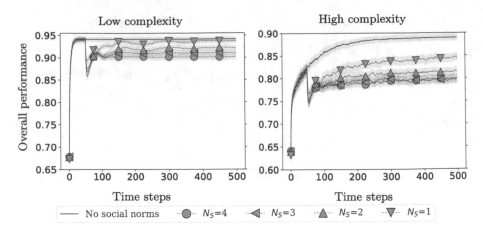

Fig. 5. Overall performance for different values of N_S (99.9% confidence interval)

4   Conclusion

A thorough examination of alternatives is costly for individuals and for this reason they turn to descriptive social norms to avoid computational costs and use them as a shortcut by complying with the perceived consensus of their colleagues. We have provided evidence that this comes at a cost for the overall performance of a team, unless tasks are highly correlated. We have analyzed the effect for different environments and levels of communication and found that while the number of common tasks increases the effect of the social norms on performance, the level of communication does not. Finally, we have confirmed that the established finding that team-based incentives increase overall performance for highly complex tasks also works in presence of social norms.

Further research possibilities include considering different network structures and observing dynamics of compliance with the descriptive norms. Another research direction is to find similar empirical studies and test our simulations against real data.

References

1. Cialdini, R., Reno, R., Kallgren, C.: A focus theory of normative conduct: recycling the concept of norms to reduce littering in public places. J. Person. Soc. Psychol. **58**, 1015–1026 (1990)
2. Fischer, P., Huddart, S.: Optimal contracting with endogenous social norms. Am. Econ. Rev. **98**(4), 1459–1475 (2008)
3. Iversion, K.E.: A Programming Language. Wiley, New York (1962)
4. Kauffman, S.A., Johnsen, S.: Coevolution to the edge of chaos: coupled fitness landscapes, poised states, and coevolutionary avalanches. J. Theoret. Biol. **149**(4), 467–505 (1991)

5. Kauffman, S.A., Weinberger, E.D.: The NK Model of rugged fitness landscapes and its application to maturation of the immune response. J. Theoret. Biol. **141**, 211–245 (1989)
6. Levinthal, D.A.: Adaptation on rugged landscapes. Manag. Sci. **43**(7), 934–950 (1997)
7. Priebe, C.S., Spink, K.S.: When in Rome: Descriptive norms and physical activity. Psychol. Sport Exerc. **12**(2), 93–98 (2011)
8. Pryor, C., Perfors, A., Howe, P.D.L.: Conformity to the descriptive norms of people with opposing political or social beliefs. PLoS ONE **14**(7), 1–16 (07 2019)
9. Rivkin, J.W.: Imitation of complex strategies. Manag. Sci. **46**(6), 824–844 (2000)
10. Verel, S., Liefooghe, A., Jourdan, L., Dhaenens, C.: On the structure of multiobjective combinatorial search space: MNK-landscapes with correlated objectives. Eur. J. Oper. Res. **227**(2), 331–342 (2013)
11. Wall, F., Leitner, S.: Agent-based computational economics in management accounting research: opportunities and difficulties. J. Manag. Acc. Res. (2020)
12. Xie, X.F., Zhang, W.J., Yang, Z.L.: Social cognitive optimization for nonlinear programming problems. In: Proceedings International Conference on Machine Learning and Cybernetics, vol. 2, pp. 779–783. IEEE (2002)

What Makes a Turing Award Winner?

Zihe Zheng, Zhongkai Shangguan(✉), and Jiebo Luo

University of Rochester, Rochester, NY 14627, USA
zzheng18@u.rochester.edu, zshangg2@ur.rochester.edu,
jluo@cs.rochester.edu

Abstract. Computer science has grown rapidly since its inception in the 1950s and the pioneers in the field are celebrated annually by the A.M. Turing Award. In this paper, we attempt to shed light on the path to influential computer scientists by examining the characteristics of the 74 Turing Award laureates. To achieve this goal, we build a comprehensive dataset of the Turing Award laureates and analyze their characteristics, including their personal information, family background and academic background. The FP-Growth algorithm is used for frequent feature mining. Logistic regression plot, pie chart and word cloud are generated accordingly for each of the interesting features to uncover insights regarding personal factors that drive influential work in the field of computer science. In particular, we show that the Turing Award laureates are most commonly white, male, married, United States citizen, and received a PhD degree. Our results also show that the age at which the laureate won the award increases over the years; most of the Turing Award laureates did not major in computer science; birth order is strongly related to the winners' success; and the number of citations is not as important as one would expect.

Keywords: Turing Award · Feature mining · Family background · Academic background

1 Introduction and Related Work

The field of Computer Science is one of the fastest growing disciplines in the world. As a very important part of the Digital Revolution [20], the development of computer science has direct impact on people's lives and the potential to fundamentally change the traditional way of life. Who are those people that have significant impact on the design and direction of computer technology that we use? Can we find some rules or best practices to promote further development of computer science? These questions prompt people to study from outside computer science itself, e.g., from the perspective of history and the sociology of science [3]. In this study, we try to discover patterns from the diverse information of Turing Award Laureates from the perspective of data.

Z. Zheng and Z. Shangguan—Contributed equally.

© Springer Nature Switzerland AG 2021
R. Thomson et al. (Eds.): SBP-BRiMS 2021, LNCS 12720, pp. 310–320, 2021.
https://doi.org/10.1007/978-3-030-80387-2_30

Turing Award, in full A.M. Turing Award, is Association for Computing Machinery (ACM)'s oldest and most prestigious award. The award is named after Alan M. Turing, a British pioneer of computer science, mathematics and cryptography [8], and is presented annually to individuals who have made lasting contributions of a technical nature to the computing community [1]. The Turing Award is recognized as the highest distinction in computer science, and is referred to as the computer science equivalent of the Nobel Prize [22].

Scientists in various fields have made many attempts to find the common features of high achieving scientists and how they developed those desirable characteristics. Most of them are from sociological perspective, focusing on family background [14], academic network of scientists [18], the years in which training was received [9], and affiliated institutions [15]. Many of these work focus on Nobel Prize winners, yet they still inspire our research and help us with feature selection. Educational background of the Nobel Prize winners was studied and it is found that undergraduates from small and elite institutes are the most likely to win a Nobel Prize [7]. Clark and Rice [6] studied Nobel Prize winners and found that eminent scientists tend to be earlier born when family size is not controlled while in smaller families the later born are more likely to be eminent in science. The best nations and institutions for revolutionary science was also studied [5]: they found that MIT, Stanford and Princeton are the top institutions and that the USA dominates the laureates of these prizes.

Compared to Nobel Prize, the work focused on the Turing Award is limited. From a sociology study, Akmut [3] concluded that place of birth, nationality, gender, social background, race and networks play a role in making Turing Award laureates. A metric named "Turing Number" [22] was proposed to measure the distance between scholars and Turing Award laureates, as a new way to construct a scientific collaboration network centered around the laureates. The Turing Award laureates also provide good performance when being used as a data source for studying and testing other models and metrics. To understand the process of scientific innovation, Liu and Xu [13] developed a chaining model that showed the academic career path of these computer scientists. The current research regarding the Turing Award are mostly focused on some of its specific aspects, yet a thorough analysis of the laureates from the aspect of data science is needed. Therefore, we hope to construct a comprehensive dataset and analyze the frequent features to pave the way for future studies regarding top computer scientists.

This study collects the relevant information of the 74 Turing Award winners over the years (until 2020) in order to have a comprehensive understanding of the winners of the Turing Award. We conduct a preliminary analysis of the winners' bibliography, family information, academic background, personal experience, and so on, hoping to provide reference of the discipline development and training direction in the field of computer science.

We make two contributions in this study. First, we construct the latest dataset that contains diverse data of all Turing Award winners (as of 2020) by integrating several web resources. Second, we provide an in-depth analysis of the characteristics of the Turing Award winners and uncover insights regarding personal factors that drive influential work in the field of computer science.

2   Methods

2.1   Data Preparation

To construct a dataset that reflects the Turing Award laureates' experiences and backgrounds, we select features in the following three aspects: personal information, family background and academic background.

Personal information includes: birth year, gender, race, and citizenship. Race is denoted by White, Jewish and Asian. Laureates with dual citizenship are counted as citizens for both of the countries.

Family background includes: number of children, marital status, parents' background, number of siblings, and order of birth. Marital status are specified as number of marriages if applicable. Parents' backgrounds are either a short paragraph or several sentences about the laureates' family. Order of birth are specified to younger (if one sibling), elder (if one sibling), youngest (if more than one sibling), middle (if more than one sibling), and eldest (if more than one sibling).

Academic background includes: highest degree, universities attended, majors studied, location of education, fields of study, citation count, PhD advisor, and university affiliation. Since honorary degrees are usually given when the recipients already have distinguished achievements, honorary degrees are not included as part of the educational background. Majors studied include undergraduate majors and graduate majors. Same majors are counted for undergraduate and graduate, respectively, while same majors for more than one graduate programs are only counted once. Double majors are included as well. Location of education is specified as countries, with no distinction between undergraduate study and graduate study locations. Fields of study are the fields in computer science that the laureates have made significant contributions to and are mentioned in the rationale. University affiliation includes both long-term affiliation as well as short-term affiliation.

The main source of our data is the ACM official website of the Turing Award [1]. On the main page of most of the laureates, their birthday, educational path, past and current affiliations, as well as a short biography are shown. Missing information are manually collected from the Wikipedia page of the laureates.

The laureates' academic background are collected from Google Scholar [10], Semantic Scholar [16] and ACM Author Profile [1]. A laureate's citation count for the most cited paper, the top five cited papers and total citations are collected from each of the three sources. While the profile for each laureate can be found on Semantic Scholar and ACM Author Profile, only 45% of the laureates are documented by Google Scholar. The family backgrounds are first collected from the biographies on the laureates' ACM main pages and then supplemented by their interview transcripts if available. 72% of the laureates' description of their family backgrounds are found. Overall, despite the missing laureates in Google

Scholar and the missing values in family background, all of the other features are successfully collected for all the 74 laureates.

2.2 Data Analysis

To see the general features of the Turing Award laureates, we implement the frequent pattern growth (FP-Growth) algorithm [12] for pattern mining. It adopts a divide-and-conquer strategy to mine the frequent itemsets without costly candidate generation. After a frequent pattern tree is generated from the whole dataset, it then recursively generates and mines each of the conditional frequent pattern trees for frequent patterns, which significantly reduces the time complexity of pattern mining [11]. In our study, we group the features for each laureate as an itemset and utilize the FP-Growth algorithm to extract the frequent patterns of these features. To avoid low support, we select features that have small variations, including sex, citizenship, race, marital status, order of birth, highest education level, schools attended, field of study, university affiliation, majors, education location, and their affiliation at the time of award (grouped). Using this subset of our dataset, the frequent patterns of features of the laureates are generated. The features are then separately analyzed. Logistic regression plot, pie chart and word cloud are generated accordingly for each of the interesting features we find.

3 Results

3.1 Turing Award General Information

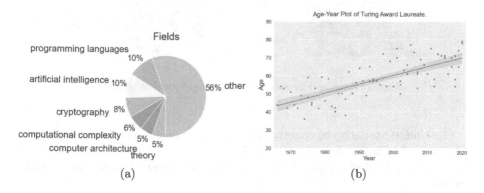

(a)   (b)

Fig. 1. (a): Turing Awards by fields; (b): Age-Year relationship of Turing Award laureates. The blue line is fitted by Logistic Regression, and the shadow part shows the 95% confidence interval for the regression. (Color figure online)

The Turing Award has been awarded 55 times with 74 winners as of 2020. Among them, 17 times were team awards (14 times with two winners, and 3 times with

Z. Zheng et al.

three winners). Solo award is the most common, accounting for 69% of the total number of awards. Figure 1(a) shows their awarded fields, where 31 people were awarded in more than one field. Artificial Intelligence, Programming Languages, and Cryptography are the most popular fields being awarded, followed by Computational Complexity and Theory. A total of 45 scientists in the above five fields have won the Turing Award.

3.2 Personal Information

The relationship between the age of the laureates and the year of the award are shown in Fig. 1(b). Logistic regression is used to fit the linear relationship between age and year of the award by minimizing the Mean Squared Error (MSE). The age and year of award are positively correlated, i.e., the age at which a laureate won the award increases over time. The average age of the scientist at the time of the award is 57.82. Donald Knuth [21] is the youngest and Alfred Aho [19] is the eldest when they were awarded, at the age of 36 and 79, respectively.

In terms of race and gender, as shown in Fig. 2, white males dominate the Turing Award. Among the 74 winners, 50 of them are white, and 22 of them are Jewish; only three females have won the award. The first woman who won the Turing Award was Frances Elisabeth Allen, in 2006, 41 years after the award was established.

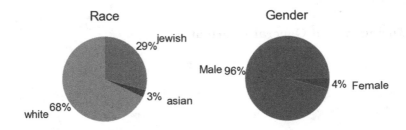

Fig. 2. Race and gender of the Turing Award laureates.

The United States is the country with the most awarded computer scientists. Among the 74 Turing Award winners, 57 are United States citizens and 8 are citizens of the United Kingdom. Followed by Israel and Canada with 5 and 6 winners each.

3.3 Family Background

We extract a winner's family information mentioned in the interview transcript and generate a word cloud based on the word frequency, as shown in Fig. 3. The most common words are "teacher", "school", "worked" and "college", which

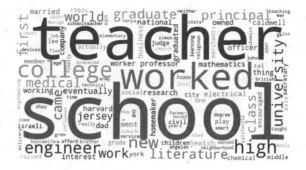

Fig. 3. Word cloud when the Turing Award laureates talk about their family. The larger, the more frequent.

indicate that the Turing award laureates are likely to have good family conditions, and that their parents' professions are mostly related to education. All winners are heterosexual and have been married at least once.

The relationship between birth order and personal achievement has always been a controversial topic. Thurstone and Jenkins [17] show that first-born and second-born children have IQ differences, and Clark and Rice [6] show that the birth order has impact on the Nobel Prize winners. The order of birth information of the Turing award laureates is shown in Fig. 4. We find that if the winner only has one sibling, then he/she is much more likely to be the younger one; if the winner has more than one sibling, then he/she is the most likely to be the eldest one. Our findings are consistent with the research by Clark and Rice [6].

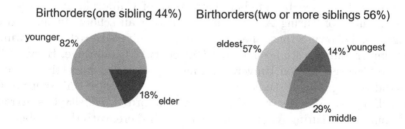

Fig. 4. Birth order of the Turing Award laureates.

3.4 Academic Background

Figure 5 shows the highest degree the winners obtained and their majors. Among all the 74 Turing Award winners, 63 obtained a PhD degree, 5 obtained a master's degree and 6 obtained a bachelor's degree. Regarding their majors, about one third of the awarded scientists studied mathematics during their higher education. We continue to show details about the winners' undergraduate major and graduate major, as shown in Fig. 6. Surprisingly, only three winners had a major in Computer Science when they were undergraduate, and they all had double majors. This could be explained by the fact that computer science is a relatively new major. The world's first computer science degree program, the Cambridge Diploma in Computer Science, began at the University of Cambridge Computer Laboratory in 1953. The first computer science department in the United States was formed at Purdue University in 1962. More than half of the scientists studied mathematics during their undergraduate studies. In their graduate studies, computer science became the most popular major, while many people continued to study mathematics.

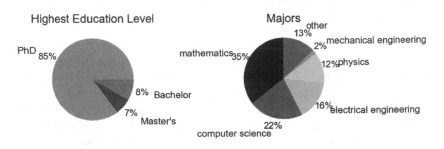

Fig. 5. Educational background.

We further explore where they received their degrees and the United States remains the most dominant country, with 56 of the winners studied in the United States. We show the top 12 universities where the Turing Award winners graduated in Table 1, ranked by number. University of California at Berkeley has the most number of Turing Award winners among all schools in undergraduate programs. Following that, the most popular universities that the winners did their undergraduate studies include: University of Cambridge, Harvard University and Carnegie Mellon University. The most popular universities that the winners did their graduate studies are: Princeton University, University of California at Berkeley, and Harvard University. Stanford University has attracted the most number of Turing Award winners to teach there, with the number of 23, although only two undergraduate and six graduate students are from Stanford University.

Citation is often a way to evaluate a scholar's academic influence. We collect the winners citation information, including the most cited paper, the sum of top 5 paper citations, and the total citations, as of November 30, 2020. (For the 2020 winners, the data is obtained on April 1, 2021). Since different databases have different standards for citation, we use Semantic Scholar, Google Scholar and ACM Author Profile as references. In order to make the result more intuitive, we sort the citations and the results are shown in Fig. 7. All the citation measures show a similar exponential distribution, regardless of the resource. Only a few of them have a very high citation count and they are all in the Artificial Intelligence

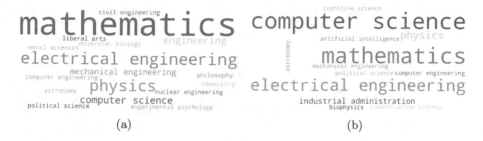

(a)   (b)

Fig. 6. Word cloud showing the distribution of the majors of the Turing Award laureates: (a) undergraduate majors, (b) graduate majors. The larger, the more frequent.

Table 1. The number of Turing Award laureates by universities.

Undergraduate universities	Count	Graduate universities	Count	Working universities	Count
University of California, Berkley	6	Princeton University	9	Stanford University	23
University of Cambridge	5	University of California, Berkley	7	Massachusetts Institute of Technology	19
Harvard University	4	Harvard University	7	University of California, Berkley	13
Carnegie Mellon University	4	Stanford University	6	Carnegie Mellon University	10
Massachusetts Institute of Technology	3	Massachusetts Institute of Technology	4	Princeton University	8
California Institute of Technology	3	Weizmann Institute of Science	2	Harvard University	7
University of Oxford	3	University of Michigan	2	New York University	5
Duke University	3	California Institute of Technology	2	University of Oxford	4
University of Chicago	3	University of Oslo	2	University of Toronto	4
Technion–Israel Institute of Technology	2	University of Toronto	2	Weizmann Institute of Science	3
Stanford University	2	University of Illinois at Urbana–Champaign	2	Cornell University	3
University of Oslo	2	Carnegie Mellon University	2	University of Edinburgh	2

area. We compare the rankings of the Turing Award winners among all computer science scholars given by Guide2Research [2], and only five of the Turing Award Laureates are listed in top 100 computer scientists. The results indicate that one may not need a very high citation to win the award. It is worth noting that 21 of the 74 laureates have an advisor-student relationship as shown in Fig. 8.

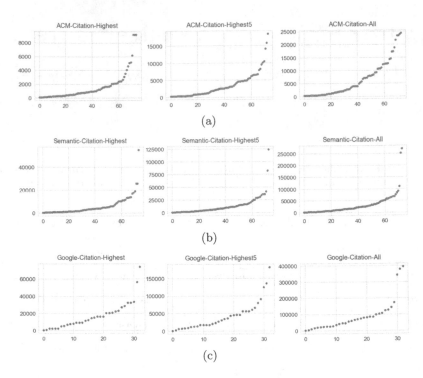

Fig. 7. Sorted citations from ACM Author Profile, Semantic Scholar, and Google Scholar: (a) ACM Author Profile, (b) Semantic Scholar, and (c) Google Scholar.

Fig. 8. Advisor-student relationship.

4   Conclusion and Future Work

In this study, we build a comprehensive dataset of the Turing Award laureates, including their personal information, family background and academic background. We then investigate potential patterns in the personal background of these Turing Award winners. Our analysis results show that most scientists who won the Turing Award have the following characteristics: white, male, married, United States citizen, and having a PhD degree. Four particular findings are interesting: the age at which a laureate won the award increases over time, most of the Turing Award laureates did not major in computer science even though the award celebrates the highest achievements in computer science, the order of birth has strong correlations with the winners' success, and the number of citations is not as important as one would expect.

We can consider three directions in our future work: 1) collecting more comprehensive data of the laureates' social networks, including their collaborators, and apply graph mining techniques [4] to mine the winners' social networks, 2) counting the average citations in different areas of computer science and normalize the citation data for each area, and 3) including top computer scientists who have not won the Turing Award in our dataset and utilizing machine learning algorithms to predict future Turing Award winners.

References

1. ACM: Alphabetical listing of a.m. turing award winners (2020). https://amturing. acm.org/alphabetical.cfm. Accessed 1 Nov 2020
2. ACM: Ranking for computer science (2020). http://www.guide2research.com/ scientists/. Accessed 9 Dec 2020
3. Akmut, C.: Social conditions of outstanding contributions to computer science: a prosopography of Turing Award laureates (1966–2016) (2018)
4. Chakrabarti, D., Faloutsos, C.: Graph mining: laws, generators, and algorithms. ACM Comput. Surv. (CSUR) **38**(1), 2-es (2006)
5. Charlton, B.G.: Which are the best nations and institutions for revolutionary science 1987–2006? Analysis using a combined metric of Nobel prizes, Fields medals, Lasker awards and Turing awards (NFLT metric). Med. Hypotheses **68**(6), 1191–1194 (2007)
6. Clark, R.D., Rice, G.A.: Family constellations and eminence: the birth orders of Nobel prize winners. J. Psychol. **110**(2), 281–287 (1982)
7. Clynes, T.: Where Nobel winners get their start. Nat. News **538**(7624), 152 (2016)
8. Cooper, S.B., Van Leeuwen, J.: Alan Turing: His Work and Impact. Elsevier, Amsterdam (2013)
9. Goldstein, J.L., Brown, M.S.: A golden era of Nobel laureates. Science **338**(6110), 1033–1034 (2012)
10. Google: Google scholar (2020). https://scholar.google.com. Accessed 2 Nov 2020
11. Han, J., Pei, J., Kamber, M.: Data Mining: Concepts and Techniques. Elsevier, Amsterdam (2011)
12. Han, J., Pei, J., Yin, Y.: Mining frequent patterns without candidate generation. ACM SIGMOD Rec. **29**(2), 1–12 (2000)

13. Liu, E., Xu, Y.: Chaining and the process of scientific innovation (2020)

14. Rothenberg, A.: Family background and genius II: Nobel laureates in science. Can. J. Psychiatry **50**(14), 918–925 (2005)

15. Schlagberger, E.M., Bornmann, L., Bauer, J.: At what institutions did Nobel laureates do their prize-winning work? An analysis of biographical information on Nobel laureates from 1994 to 2014. Scientometrics **109**(2), 723–767 (2016). https://doi.org/10.1007/s11192-016-2059-2

16. SemanticScholar: Semantic scholar (2020). https://www.semanticscholar.org. Accessed 4 Nov 2020

17. Thurstone, L.L., Jenkins, R.L.: Birth order and intelligence. J. Educ. Psychol. **20**(9), 641 (1929)

18. Wagner, C.S., Horlings, E., Whetsell, T.A., Mattsson, P., Nordqvist, K.: Do Nobel Laureates create prize-winning networks? An analysis of collaborative research in physiology or medicine. PLoS ONE **10**(7), e0134164 (2015)

19. Wikipedia: Alfred aho (2020). https://en.wikipedia.org/wiki/Alfred_Aho. Accessed 2 Dec 2020

20. Wikipedia: Digital revolution (2020). https://en.wikipedia.org/wiki/Digital_Revolution. Accessed 9 Dec 2020

21. Wikipedia: Donald knuth (2020). https://en.wikipedia.org/wiki/Donald_Knuth. Accessed 2 Dec 2020

22. Xia, F., Liu, J., Ren, J., Wang, W., Kong, X.: Turing number: how far are you to A.M. Turing award? ACM SIGWEB Newslett. (Autumn), 1–8 (2020). Article No. 5. https://doi.org/10.1145/3427478.3427483

Kinetic Action and Radicalization: A Case Study of Pakistan

Brandon Shapiro[1] (✉) and Andrew Crooks[2]

[1] George Mason University, 4400 University Dr, Fairfax, VA 22030, USA
bshapir3@gmu.edu
[2] University at Buffalo, Buffalo, NY, USA
atcrooks@buffalo.edu

Abstract. Drone strikes have been ongoing and there is a debate about their benefits. One major question is what is their role with respect to radicalization. This paper presents a data-driven approach to explore the relationship between drone strikes in Pakistan and subsequent responses, often in the form of terrorist attacks carried out by those in the communities targeted by these counterterrorism measures. Our analysis of news reports which discussed drone strikes and radicalization suggests that government-sanctioned drone strikes in Pakistan appear to drive terrorist events with a distributed lag that can be determined analytically. We then utilize these news reports to inform and calibrate an agent-based model which is grounded in radicalization and opinion dynamics theory. In doing so, we were able to simulate terrorist attacks that approximated the rate and magnitude observed in Pakistan from 2007 through 2018. We argue that this research effort advances the field of radicalization and lays the foundation for further work in the area of data-driven modeling and kinetic actions.

Keywords: Radicalization · Data-driven modeling · Drone strikes · Terrorism · Pakistan · Agent-based modeling

1 Introduction

Since the events of September 11, 2001, kinetic actions in the form of drone strikes have become a tool for the United States (U.S.) as it fights a global war against terrorist organizations. The U.S. has been conducting drone strike campaigns in Pakistan, Yemen, Somalia, Afghanistan, and Libya since 2002 (Bureau of Investigative Journalism 2020 [BIJ]), and both the local and national media outlets have reported on these operations. Advantages and disadvantages of this particular counterterrorism option continue to be debated, including whether drone strikes reduce the number of terrorist attacks or lead to additional violence (e.g., Plaw et al. 2016; Shah 2018). Though the majority of the U.S. population appears to support the use of drone strikes to eliminate suspected terrorists overseas, much of the international landscape holds an opposing view (e.g., Pew Research Center 2014).

These controversies and ongoing public discourse over the use of drone strikes have led to a surge in the open-source literature within the academic community.

© Springer Nature Switzerland AG 2021
R. Thomson et al. (Eds.): SBP-BRiMS 2021, LNCS 12720, pp. 321–330, 2021.
https://doi.org/10.1007/978-3-030-80387-2_31

Research in the areas of radicalization and terrorism engage heavily with opinion dynamics, a field that has embraced computational methods and agent-based models to characterize and explore complex adaptive systems and capture emergent phenomena by simulating individual actions of heterogenous agents (e.g., Crooks et al. 2019). While the fields of radicalization and opinion dynamics have employed computational tools, generally speaking they do not appear have leveraged an abundance of empirical data to inform their modeling efforts. To address this noted gap, in this paper we present an agent-based model to explore the relationship between drone strikes and terrorist attacks that have occurred within Pakistan.

Agent-based modeling can be applied to complex systems to address multiscale processes (Ozik et al. 2008). The ability for agent-based models to execute large-scale scenarios against representative populations of agents make them an efficient and effective tool against analytically-challenging endeavors. Agent-based models are also particularly well-suited for exploring social and behavioral dynamics that might be difficult to observe in the real world (e.g., radicalization). While it is easier to think of an ideological progression toward radicalization and the use of violence as easy-to-define, non-overlapping categories, in reality, individuals do not follow a sequential or linear advancement (Jensen et al. 2020). Rather, the radicalization trajectory from inception of an idea to execution of violent act is likely characterized as a set of "complex pathways" (Defense Science Board 2015). Therefore, an agent-based model which captures these non-linear behaviors and dependencies among the various elements contained within the system could serve as a helpful tool for senior-level decisionmakers within the U.S. Government. Instead of focusing on identifying trigger points that might prompt individuals to undergo a phase change and move along the radicalization spectrum, as previous work within this field have modeled (e.g., Cioffi-Revilla and Harrison 2011; Galam and Javarone 2016; Genkin and Gutfraind 2011; Shults and Gore 2020), we have employed a bottom-up modeling approach to characterize Pakistan's macro-level system as a whole—that is, the relationship between drone strikes and terrorist attacks—using a simple set of rules governing each of the agents (Axtell and Epstein 1994).

Leveraging empirical data to calibrate/parameterize an agent-based model—which has been built upon radicalization and opinion dynamics theories—in order to define and characterize a real-world, observed relationship between drone strikes and terrorist attacks that occurred in Pakistan appears to be the first. As such, this paper provides a unique contribution to the radicalization and data-driven modeling fields. In the remainder of this paper, Sect. 2 presents a brief background of radicalization and the employment of computational models in this area of research. We then turn to methodology, which describes the process of translating the relationship between drone strikes and terrorist attacks carried out in Pakistan by a network of individuals into an agent-based model (Sect. 3). This is followed with the presentation of some initial results from the agent-based model (Sect. 4) and a summary of the paper and proposed areas of further work is given in Sect. 5.

2   Background

There does not appear to be an agreed-upon or universally-accepted working definition of radicalization within either the U.S. Government or among those in the academic community. Despite this disagreement, many describe radicalization as a process. Silber and Bhatt (2007), for example, defined the trajectory as a four-phase process (i.e., pre-radicalization, self-identification, indoctrination, and jihadization). Other academic frameworks characterized the progression of individuals to commit terrorist acts as a narrowing staircase (Moghaddam 2005), influenced by organizations that served as a conveyor belt which conditioned them for recruitment by more extreme terrorist organizations (Baran 2005), or even a metaphorical pyramid where ideas become more extreme and the tendency for violent actions become more likely as individuals ascend toward the top of the pyramid (McCauley and Moskalenko 2008, 2017).

Though many within the academic community have presented a variety of diverse frameworks in an attempt to explain the radicalization journey of an individual, several of these theories exhibit overlapping assumptions and themes, such as individuals with feelings of insignificance who looked for opportunities to shed their current social identity for a new one (Doosje et al. 2016; Webber and Kruglanski 2016, 2018). As a result, such individuals might then connect with small, tightly-knit group, thereby isolating themselves from others. Such interactions over time potentially serve as a reinforcing mechanism, or echo chamber, that fosters radical beliefs (della Porta 2018; Orsini 2020).

Motivating factors or trigger points, the number of steps, and length of time along this radicalization journey, however, continue to be debated. Despite an ongoing dialogue, most, if not all, agree that that the overall radicalization trajectory has shortened due to the technology boom. The Internet and social media platforms have made content available to the masses instantly (e.g., Klausen et al. 2020).

As noted earlier, previous work has employed agent-based models in an attempt to identify and address potential trigger points or temporal scales (e.g., Cioffi-Revilla and Harrison 2011; Galam and Javarone 2016; Genkin and Gutfraind 2011; Moon and Carley 2007; Shults and Gore 2020). However, we argue that these models fall short because, in order to successfully implement an agent-based model, several items must be addressed, including time scales, calibration, internal model verification, and validation. While prior research has shed light onto various aspects of radicalization, they did not include one or more of the following: the necessary elements listed above, a description of the interaction network, or empirical data to inform and parameterize the computational simulation. Our agent-based model, which is built upon theory (see Sect. 3) and our exploratory analyses of news reporting, includes all of the elements described above, and we turn to this next.

3 Methodology

Within this section, we introduce our radicalization model for Pakistan. Important aspects relating to this system, the agents, and environment are introduced below while a description of the graphical user interface is in Fig. 1. For those interested, the model, which we have developed in NetLogo 6.0.4 (Wilensky 1999) along with a detailed Overview, Design concepts, and Details (ODD) framework protocol for describing individual-based models and ABMs (Grimm et al. 2006) can be found at https://bit.ly/3qLJynv. The rationale for providing the ODD and for sharing the model is that it allows broader dissemination of the model and its methodology. Furthermore, it provides additional information (and in greater detail) on the model logic, variables, and agent decisions than what is allowed in this paper due to page count limitations. Providing this documentation, model source code, and data will allow for replication of the results presented here and for others to extend the model if they so desire.

We use the BIJ data as our source for Pakistan drone strikes (i.e., kinetic actions) and the National Consortium for the Study of Terrorism and Responses to Terrorism (START) University of Maryland (2019) Global Terrorism Database (GTD) as our source for terrorist incidents. Due to the secrecy surrounding the U.S. drone strike campaign carried out in Pakistan, the BIJ relied on open-source reporting, including news articles reporting on suspected U.S. drone strikes near particular locations, and compiled the event information into a publicly-available dataset. For terrorist attacks in Pakistan, GTD was compiled using a variety of publicly-available, open-source materials. Depending on availability of information, each event could have had over one hundred attributes (e.g., date, location, claims of responsibility, intended target, and source).

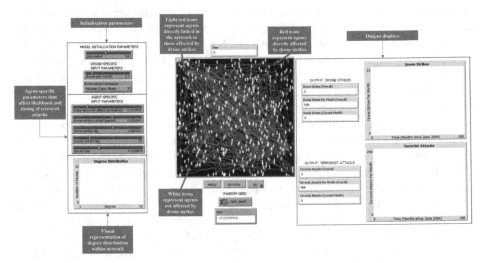

Fig. 1. Pakistan radicalization model's graphical user interface.

Agent-based modeling focuses on individuals—in this case, those who were targeted by U.S. drone strikes in Pakistan during the period 2004 through 2018—making decisions in response to a stimulus introduced within the simulated environment. However, capturing this decision-making process is difficult (Heppenstall et al. 2020). To ensure a realistic model of heterogenous behavior, we employed a network of agents who communicated their respective levels of anger in a manner inspired by the radicalization and opinion dynamics literature (e.g., della Porta 2018; Orsini 2020; Webber and Kruglanski 2018) and who carried out their terrorist acts consistent with our review and analyses of the publicly-available, open-source empirical data.

As highlighted in Sect. 2, radicalization frameworks differ in the specific details, but all seem to agree that tipping points (e.g., significance loss) prompted phase changes for individuals. For some, loss of life caused by a drone strike served as that trigger.

Needing to feel important and respected, those individuals then sought to restore their significance. Oftentimes, this took the form of terrorist attacks, particularly for those who subscribed to radical ideological narratives influenced by friends, family, and the Internet (e.g., Webber and Kruglanski 2018). Within these tight-knit networks, people regularly exchanged viewpoints and updated their opinions to reflect the information disseminated and consensus of the group (e.g., della Porta 2018; Orsini 2020).

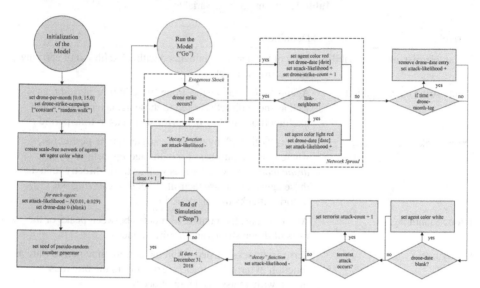

Fig. 2. The agent-based model flow diagram.

Consistent with the radicalization and opinion dynamics theory discussed in Sect. 2 and above in this section, Fig. 2 provides an overview of the forcing function, decision-making processes, and subsequent actions that agents within the simulated environment of our Pakistan radicalization model employs at each time step. U.S. drone strikes conducted in Pakistan are exogenous events to the model, and serve as a forcing function by injecting a traumatic grievance into the environment (e.g., McCauley and Moskalenko 2008; Webber and Kruglanski 2016). This, in turn, causes members of the targeted population to increase their anger toward the U.S. and its Western allies. As the level of anger increases among members of this population of agents, the likelihood that they choose to carry out a terrorist attack also increases (e.g., United States House of Representatives 2009). This likelihood to carry out a terrorist attack (i.e., anger) lingers for an amount of time beyond the drone strike, and is represented by a distributed lag model with a decay rate associated with that relationship.

Specifically, agents who experience a drone strike communicate that anger—which is proxied by an increase in the likelihood of carrying out a terrorist attack—with those directly linked in their network. Recipients of the information update their respective opinions based on observations of their immediate neighbors. Following the kinetic action, the anger of the affected agents begins to decrease, as defined by the decay function, which is consistent with empirical studies (e.g., Jaeger and Siddique 2018).

Table 1 defines the agent-specific variables incorporated into this simulation. The temporal scale for each time step is one month since the data for both the Pakistan drone strikes and terrorist attacks within Pakistan are aggregated to that level of detail. To match the timeline of the drone strike and terrorist attack data collected for this research effort, the model runs from June 2004 through December 2018 (see Table 2).

Table 1. Agent-specific variables.

Variable	Definition
Initial terrorist attack probability	Each of the agents is initialized with a likelihood of carrying out a terrorist attack
Drone strike current period	Agents "affected" by the drone strike in time period t, as well as those directly linked to those agents, increase their likelihood of carrying out terrorist attacks in time period t
Drone strike lag	Agents "affected" by the drone strike in time period t-m, where m is the number of months ago (as defined by the *drone month lag* variable), as well as those directly linked to those agents, increase their likelihood of carrying out terrorist attacks in time period t
Drone month lag	Approximates the number of months between peaks of the rates of drone strikes and terrorist attacks
Decay rate	Terrorist attacks carried out in the current time period t are a function of terrorist attacks carried out in the previous month with an assumed rate of decay

4   Results

Before presenting the model results, we first describe how we have verified the model, meaning the process we have used to ensure the model matches its design. We achieved model verification using iterative design review (i.e., code walkthroughs) and parameter testing via sensitivity analysis. For example, parametric analysis was conducted using NetLogo's Behavior Space tool where we tested the various parameter settings in combination to ensure the relationships among the output variables were reasonable and consistent with both theory and our review of the empirical data.

The values specified in Table 2 were used in the baseline run of the simulation to ensure the reported output of the rate of terrorist attacks qualitatively agreed with the macro-level behaviors exhibited in Pakistan (Axtell and Epstein 1994). Figure 3 shows the U.S. drone strike campaign in Pakistan (solid red line) and average rate of terrorist attacks (solid blue line) outputted from one thousand repeated simulation runs of the computational agent-based model using the baseline parameter settings identified in Table 2. For ease of comparison, the actual terrorist attacks in Pakistan are overlaid onto Fig. 3, and are represented by the dashed gray line.

Agent-based models typically require a "burn-in" period where early observations are ignored and discarded from the analysis due to initial conditions biasing model dynamics (e.g., Caiani et al. 2016). Non-stable behaviors and processes exhibited at the beginning of the simulation required a "burn-in" period of thirty-five months to allow the terrorist attack curve to approach the probable region of the sample space.

Table 2. Input parameter settings used to calibrate/parameterize Pakistan radicalization model.

Variable	Parameter setting	Source
Time duration per run	June 2004 to December 2018	BIJ (2020); GTD (2019)
Population	500 agents (i.e., 1 agent represented 6,000 people in FATA region)	Marten et al. (2008); Nawaz (2009); Parametric exploration
Network structure	Preferential attachment (i.e., scale-free)	Sageman (2004)
Rate of drone strikes	Varies by month	BIJ (2020)
Initial terrorist attack probability	~N(0.010933, 0.029)	Parametric exploration
Drone strike current period	~N(0.430692, 0.116)	Parametric exploration
Drone strike lag	~N(1.582621, 0.510)	Parametric exploration
Drone month lag	25	Parametric exploration
Decay rate	~N(0.918805, 0.024)	Parametric exploration

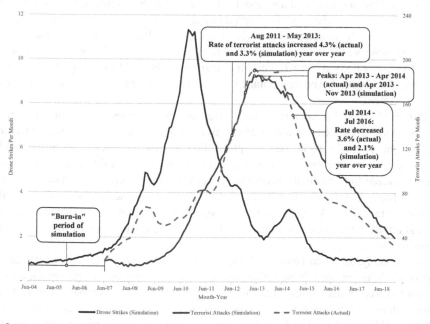

Fig. 3. Terrorist attacks simulated by Pakistan radicalization model qualitatively agree with real-world system.

From June 2007 through the initial peak of the terrorist attack curve, our simulation estimated 81 percent of the terrorist attacks that occurred in Pakistan during that time, and from July 2014 through December 2018, our simulation overestimated the number of terrorist attacks by 27 percent. Over the entire duration June 2007 through December 2018, however, our model accounted for 99.95 percent of the 12,450 terrorist attacks that actually occurred in Pakistan. By introducing shocks into the system at a rate and magnitude equivalent to the U.S. drone strike campaign in Pakistan from June 2004 through December 2018, we can conclude that the simulation of agent interactions occurring at the local level ("bottom-up approach") simulated a terrorist attack series which approximated and qualitatively agreed with the observed, real-world data. Sensitivity analyses revealed that, though the magnitude of the peaks for the terrorist attack curves changed, the shapes of those curves were somewhat insensitive to changes in the input parameters.

5 Summary and Further Work

The field of radicalization has employed computational models, but much of the prior work has focused on attempting to identify and explain the trigger points which might cause individuals to undergo phase changes as they progress along the ideology spectrum and adopt increasingly radical thoughts and viewpoints. We begin to address this noted gap by developing and presenting a simple agent-based model informed by theory and calibrated using empirical data to explore the relationship between kinetic actions (i.e., drone strikes) and terrorist attacks in Pakistan from 2004 through 2018. Rather than try to pinpoint and define the motivating factors which might influence somebody down a path toward radicalization, we set out to develop a data-driven agent-based model that incorporated a distributed lag model to characterize the interdependencies between drone strikes and terrorist attacks observed in Pakistan. Based on parametric and validation tests, the model simulates a terrorist attack curve which approximates the rate and magnitude observed in Pakistan from 2007 through 2018. Though not the final word on the subject, this body of research continues to move the field in the right direction by implementing a data-driven, computational model to explore radicalization as a complex adaptive system.

Using this model as a starting point, we see several potential avenues for future applications to examine the relationships between drone strikes and subsequent terrorist attacks. One example would be to explore the type and severity of terrorist attacks. Another example could be applied to other case studies to determine whether the model is generalizable for underlying structures present in those countries. Somalia, a country in which the U.S. has an active drone strike campaign, is one candidate to apply this computational ABM. A logical next step would be to compare the drone strike and terrorist attack curves outputted by the simulation for Pakistan against the comparable curves for Somalia.

Additionally, the agent-based model applied to Pakistan only focused on the temporal autocorrelations between drone strikes and the observed lagged effects seen with subsequent terrorist attacks. Using geospatial analysis to explore the spatial autocorrelation between these two events, however, would uncover patterns that emerged in space (i.e., Tobler's (1970) First Law of Geography) by characterizing the relationships between the geographic entities of drone strikes and terrorist attacks based on their proximity to each other. Specifically, future research could extend our computational model to incorporate both the temporal and location elements into a spatial autoregressive model by leveraging the Pakistan drone strike and terrorist attack datasets which we have made available at https://bit.ly/3qLJynv.

References

Axtell, R.L., Epstein, J.M.: Agent-based modeling: understanding our creations. Bull. Santa Fe Inst. **9**(4), 28–32 (1994)

Baran, Z.: Fighting the War of Ideas. Foreign Aff. **84**(6), 68–78 (2005)

Bureau of Investigative Journalism. https://www.thebureauinvestigates.com/projects/drone-war. Accessed 20 June 2020.

Caiani, A., Godin, A., Caverzasi, E., Gallegati, M., Kinsella, S., Stiglitz, J.E.: Agent based-stock flow consistent macroeconomics: towards a benchmark model. J. Econ. Dyn. Control **69**, 375–408 (2016)

Cioffi-Revilla, C., Harrison, J.F.: Pandemonium in silico: individual radicalization for agent-based modeling. Paper Presented at the Annual Convention of the International Studies Association, Montreal, Quebec, Canada (2011)

Crooks, A., Malleson, N., Manley, E., Heppenstall, A.: Agent-Based Modelling and Geographical Information Systems: A Primer. SAGE Publications Inc., Thousand Oaks (2019)

Defense Science Board: Task Force on Department of Defense Strategy to Counter Violent Extremism Outside of the United States. Department of Defense (2015)

della Porta, D.: Radicalization: A Relational Perspective. Annual Review of Political Science, 21, 461–474 (2018).

Doosje, B., Moghaddam, F.M., Kruglanski, A.W., de Wolf, A., Mann, L., Feddes, A.R.: Terrorism, radicalization and de-radicalization. Curr. Opin. Psychol. **11**, 79–84 (2016)

Galam, S., Javarone, M.A.: Modeling radicalization phenomena in heterogeneous populations. PLoS ONE **11**(5), 1–15 (2016)

Genkin, M., Gutfraind, A.: How do terrorist cells self-assemble: insights from an agent-based model of radicalization. SSRN (2011)

Grimm, V., et al.: A standard protocol for describing individual-based and agent-based models. Ecol. Model. **198**(1–2), 115–126 (2006)

Heppenstall, A., Crooks, A., Malleson, N., Manley, E., Ge, J., Batty, M.: Future developments in geographical agent-based models: challenges and opportunities. Geograph. Anal. **53**(1), 76–91 (2021)

Jaeger, D.A., Siddique, Z.: Are drone strikes effective in Afghanistan and Pakistan? On the dyanmics of violence between the United States and the Taliban. CESifo Econ. Stud. **64**(4), 667–697 (2018)

Jensen, M.A., Seate, A.A., James, P.A.: Radicalization to violence: a pathway approach to studying extremism. Terrorism Polit. Violence **32**(5), 1067–1090 (2020)

Klausen, J., Libretti, R., Hung, B.W.K., Jayasumana, A.P.: Radicalization trajectories: an evidence-based computational approach to dynamic risk assesssment of "Homegrown" Jihadists. Stud. Conflict & Terrorism **43**(7), 588–615 (2020)

Marten, K., Johnson, T.H., Mason, C.: Misunderstanding Pakistan's federally administered Tribal area? Int. Secur. **33**(3), 180–189 (2008)

McCauley, C., Moskalenko, S.: Mechanisms of political radicalization: pathways toward terrorism. Terrorism Polit. Violence **20**(3), 415–433 (2008)

McCauley, C., Moskalenko, S.: Understanding political radicalization: the two-pyramids model. Am. Psychol. **72**(3), 205–216 (2017)

Moghaddam, F.M.: The staircase to terrorism: a psychological exploration. Am. Psychol. **60**(2), 161–169 (2005)

Moon, I.-C., Carley, K.M.: Modeling and simulating terrorist networks in social and geospatial dimensions. IEEE Intell. Syst. **22**(5), 40–49 (2007)

National Consortium for the Study of Terrorism and Responses to Terrorism (START). University of Maryland. Global Terrorism Database. https://www.start.umd.edu/gtd. Accessed 26 Oct 2020

Nawaz, S.: FATA - a most dangerous place: meeting the challenge of militancy and terror in the federally administered Tribal areas of Pakistan. Center for Strategic & International Studies (CSIS) (2009)

Orsini, A.: What everybody should know about radicalization and the DRIA model. Stud. Conflict Terrorism 1–33 (2020). https://doi.org/10.1080/1057610X.2020.1738669

Ozik, J., Sallach, D.L., Macal, C.M.: Modeling dynamic multiscale social processes in agent-based models. IEEE Intell. Syst. **23**(4), 36–42 (2008)

Pew Research Center: Global Opposition to U.S. Surveillance and Drones, But Limited Harm to America's Image. Pew Research Center (2014)

Plaw, A., Fricker, M.S., Colon, C.R.: The Drone Debate: A Primer on the U.S. Use of Unmanned Aircraft Outside Conventional Battlefields. Rowman & Littlefield, New York (2016)

Sageman, M.: Understanding Terror Networks. University of Pennsylvania Press, Philadelphia (2004)

Shah, A.: Do U.S. drone strikes cause blowback? Evidence from Pakistan and beyond. Int. Secur. **42**(4), 47–84 (2018)

Shults, F.L., Gore, R.: Modeling radicalization and violent extremism. In: Verhagen, H., Borit, M., Bravo, G., Wijermans, N. (eds.) Advances in Social Simulation. SPC, pp. 405–410. Springer, Cham (2020). https://doi.org/10.1007/978-3-030-34127-5_41

Silber, M.D., Bhatt, A.: Radicalization in the West: The Homegrown Threat. New York Police Department Intelligence Division (2007)

Tobler, W.R.: A computer movie simulating urban growth in the detroit region. Econ. Geogr. **46**(Supplement 1), 234–240 (1970)

United States House of Representatives: Reassessing the Evolving al-Qaeda Threat to the Homeland, Hearing before the Subcommittee on Intelligence, Information Sharing, and Terrorism Risk Assessment of the Committee on Homeland Security, 111th Congress, First Session (2009)

Webber, D., Kruglanski, A.W.: Psychological factors in radicalization: a "3N" approach. In: LaFree, G., Freilich, J.D. (eds.) The Handbook of the Criminology of Terrorism, pp. 33–46. Wiley-Blackwell, Malden (2016)

Webber, D., Kruglanski, A.W.: The social psychological makings of a terrorist. Curr. Opin. Psychol. **19**, 131–134 (2018)

Wilensky, U.: NetLogo. Center for Connected Learning and Computer-Based Modeling, Northwestern University, Evanston (1999). http://ccl.northwestern.edu/netlogo/

Multi-agent Naïve Utility Calculus: Intent Recognition in the Stag-Hunt Game

Lux Miranda[✉][iD] and Ozlem Ozmen Garibay[iD]

Human-Centered Artificial Intelligence Research Laboratory,
University of Central Florida, Orlando, FL, USA
lux@knights.ucf.edu

Abstract. The human ability to utilize social and behavioral cues to infer each other's intents, infer motivations, and predict future actions is a central process to human social life. This ability represents a facet of human cognition that artificial intelligence has yet to fully mimic and master. Artificial agents with greater social intelligence have wide-ranging applications from enabling the collaboration of human-AI teams to more accurately modelling human behavior in complex systems. Here, we show that the Naïve Utility Calculus generative model is capable of competing with leading models in intent recognition and action prediction when observing *stag-hunt*, a simple multiplayer game where agents must infer each other's intentions to maximize rewards. Moreover, we show the model is the first with the capacity to out-compete human observers in intent recognition after the first round of observation.

Keywords: Artificial social intelligence · Action understanding · Social cognition

1 Introduction

A fundamental process in our everyday life is our ability to observe peoples' actions, infer their beliefs, desires, and intentions, and predict what they will do next. A great many algorithms have been created, studied, and used to emulate this kind of intent recognition in a variety of domains and contexts [2,20]; this work principally focuses on evaluating a novel computational framework (the Naïve Utility Calculus) in inferring cooperative intent among multiple agents in a shared environment.

One application of immediate interest for this particular inference problem is its use in human-AI teaming. Algorithmic intent recognition of cooperation between humans is critical to developing AI that can enhance teams' coordination and efficacy in performing complex tasks such as urban search-and-rescue or military operations [1,4]. Furthermore, there has been a shift in recent years in scholars beginning to view cognitive science and generative social science as interdependent fields [13], and many models have been advanced by using more complex and realistic cognitive architectures for agents [8,17]. Thus, improving

© Springer Nature Switzerland AG 2021
R. Thomson et al. (Eds.): SBP-BRiMS 2021, LNCS 12720, pp. 331–340, 2021.
https://doi.org/10.1007/978-3-030-80387-2_32

these cognitive architectures has consequences for, for example, detecting and preventing threats to public safety [2] such as the spread of disinformation on social media [7,16], or informing policy to better deal with the effects of the global climate crisis [3,6].

The Naïve Utility Calculus (NUC) is a recently formalized framework for action-understanding [11]. While, prior to this work, the NUC displayed much promise as a general model for action-understanding, it had only been put to use in a single-agent setting with limited inferences on social behavior.

Thus, this work is presented with the primary purpose of testing the Naïve Utility Calculus in a multi-agent setting with greater requirements on social inference. To do this, we utilize a well-studied cooperative action game known as *stag-hunt*. Stag-hunt is a multiplayer game where agents work to maximize their rewards by choosing to pursue and capture either a high-reward stag or a low-reward hare. While low-reward hares can be captured individually, agents must work together in order to capture a high-reward stag.

While stag-hunt was introduced by Skyrms [19] and famously used via a Minecraft implementation known as *Pig Chase* by the Microsoft Malmo Collaborative AI Challenge [12], this work primarily draws on a version of stag-hunt used to more directly test general models of artificial theory of mind. This version, introduced by Shum et al. [18], includes data from human subjects performing the same tasks as their computational model, which relies on Bayesian inference over a generative model encoding relations known as Composable Team Hierarchies (CTHs). Further utilized by Rabkina and Forbus [15] with the introduction of a model known as Analogical Reasoning, this version of stag-hunt allows for a streamlined, simple experiment with directly comparable results between competing models.

We find the Naïve Utility Calculus model is, in terms of both accurately recognizing intent and predicting actions, comparable to these prior two models—with the exception of intent recognition after the first timestep. Intriguingly, after observation of this timestep, NUC significantly outperforms both prior models *and* the human subjects in intent recognition.

2 Background

2.1 The Naïve Utility Calculus

The Naïve Utility Calculus (NUC) operates under the assumption that agents tend to choose their actions in order to maximize some notion of subjective utility [11]. Research in childhood psychology indicates that this paradigm is likely a good approximation of how individuals reason about the behavior of others [9,10].

Within NUC, agents are treated as boundedly rational. The calculus itself is implemented as a generative model of each agents' mental processes, with cost and reward functions being used to formulate a plan of actions for the agent to take. Observers, however, are not aware of these underlying cost and reward functions. Thus, given only the actions that an agent has taken, Bayesian inference may be run over the model to produce inferred estimates of the agent's cost and reward functions. These estimated functions may then be used as input to the generative model, producing a predicted set of actions the agent will take [11].

In Jara-Ettinger et al. [11], the authors introduced a formal computational model of the Naïve Utility Calculus and extensively tested it in a series of single-agent experiments. In each of these experiments, both the model and human observers were presented with images of an astronaut exploring an extra-terrestrial grid world with various types of terrain, collectible "care packages," and goal locations. The experiments found that, in this simple setting, the model was sufficient for matching the abilities of human subjects to estimate the utility function of a single agent in a variety of scenarios. However, the implications of social reasoning for these experiments are limited—only a semi-social scenario is created during a single experiment where "collectible packages" are replaced with "rescuable astronauts." However, "other agents" were merely treated as part of the environment—the same as care packages—and thus did not constitute a true multi-agent environment.

In this work, we pit the Naïve Utility Calculus up against two prior models tested in the multi-agent stag-hunt game. Analogical Reasoning, the most recent of these models, is likewise a computational formalization of qualitative theory from research in childhood psychology, but it instead formalizes the separate cognitive process of analogical thinking [14,15]. The second model is a hierarchical Bayesian model introduced by Shum et al. [18] that explicitly encodes causal models known as Composable Team Hierarchies (CTHs) to aid in cooperation inferences.

2.2 Analogical Reasoning

The Analogical Reasoning model [15], similar to NUC, is a computational formulation inspired by findings in cognitive psychology. It takes after the process of human reasoning known by the same name. The primary reasoning engine used by Rabkina and Forbus [15] is a modern iteration of an analogical reasoning algorithm known as the Structure-Mapping Engine [5]. Cases in stag-hunt are represented in the form of how the spatial relationships between agents change over each timestep in addition to several non-spatial events such as a target being captured. Analogical Reasoning is a traditional machine learning method; the authors trained and tested their model via leave-one-out cross validation over the nine scenarios (i.e., for each scenario, the model was trained on the eight other scenarios and tested on the one).

2.3 Composable Team Hierarchies

Like NUC, the Composable Team Hierarchies (CTH) model presented in Shum et al. [18] is a generative model over which Bayesian inference may be performed. Instead of encoding the internal mental state of agents, however, the

CTH model assumes that agent actions are based around a certain team hierarchy and encodes a causal model of how a given team structure would dictate agents' future actions. Bayesian inference over the model accepts observations of agent actions to infer a team hierarchy (e.g., Players A and C are teaming up against Player B), and that inferred team hierarchy is then used to predict agent actions (e.g., Players A and C will pursue a stag, Player B will pursue a hare).

3   Experiment

We test the abilities of the Naïve Utility Calculus using observations of the multi-agent stag-hunt game à la Rabkina and Forbus [15] and Shum et al. [18].

The game operates on the premise that a group of hunters must each choose to attempt to capture either a hare or a stag. Hares provide a low reward and can be captured by a single hunter. Stags provide a much higher reward but require a team of two or more hunters to be captured. There is no penalty for not capturing a target.

The version of the game used here places three hunters, two stags, and two hares into a 5×7 grid world. In this grid world, there are traversable white tiles ("floors") and non-traversable black tiles ("walls"). At each timestep, hunters may move up, down, left, or right, but not diagonally. Hares are stagnant, and the movement of stags is pre-determined as part of the map structure. In general, stags attempt to move away from pursuing hunters.

This configuration is used to construct nine unique scenarios that each encode different possibilities of agent cooperation and non-cooperation. For each scenario, three timesteps of agent movement are encoded. Successful stag captures occur in five scenarios (Fig. 1a, c, d, g, i), indicating cooperation, while only hares are captured in the other four scenarios (Fig. 1b, e, f, h), indicating no cooperation.

These board states are used on two fronts: Firstly, the game encodes sufficient information for an observer to deduce each agent's goal by the third time step, and, consequently, information to see which agents are cooperating to capture a stag (and which are not). Indeed, in Shum and colleagues' trials with real human observers, cooperation inferences were made correctly 100% of the time by the third timestep. The challenge lies in inferring each agent's goal before this final timestep.

Secondly is the prediction of agents' future actions. This is not a task fully encompassed by either Shum and colleagues' model or human subjects, thus our accuracy metric for this task is only compared against Rabkina and Forbus's model.

4   Methods

We utilize the same existing implementation of the Naïve Utility Calculus used by Jara-Ettinger et al. [11]. This implementation is in the form of a Python 2.7

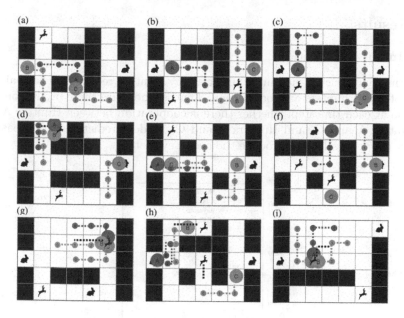

Fig. 1. The nine scenarios of the stag-hunt game. Used with permission from Rabkina and Forbus [15], originally adapted from Shum et al. [18].

package known as Bishop (https://github.com/julianje/Bishop/). In this study, we have adapted Bishop to use in the multi-agent setting without any modifications to the base package. This was possible as Bishop is uniquely suited for studying this version of stag-hunt, as it is designed specifically for observations of a single agent in a grid world with various objectives that may be obtained.

To adapt the usage of Bishop into a multi-agent setting, we simply encode the stag-hunt game into a format that Bishop may read and run the program on each agent individually. Of note is that, although, in theory, each player incorporates information on the other hunters to make their decisions, an outside observer of the game does not necessarily need to incorporate the state of other players into their inferences on each individual hunter. For example, if an observer is attempting to determine whether Hunter B is trying to capture a hare or a stag, they do not necessarily need to know the locations of Hunter A or Hunter C, as (at least in our nine scenarios) only Hunter B's movements are sufficient information to make the inference. Thus, in service to the limitations of the existing implementation of Bishop, a separate "map" is created for each of the three hunters between the nine scenarios, and Bishop's inference functions are run on each of these maps individually.

Lastly, we leave all specifiable hyperparameters within Bishop at their default values. All reported results below are from 100 runs of NUC at 500 samples per run. All code and data is available via DOI: 10.5281/zenodo.4430598.

5 Results

5.1 Intent Recognition

Our primary metric for measuring the accuracy of intent recognition is a pairwise count of which hunters are cooperating. That is, there are three predictions per scenario at each timestep: whether there is cooperation between Hunters A and B, Hunters A and C, and Hunters B and C. Model predictions are measured against the true values to produce the accuracy metric.

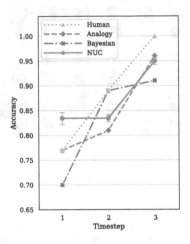

Fig. 2. Pairwise cooperation inference accuracy across each model and the human subjects. Error bars on the NUC data indicate one standard error across 100 runs with 500 samples each. Note that accuracies for human and CTH (Bayesian) model data are based on good-faith estimates from the figures of Shum et al. [18] and Rabkina and Forbus [15]

The NUC's predictions are summarized in Fig. 2. Note that the precise values for the Bayesian and Human metrics are good-faith estimates from Shum and colleagues' figures determined by Rabkina and Forbus.

We find that, after the first timestep, the NUC outperforms human subjects and all other models by a minimum of eight percentage points with a very high margin of certainty. The reason for this is uncertain; one possible hypothesis to explain this is that human observers may have less confidence in their judgement when possessing only limited information, and are therefore perhaps not achieving the maximum possible inference accuracy.

After the second timestep, NUC performs better than Analogical Reasoning but not as well as humans or the Bayesian model. After the final timestep, NUC outperforms the Bayesian model and is roughly on par with Analogical Reasoning.

5.2 Future Action Prediction

We replicate Rabkina and Forbus's [15] metric of measuring prediction accuracy. Due to the limitations of the current implementation of Bishop, we do not calculate action prediction accuracy in precisely the same way—while Rabkina and Forbus incorporate predicting the actions of stags into their accuracy metric, here stags are necessarily treated as parts of the environment rather than agents with predictable behavior. Thus, the metric is best used as an approximate comparison between the two models rather than a precise one.

For each agent, NUC first performs intent recognition and then uses the inferred intent to predict its next moves. The agents' movements are measured relative to the other two agents, the two stags, and the two hares, with three possible states for the relative movements: toward, away, and stationary. For example, an action prediction for Agent B's next move might look like: Away from Agent A, Toward Agent C, Stationary to Stag 1, Toward Stag 2, Away from Hare 1, Away from Hare 2. Since there are three possible states to infer, we would expect a model guessing randomly to be correct in its prediction roughly a third of the time. Thus, the baseline accuracy for this metric is 33.33%.

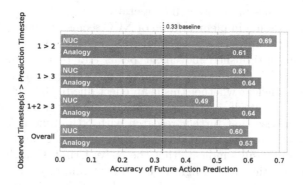

Fig. 3. Relational action prediction accuracy across each timestep case.

From timestep 1, we make these six predictions for each agent into timestep 2 and timestep 3. Additional predictions are made given the information from timestep 2 into timestep 3. Results are summarized in Fig. 3. We find that, with the exception of the first timestep, NUC roughly underperforms in comparison to Analogical Reasoning by several percentage points, though this margin might be in part due to the exclusion of stags in our prediction due to the limitations of Bishop mentioned above.

However, there is a notable case of underperformance when predicting timestep 3 movements having observed timestep 1 and timestep 2 $(1 + 2 > 3)$. We hypothesize that NUC seems to suffer from a "status quo" bias. NUC tends to fare especially poorly in cases where the first two agent movements are the same, but the agent then turns into a separate direction, forming an "L" movement pattern. We explore this quantitatively by considering raw accuracy of movement prediction on a five-fold accuracy metric predicting whether an agent will remain stationary or move up, left, down, or right. This metric, thus, has a baseline random accuracy of 20%. We find that, in the prediction of the $1+2 > 3$ case, NUC has a raw accuracy of just 22%, with an 8% accuracy in the "L" cases and 33% accuracy in the non-"L" cases. This trend may be due to the limited nature of this version of the stag-hunt game with its small board and very few number of timesteps, but it may be a worthy line of inquiry to see if a similar pattern holds in other action-understanding problems.

6   Conclusion

In this study, we adapted the recently formalized computational model of Naive Utility Calculus (NUC) to the multi-agent stag-hunt game to test the model's abilities to infer agents' intent to cooperate (or not cooperate) and predict their future actions. To this end, we found the NUC's ability is comparable to leading action-understanding models which have been tested in this same version of the game. Moreover, we found that, when having only observed the first round of the game, NUC is able to outperform human observers by a significant margin in inferring which pairs of agents will cooperate.

However, the existing implementation of NUC seems to suffer from a "status-quo bias;" when attempting to predict future agent actions, the NUC strongly favored continuing established movement paths and poorly predicted sudden changes to such paths, even if such a change would bring an agent closer to its inferred goal. It is unclear whether this is an effect due to the current implementation of NUC, NUC itself, or the limited and simple nature of the stag-hunt experiment. We advise that further research with NUC take note of this effect to see if it persists across different action-understanding scenarios.

Nonetheless, our results indicate that the Naive Utility Calculus is worthy of continued study. Particularly in light of the (albeit limited) capacity to outperform humans in some cases, the principles on which NUC is based may represent a potential avenue for advancing platforms for artificial intent recognition.

Acknowledgements. This material is based upon work supported by the Defense Advanced Research Projects Agency (DARPA) under Contract No. HR001120C0036. Any opinions, findings and conclusions or recommendations expressed in this material are those of the author(s) and do not necessarily reflect the views of the Defense Advanced Research Projects Agency (DARPA).

References

1. Barnes, M., Chen, J., Schaefer, K.E., Kelley, T., Giammanco, C., Hill, S.: Five requisites for human-agent decision sharing in military environments. In: Savage-Knepshield, P., Chen, J. (eds.) Advances in Human Factors in Robots and Unmanned Systems, vol. 499, pp. 39–48. Springer, Cham (2017). https://doi.org/10.1007/978-3-319-41959-6_4 ISBN: 978-3-319-41959-6

2. Demiris, Y.: Prediction of intent in robotics and multi-agent systems. Cogn. Process. **8**(3), 151–158 (2007). https://doi.org/10.1007/s10339-007-0168-9. ISSN: 1612-4782
3. Elsawah, S., et al.: Eight grand challenges in socio-environmental systems modeling. Soc.-Environ. Syst. Model. **2**, 16226 (2020). https://doi.org/10.18174/sesmo. 2020a16226
4. Fiore, S.M., Wiltshire, T.J.: Technology as teammate: examining the role of external cognition in support of team cognitive processes. Front. Psychol. **7**, 1531 (2016). https://doi.org/10.3389/fpsyg.2016.01531. ISSN: 1664-1078
5. Forbus, K.D., Ferguson, R.W., Lovett, A., Gentner, D.: Extending SME to handle large-scale cognitive modeling. Cogn. Sci. **41**(5), 1152–1201 (2017). https://doi. org/10.1111/cogs.12377. ISSN: 1551-6709
6. Freeman, J., Baggio, J.A., Coyle, T.R.: Social and general intelligence improves collective action in a common pool resource system. Proc. Natl. Acad. Sci. U.S.A. **117**(14), 7712–7718 (2020). https://doi.org/10.1073/pnas.1915824117. ISSN: 1091-6490
7. Garibay, I., et al.: Deep agent: studying the dynamics of information spread and evolution in social networks. arXiv preprint arXiv:2003.11611 (2020)
8. Gunaratne, C., Rand, W., Garibay, I.: Inferring mechanisms of response prioritization on social media under information overload. Sci. Rep. **11**(1), 1–12 (2021). https://doi.org/10.1038/s41598-020-79897-5. ISSN: 2045-2322
9. Jara-ettinger, J., Gweon, H., Schulz, L.E., Tenenbaum, J.B.: The Naïve utility calculus: computational principles underlying commonsense psychology. Trends Cogn. Sci. **20**(8), 589–604 (2016). https://doi.org/10.1016/j.tics.2016.05.011. ISSN: 1364-6613
10. Jara-ettinger, J., Gweon, H., Tenenbaum, J.B., Schulz, L.E.: Children's understanding of the costs and rewards underlying rational action. Cognition **140**, 14–23 (2015). https://doi.org/10.1016/j.cognition.2015.03.006. ISSN: 0010-0277
11. Jara-Ettinger, J., Schulz, L.E., Tenenbaum, J.B.: The Naïve utility calculus as a unified, quantitative framework for action understanding. Cogn. Psychol. **123**, 101334 (2020). https://doi.org/10.1016/j.cogpsych.2020.101334. ISSN: 0010-0285
12. Johnson, M., Hofmann, K., Hutton, T., Bignell, D.: The Malmo platform for artificial intelligence experimentation. In: IJCAI International Joint Conference on Artificial Intelligence 2016, pp. 4246–4247 (2016). ISSN: 1045-0823
13. Orr, M.G., Lebiere, C., Stocco, A., Pirolli, P., Pires, B., Kennedy, W.G.: Multi-scale resolution of cognitive architectures: *a paradigm for simulating minds and society*. In: Thomson, R., Dancy, C., Hyder, A., Bisgin, H. (eds.) SBP-BRiMS 2018. LNCS, vol. 10899, pp. 3–15. Springer, Cham (2018). https://doi.org/10.1007/978-3-319-93372-6_1
14. Rabkina, I.: Analogical theory of mind: computational model and applications. Ph.D. thesis, Northwestern University (2020). https://search.proquest. com/openview/9b5e17f0c672eeed61afad5273bb39df/1?pq-origsite=gscholar& cbl=18750&diss=y
15. Rabkina, I., Forbus, K.D.: Analogical reasoning for intent recognition and action prediction in multi-agent systems. In: Proceedings of the 7th Annual Conference on Advances in Cognitive Systems (2019)
16. Rajabi, A., Gunaratne, C., Mantzaris, A.V., Garibay, I.: On countering disinformation with caution: effective inoculation strategies and others that backfire into community hyper-polarization. In: Thomson, R., Bisgin, H., Dancy, C., Hyder, A., Hussain, M. (eds.) SBP-BRiMS 2020. LNCS, vol. 12268, pp. 130–139. Springer, Cham (2020). https://doi.org/10.1007/978-3-030-61255-9_13

17. Schlüter, M., et al.: A framework for mapping and comparing behavioural theories in models of social-ecological systems. Ecol. Econ. **131**, 21–35 (2017). https://doi.org/10.1016/j.ecolecon.2016.08.008. https://www.sciencedirect.com/science/article/pii/S0921800915306133. ISSN: 0921-8009
18. Shum, M., Kleiman-Weiner, M., Littman, M.L., Tenenbaum, J.B.: Theory of minds: understanding behavior in groups through inverse planning. In: 33rd AAAI Conference on Artificial Intelligence, AAAI 2019, 31st Innovative Applications of Artificial Intelligence Conference, IAAI 2019 and the 9th AAAI Symposium on Educational Advances in Artificial Intelligence, EAAI 2019, pp. 6163–6170 (2019). https://doi.org/10.1609/aaai.v33i01.33016163. ISSN: 2159-5399
19. Skyrms, B.: The Stag Hunt and the Evolution of Social Structure, pp. 1–149 (2003). https://doi.org/10.1017/CBO9781139165228
20. Sukthankar, G., Geib, C., Bui, H.H., Pynadath, D., Goldman, R.P.: Plan, Activity, and Intent Recognition: Theory and Practice. Newnes (2014)

Author Index

Printed in the United States
by Baker & Taylor Publisher Services

Printed in the United States
by Baker & Taylor Publisher Services